ENZYMES
IN ANESTHESIOLOGY

ENZYMES IN ANESTHESIOLOGY

Edited by Francis F. Foldes

With contributions by
A. A. Aszalos · F. F. Foldes · L. C. Mark
S. H. Ngai · R. W. Patterson · J. M. Perel
S. F. Sullivan · L. Triner · E. K. Zsigmond

Springer-Verlag
New York Heidelberg Berlin

Francis F. Foldes, M.D.
Montefiore Hospital and Medical Center
111 E. 210 St.
Bronx, New York 10467

Library of Congress Cataloging in Publication Data

Main entry under title:

Enzymes in anesthesiology.

Includes bibliographical references and index. 1. Anesthesiology. 2. Enzymes. I. Aszalos, A. A. II. Foldes, Francis F.
RD82.E58   615′.781   77-27631

© 1978 by Springer-Verlag New York Inc.
Softcover reprint of hardcover 1st edition 1978

9 8 7 6 5 4 3 2 1

ISBN-13: 978-1-4612-6250-3   e-ISBN-13: 978-1-4612-6248-0
DOI: 10.1007/ 978-1-4612-6248-0

Foreword

It is a pleasure to have the privilege of writing the foreword for a book edited by Dr. Francis F. Foldes. Dr. Foldes has collected in one convenient place a discussion and description of enzyme systems of use to the anesthesiologist and to those other individuals, such as undergraduate and graduate students in related basic sciences, who will profit by and can make use of this body of information.

The practicing anesthesiologist and those who work in related fields have become increasingly aware of the need to understand enzyme activities which influence the uptake, distribution, and excretion of those substances that are used in the anesthetic management of surgical patients. A variety of such activities is obvious when one considers that such diverse substances as analgesic drugs, tranquilizers, hypnotics, anesthetic agents, and muscle relaxants are strongly affected by these systems and have an influence over the basic understanding of how these drugs operate and act in the body, as well as providing a safety measure so necessary to the proper conduct of clinical anesthesia.

The editor and his colleagues have rendered us a great service in collecting information that deals with the basic activity of enzymes including their structure, their kinetics, and to the degree that knowledge permits, mechanism of actions.

Included in this small and good book, in addition to basic enzymology of practical use to the individuals mentioned, are the discussions of enzyme systems dealing with nerve transmitters, with enzymes that markedly affect in both directions the activity of nonvolatile substances, and a group of enzymes that have not been ordinarily thought to be directly applicable to clinical anesthesiology, but on simple reflection are of great importance.

I feel particularly pleased to note that I have had the opportunity of association in a direct sense with all but two of the co-authors of this splendid book.

It promises to be a collection that will be of considerable conceptual value and, probably even more important, of practical use as a reference for anesthesiologists, medical, and graduate students in the basics of the sciences as well. I hope the spirit and intention of the editor and the co-authors will find the receptivity their work deserves among these groups.

E. M. Papper, M.D.
Vice President for Medical Affairs
and Dean, School of Medicine
University of Miami

Preface

Enzymes are essential components of biochemical reactions and are indispensable for the smooth functioning of the physiologic processes based on these reactions. The field of enzymology is one of the most rapidly expanding sciences. Day by day the wealth of information available on enzyme properties and functions increases by leaps and bounds. The possibilities for interactions between enzymes and anesthetic agents are almost limitless. To consider even a small fraction of all the relevant possibilities would be impossible.

Perhaps the most difficult task that confronted us was the selection of the enzyme systems to be discussed from the list of over 4000 entries in the *Enzyme Nomenclature* (Florkin and Stotz 1973). Our choice was guided by three principles: The first was to select enzyme systems of theoretical and practical importance for anesthesiologists. Our second goal was to furnish enough, up-to-date, information on general enzymology to facilitate the understanding of the kinetics of the selected enzyme systems. Our third objective was to demonstrate an avenue of approach to the investigation of other enzyme systems not discussed in this monograph. To achieve our second and third objectives we have devoted a considerable part of the monograph to basic enzymology. This section is comprehensive enough to stand on its own as an introductory text in general enzymology.

Although the enzymes discussed were selected with primary consideration for the interests of anesthesiologists, this monograph contains much information in the discussion of the specific enzyme systems to appeal to a much larger readership. It should be useful for undergraduate and graduate students in biology, biochemistry, physiology, pharmacology, and for medical students. This volume could also serve as a starting point for those wishing to embark on research on any of the enzyme systems discussed. This applies especially to geneticists and pharmacogeneticists interested in abnormal cholinesterases and malignant hyperthermia.

Monographs are seldom read from cover to cover, but more often than not are used as reference sources. With this in mind it was attempted to make each chapter self sufficient. To facilitate this, we have compiled a glossary that will help the reader to comprehend any chapter, with frequent references to other chapters, or to other sources of information.

Francis F. Foldes

Contents

Part II.   Specific Enzymes

Contributors

Adorjan A. Aszalos, Ph.D.
Head, Biochemistry Section
Frederick Cancer Research Center
National Cancer Institute
Frederick, Maryland

Francis F. Foldes, M.D.
Professor Emeritus of Anesthesiology
Albert Einstein College of Medicine
and Consultant in Anesthesiology
Montefiore Hospital and Medical Center
New York, New York

Lester C. Mark, M.D.
Professor of Anesthesiology
College of Physicians and Surgeons
Columbia University
New York, New York

S. H. Ngai, M.D.
Professor of Anesthesiology
and Pharmacology
College of Physicians and Surgeons
Columbia University
New York, New York

Richard W. Patterson, M.D.
Professor of Anesthesiology
University of California
School of Medicine
Los Angeles, California

James M. Perel, Ph.D.
Associate Professor of Clinical
Psychopharmacology
College of Physicians and Surgeons
Columbia University
New York, New York

Stuart F. Sullivan, M.D.
Professor of Anesthesiology
University of California
School of Medicine
Los Angeles, California

Lubos Triner, M.D., Ph.D.
Associate Professor of Anesthesiology
College of Physicians and Surgeons
Columbia University
New York, New York

Elemer K. Zsigmond, M.D.
Professor of Anesthesiology
University of Michigan
School of Medicine
Ann Arbor, Michigan

Glossary

AADC:	Aromatic L-amino acid decarboxylase
AcCoa:	Acetylcoenzyme-A
ACh:	Acetylcholine
AChE:	Acetylcholinesterase
ACP:	Acyl carrier protein
ACTH:	Adrenocorticotropic hormone
ADH:	Antidiuretic hormone
ADP:	Adenosine diphosphate
AK:	Adenylate-kinase; myokinase; EC 2.7.4.3.
AMP:	Adenosine monophosphate
5'-AMP:	5'-Adenosine monophosphate
ATP:	Adenosine triphosphate
ATPase:	Adenosine-triphosphatase; ATP-phosphohydrolase; EC 3.6.1.4.
BuChE:	Butyrylcholinesterase
cAMP:	Cyclic adenosine 3', 5'-monophosphate
CD:	Circular dichroism
CDP:	Cytidine diphosphate
ChAc:	Choline acetylase
ChE:	Cholinesterase
CM Sephadex:	Carboxymethyl Sephadex
CMP:	Cytidine monophosphate
CNS:	Central nervous system
CoA:	Coenzyme A
COMT:	Catechol-O-methyltransferase
CoQ:	Coenzyme Q
CP:	Creatine phosphate

CPK:	Creatine-phosphokinase; ATP: creatine phosphotransferase; EC 2.7.3.2.
CTP:	Cytidine triphosphate
DβH:	Dopamine-β-hydroxylase
dcMP:	2′Deoxycytidine monophosphate
dcTP:	2′Deoxycytidine triphosphate
DEAE Sephadex:	Diethylaminoethyl Sephadex
DFP:	Diisopropyl fluorophosphate
dG:	2′Deoxyguanosine
dGTP:	2′Deoxyguanosine triphosphate
DNA:	Deoxyribonucleic acid
DOCA:	Deoxycorticosterone
DOPA:	3,4-Dihydroxyphenylalanine
DOPAC:	3,4-Hydroxyphenylacetic acid
DTT:	Dithiothreitol, Cleland reagent
EC:	International Union of Biochemistry enzyme numbering system
EDTA:	Ethylene diamine tetraacetic acid
EPR:	Electron paramagnetic resonance
ETP:	Electron transport particles, submitochondrial particles
F:	Folic acid
FH$_2$:	Dihydrofolic acid
FH$_4$:	Tetrahydrofolic acid
FMN:	Flavin mononucleotide
FP$_1$:	Ferroflavoprotein, NADH dehydrogenase
G6-P:	Glucose6-phosphate
G-6-PDH:	Glucose-6-phosphate dehydrogenase; D-glucose6-phosphate: NADP oxidoreductase; EC 1.1.1.49.
GDP:	Guanosine diphosphate
GMP:	Guanosine monophosphate
GTP:	Guanosine triphosphate
5-HIAA:	5-Hydroxyindoleacetic acid
HK:	Hexokinase; ATP: D-hexose-6-phosphotransferase; EC 2.7.1.1.
5-HT:	5-Hydroxytryptamine, serotonin
HVA:	Homovanillic acid
ICSH:	Interstitial cell-stimulating hormone
IMP:	Inosine monophosphate
K_m:	Dissociation constant of an enzyme–substrate complex (index of affinity)
LDH:	Lactate dehydrogenase; L-lactate: NAD-oxidoreductase; EC 1.1.1.27.
LH:	Luteinizing hormone
LSD:	Lysergic acid diethylamide
M:	Mole
MAC:	Minimum alveolar concentration of anesthetic to prevent gross movement in response to surgical stimulation in 50 percent of patients

MAO:	Monoamine oxidases
MAOI:	Monoamine oxidase inhibitors
MH:	Malignant hyperthermia
mU:	One milliunit equals 1 nmole of substrate converted in 1 min at 25°C
NAD:	Nicotine adenine dinucleotide
NADH:	Nicotine adenine dinucleotide, reduced form
NADH$_2$:	Reduced nicotineamide-adenine dinucleotide
NADP	Nicotine adenine dinucleotide phosphate
NADPH:	Nicotine adenine dinucleotide phosphate, reduced form
NADPH$_2$:	Reduced nicotineamide-adenine dinucleotide phosphate
NBT:	Nitrobluetetrazolium
n.m.:	Neuromuscular
NREM:	Nonrapid eye movement sleep
ORD:	Optical rotatory dispersion
2-PAM:	Pyridine-2-aldoxine methiodide
4-PAM:	Pyridine-4-aldoxine methiodide
PCP:	Parachlorophenylalanine
PEP:	Phosphoenolpyruvate
PHS:	Phenazine methosulfate
Pi:	Inorganic phosphate
PK:	Pyruvate kinase; ATP: pyruvate phosphotransferase; EC 2.7.1.40.
PNMT:	Phenylethanolamine-N-methyltransferase
REM:	Rapid eye movement sleep
RNA:	Ribonucleic acid
S:	Svedberg unit (sec)
SGOT:	Serum glutamic-oxalacetic transaminase; aspartate aminotransferase; L-aspartate: 2-oxaloglutarate aminotransferase; EC 2.6.1.1.
TH:	Tyrosine hydroxylase
TRIS:	2-Amino-2-hydroxymethyl-1,3-propanediol (trometamol)
tRNA:	Transfer ribonucleic acid
TSH:	Thyroid-stimulatory hormone
UDP:	Uridine diphosphate
UMP:	Uridine monophosphate
UTP:	Uridine triphosphate
V_{max}:	Maximum velocity of the enzyme-catalyzed reaction
V_{O_2}:	Oxygen consumption
VMA:	3-methoxy-4-hydroxy mandelic acid

PART I

BASIC CONSIDERATIONS

Chapter 1

Structure of Enzymes

A. A. Aszalos

Introduction

The biologic functions of living organisms depend on complex systems of biochemical reactions catalyzed by specific proteins, the enzymes. Any alteration of the enzyme pattern may have far-reaching consequences for these organisms. The study of these specific proteins requires the combined skills of many scientists, such as biologists, physical chemists, physicists, biochemists and chemists. Because of the importance of the biochemical reactions carried out by enzymes, and because of the fascinating aspects of the research related to it, enzymology has become one of the most rapidly expanding sciences.

The primary purpose of this book is to discuss the role of enzymes in anesthesiology. To be able to accomplish this, it is necessary to consider first some basic aspects of enzymology. Because of limitations of space, this summary will be brief and far from complete, but will hopefully supply the background necessary for the understanding of specific topics to be discussed in the second part of this monograph. Subjects such as evolutionary problems, synthesis of enzymes in cells or *in vitro*, and certain physical studies will be treated briefly. Those interested in obtaining more information on certain topics are referred to textbooks and other publications cited in the following chapters.

There are basic differences between catalyzed organic reactions in homogeneous solutions and enzyme-catalyzed reactions. In the first case, the chemical structures of all components are known, but in most enzyme-catalyzed reactions the structure of the catalyst, the enzyme, is not known. In this chapter the structural aspects of the chemistry and physical chemistry of enzymes are discussed briefly. This topic may be considered most conveniently under the following headings: (a) molecular weight of and number of peptide chains in enzyme structures; (b) primary structure; (c) secondary structure; (d) tertiary structure; and (e) quaternary structure. Catalysis, kinetics, and coenzymes are considered under separate headings.

Molecular Weight of Enzymes and Number of Peptide Chains in Them

The molecular weight of an enzyme can be determined by a number of chemical and physicochemical methods. The chemical methods are based on the determination of a component, such as a prosthetic group or an amino acid in the molecule. Knowledge of the number of protein chains present in the enzyme is a prerequisite of these determinations, because a native enzyme may contain several protein chains and, therefore, several terminal amino acids and prosthetic groups.

Physicochemical methods fall into two main categories. "Number-average methods" depend on the number of particles present per unit volume. An example of such a method is the measurement of osmotic pressure. By this method, the molecular weight can be calculated from the osmotic pressure, the concentration, and known physical constants (23). If the solution is polydispersed, the molecular weight determined by this method is usually lower than one obtained by the use of a "weight-average method," such as sedimentation analysis. With this technique, molecular weight can be calculated from sedimentation rates (expressed as the sedimentation coefficient) obtained by ultracentrifugation, known physical constants, and the diffusion coefficient of the enzyme (15).

A very versatile method of molecular-weight determination is density-gradient centrifugation, in which the enzyme solution, together with marker macromolecules, is layered on top of density-gradient sucrose in a centrifuge tube. Centrifugation causes the macromolecules to sediment down the gradient. The sedimentation coefficient of the enzyme can be calculated from the position of the enzyme relative to those of the marker macromolecules. In the equilibrium-density-gradient procedures, a cesium chloride gradient is formed during centrifugation. Each macromolecule will come into equilibrium at a position in which its density equals that of the cesium chloride solution, which in turn can be calculated. A detailed description of these methods is given by Vinograd and Brunes (56).

A simple and elegant method for the determination of molecular weights of enzymes is molecular-exclusion chromatography, also called gel filtration. In this technique, a porous polymer, like Sephadex, retains molecules according to their ability to penetrate into the pores. From the relative positions of the enzyme and marker molecules in the eluate, the molecular weight can be calculated by a simple graphical method (1).

The molecular weight of enzymes measured by the above or other methods varies from less than 1×10^4 to about 1×10^6. The smallest active enzyme found in a living organism has a molecular weight of about 9×10^3.

Sometimes the determined molecular weight is not the molecular weight of the native enzyme, but that of active subunits or of inactive individual protein chains called protomers, or of aggregates of the native enzyme. The equilibrium between the aggregate and the monomeric enzyme may show up in the ultracentrifuge either as two peaks or as an asymmetry of a single peak. The state of aggregation may depend on the concentration of the enzyme. A well-known enzyme that aggregates is acetylcoenzyme-A (AcCoA) carboxylase, which catalyzes the uptake of CO_2 to form malonylcoenzyme-A, an important precursor in the biosynthesis of fatty acids. This enzyme, if prepared from chicken liver, is obtained in a slightly active monomeric form and a fully active polymeric form. The monomer has a molecular weight of 4×10^5 and a sedimentation coefficient of 20 S and contains four equal protomers. The monomer aggregates in the presence of its positive modulator, isocitrate, and forms an aggregate fiber with a particle weight of 4×10^6. Another example of an enzyme that forms aggregates is bacterial isocitric dehydrogenase (34). This enzyme has a monomeric molecular weight of 1.1×10^5 and aggregates at acid pH into particles with molecular weights of 2×10^5, 5×10^5, and even higher. The particles formed can be disaggregated to monomers by the action of isocitrate, citrate, and Mg^{2+}.

Still another example of an enzyme that forms aggregates is L-asparaginase, which aggregates at higher enzyme concentrations (26, 28) and at low ionic strength to dimers and tetramers with particle weights of 2.6×10^5 and 5.2×10^5, respectively. α-Chymotrypsin also exists, at low concentrations, as a monomer, but forms a dimer at higher concentrations (39).

Many enzymes, especially regulatory enzymes, consist of a number of protomers. In some cases, these enzymes dissociate spontaneously on dilution, or on the removal of their prosthetic groups. Other enzymes require more drastic treatment (e.g., with concentrated urea, dilute acid, or detergents) to separate the protomers. An example of an enzyme that behaves like this is aldolase. This enzyme, isolated from skeletal muscle, has a molecular weight of 1.5×10^5 and contains four protomers. Acidification of its solution causes it to dissociate to inactive protomers. Upon neutralization of the solution, the protomers reassociate to the active enzyme. Another example is glutamate dehydrogenase of liver, which dissociates to protomers when diluted (17). L-Asparaginase of *Escherichia coli* dissociates to a dimeric form from its original tetrameric form when treated with sodium dodecyl sulfate, a detergent. Addition of salts like dipotassium hydrogen phosphate or sodium citrate reverses the action of the detergent and restores the tetrameric form and the full activity of the enzyme (2). An interesting enzyme that dissociates to two subunits is phosphorylase. The dissociation of the active tetramer to the slightly active dimer is due to the action of another enzyme, phosphorylase phosphatase. This enzyme dephosphorylates the phosphorylase. Rephosphorylation and the restoration of the tetrameric form are brought about by the action of the enzyme phosphorylase kinase.

Some enzymes are present in large complexes. For example, the biosynthesis of fatty acids in the cytoplasm is catalyzed by the fatty acid synthetase complex (32). This complex contains seven enzymes and has a particle weight of 2.3×10^6. Individual enzymes of the complex, isolated from yeast, are inactive, but are active if isolated from *E. coli*. Other examples are the pyruvate dehydrogenase complex, which catalyzes the formation of AcCoA from pyruvic acid, and the α-ketoglutarate dehydrogenase complex, which catalyzes the formation of succinylcoenzyme-A from α-ketoglutaric acid. The latter complex has a particle weight of 2.1×10^6.

Some enzymes have different molecular weights if isolated from different sources. For example, aldolase isolated from skeletal muscle has a molecular weight of 1.5×10^5, but that isolated from bacteria has a molecular weight of 6.5×10^4. Similarly, alcohol dehydrogenase of yeast has a molecular weight of 1.5×10^5 and that of liver has a molecular weight of 7.3×10^4.

For further details of molecular-weight determination of enzymes, the reader should consult the excellent articles (33, 47, 49) on this subject.

Primary Structure of Enzymes

The primary structure of enzymes includes the covalently bonded amino acid sequences and the cofactors. It is obvious from the preceding section that protomers have to be treated individually. Therefore, the primary structure of the amino acid sequence of a single protein molecule refers to that of the apoenzyme.* The nature of the associated cofactors of certain enzymes will be discussed under a separate heading.

Amino acids are bound to each other by peptide ($=$N—CO—) linkages. The sequence of amino acids is very specific and important from the point of view of biologic activity. Aggregation, adsorption to membranes, conformational stability, solubility, affinity to allosteric sites, and ionic interactions are specifically determined by the amino acid sequence. These factors, which have a marked influence on the catalytic activities of the enzymes, are discussed in greater detail in subsequent paragraphs.

Determinations of the amino acid sequence of the protein chains of enzymes usually starts with the oxidative cleavage of the disulfide bonds. Half-cysteine residues are obtained as cysteic acids after hydrolytic or proteolytic cleavage.

The next step is cleavage by trypsin. This enzyme attacks the carboxy endings of arginine and lysine residues exclusively, yielding a number of

---

*An apoenzyme is the enzyme without its cofactor.

small peptides that can be separated chromatographically or otherwise. The number of these peptides can be calculated from the number of arginine and lysine molecules obtained after total hydrolysis with 6 N hydrochloric acid in vacuum-sealed vials at 105°C for 17 hr. The separated smaller peptides can be further analyzed for their amino acid sequence in various ways. For example, they can be treated with dinitrofluorobenzene, which attacks free amino groups and pinpoints the amino terminal residues in the tryptic peptides after complete acid hydrolysis.

Partial acid hydrolysis of these tryptic peptides, followed by isolation of their products and determination of their amino acid composition, can establish the sequence of amino acids within these peptides. The basis of the sequence analysis is the matching of amino acids in the overlapping peptide chains.

Tryptic peptides and, theoretically, even larger proteins can be analyzed for their amino acid sequence by stepwise chemical analysis. The most useful sequence analysis is based on the reaction with phenylisothiocyanate (16). This method reveals the amino acid sequence starting from the amino terminus. The sequence of amino acids in peptides can also be determined starting from the carboxy terminus, by the use of the enzyme carboxypeptidase. The time-dependent liberation of free amino acids from peptides can be followed by chromatography, which permits a deduction of the amino acid sequence.

Formation of other peptides from the original protein by cleavage at specific sites, for example by chymotrypsin and pepsin, is also possible. The former enzyme hydrolyzes peptide bonds in which the carbonyl function is contributed by phenylalanine, tyrosine, or tryptrophan. The latter enzyme cleaves, in addition to these points, at bonds where leucine, aspartic, and glutamic acids are involved. The small peptides obtained may be sequenced by any of the methods cited above. Overlapping amino acid sequences in these peptides and in the tryptic peptides may reveal the entire amino acid sequence of the original protein. Details of the analytical techniques involved in these procedures are described in various publications (3, 25, 41, 51).

The amino acid sequence of a relatively small enzyme ribonuclease (molecular weight 1.27×10^4) is illustrated in Figure 1-1.

Besides the determination of the primary structure of the entire enzyme, identification of the amino acids present at the active site is also important for an understanding of the catalyzed reaction. This identification can be accomplished by labeling the active site with the substrate or with a related compound. Such labeling can be achieved easily with enzymes whose catalysis involves covalent enzyme–substrate intermediates, such as trypsin (5), aldolases (20), and phosphoglyceric acid mutase (46). The enzyme–substrate complexes of these enzymes are sufficiently stable to permit their isolation. Chemical or enzymatic degradation of such an enzyme–substrate complex yields one or two amino acids of the active site covalently bound to the

Lys·Glu·Thr·Ala·Ala·Ala·Lys·Phe·Glu·Arg· 10
Gln·His·Met·Asp·Ser·Ser·Thr·Ser·Ala·Ala· 20
Ser·Ser·Ser·Asn·Tyr·Cys·Asn·Gln·Met·Met· 30
Lys·Ser·Arg·Asn·Leu·Thr·Lys·Asp·Arg·Cys· 40
Lys·Pro·Val·Asn·Thr·Phe·Val·His·Glu·Ser· 50
Leu·Ala·Asp·Val·Gln·Ala·Val·Cys·Ser·Gln· 60
Lys·Asn·Val·Ala·Cys·Lys·Asn·Gly·Gln·Thr· 70
Asn·Cys·Tyr·Gln·Ser·Tyr·Ser·Thr·Met·Ser· 80
Ile·Thr·Asp·Cys·Arg·Glu·Thr·Gly·Ser·Ser· 90
Lys·Tyr·Pro·Asn·Cys·Ala·Tyr·Lys·Thr·Thr· 100
Gln·Ala·Asn·Lys·His·Ile·Ile·Val·Ala·Cys· 110
Glu·Gly·Asn·Pro·Tyr·Val·Pro·Val·His·Phe· 120
Asp·Ala·Ser·Val 124

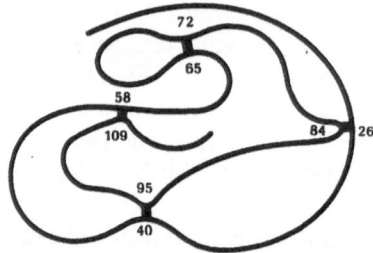

Figure 1-1. Primary structure of bovine ribonuclease.

substrate. If the enzyme–substrate complex is not stable enough, it may be stabilized during its formation. For example, aldolase (38) forms a Schiff base* composed of dihydroxyacetonephosphate and the ε-amino group of a lysine residue. After reduction of the Schiff base, a stable product is formed, and can be isolated (6).

A more selective method is affinity labeling (60). With this technique, substrate-like molecules are covalently bound to the active site via a reactive group. This type of reaction, which is the basis of so-called irreversible enzyme inhibition, pinpoints one of the amino acids involved in the catalysis. The other residues, however, which are not immediate neighbors, have to be determined by other techniques.

Indirect methods for determining the amino acids participating in the catalysis involve measurement of the influence of such factors as pH and temperature on kinetic parameters, e.g., maximum velocity. From these measurements, one may deduce what types of amino acids participate in the catalytic reaction. One of the most elegant methods of determining the active site of enzymes is the measurement of wide-angle X-ray diffraction patterns of enzyme–substrate or enzyme–inhibitor complexes.

---

*Schiff bases are compounds in which a primary amine reacts with an aldehyde or a ketone, with the elimination of a molecule of water.

The amino acid sequences surrounding the central serine molecule of the active site of several enzymes are presented below.

Trypsin: -glycine-glycine-asparagine-serine-glycine-proline-valine
Chymotrypsin: -glycine-asparagine-serine-glycine
Cholinesterase: -glycine-glutamine-serine-alanine-glycine
Thrombin: -asparagine-serine-glycine

Further details on these topics are available in the literature (30, 53).

Minor changes in the primary structure of proteins may induce significant alteration of their biologic activities. For example, enzyme protomers coded for the same enzymatic activity, but varying in their amino acid composition, may associate in different combinations to form natural enzymes containing several protomers. The natural enzymes so formed catalyze identical reactions, but differ in catalytic and physical properties. These enzymes are called isoenzymes, and they provide living organisms with an ability to catalyze the same reaction under different physiologic conditions (55).

Alterations in enzyme structure may have serious pathologic implications. Change of a single amino acid in the β-chain of hemoglobin alters its oxygen uptake ability and causes sickle-cell anemia (58).

Secondary Structure of Enzymes

X-ray crystallographic studies of some proteins revealed the presence of a periodically recurring unit of 5.4 Å size, along their long axes. Further investigation of the arrangement of the amino acid sequences in proteins revealed the possible existence of four major secondary structures. It was found that these secondary structures develop because the carbonyl oxygen and the amide hydrogens of the peptide bond have the ability to form hydrogen bonds. The hydrogen bonds are formed between the groups within the protein molecule, rather than between these groups and the solvent. Structures resulting from such interactions constitute the secondary structure arrangements of proteins.

Hydrogen bonding can occur between the hydrogen atoms attached to the nitrogen of amino acids and the oxygen of the carbonyl group of the fourth amino acid of the protein chain. Since the peptide bond is planar, the polypeptide chain has only two degrees of freedom. These restrictions result in a structural arrangement that is called an α-helix. Peptide chains may be expected to assume the α-helical configuration spontaneously for two reasons: it is the most stable of all possible conformations and has the least free energy, provided that the side chains of the amino acids have no opposing interactions. The α-helical arrangement of peptide chains is shown in Figure 1-2.

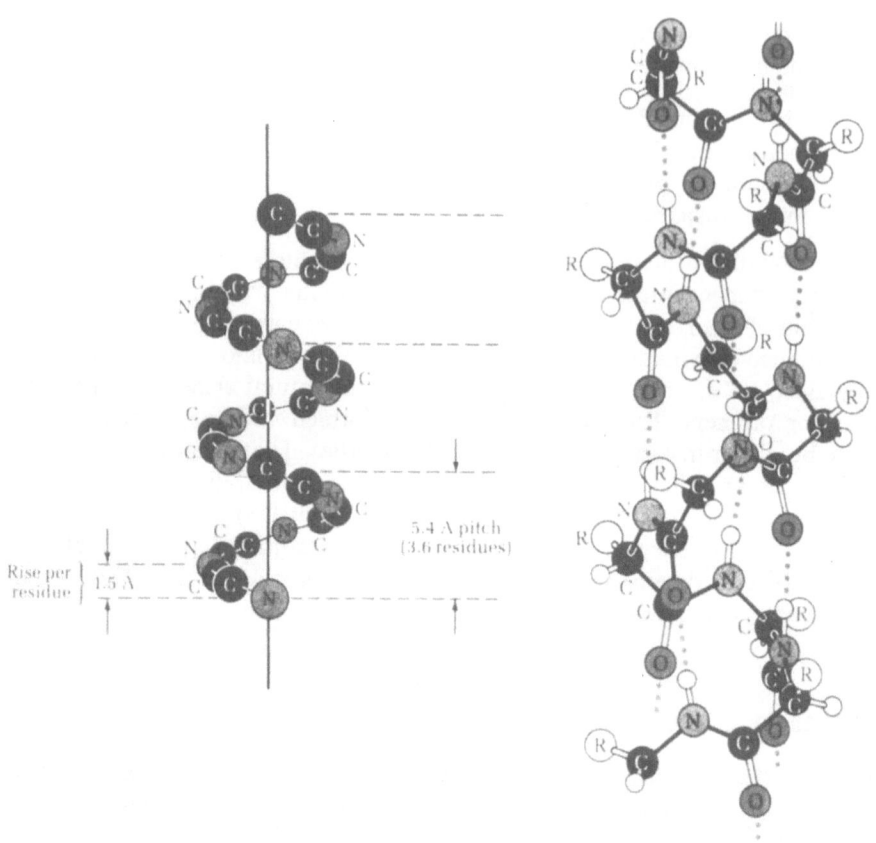

Figure 1-2. Models of α-helical arrangement of proteins.

The α-helical arrangement confers on a protein rigidity, a relatively sharp melting point, and characteristic optical properties. The optical rotation of α-helical proteins is not merely the sum of the rotation of the separate amino acids. In addition to these rotations, the α-helical configuration introduces another dextrorotatory component. If the α-helical configuration is broken without the destruction of any peptide bonds, the rotation will be a simple additive function of the residues. Therefore, the optical rotation can help to determine the helical content of a protein. Optical behavior of structured proteins containing α-helices can be studied by optical rotatory dispersion (ORD). With this technique, optical rotation is measured as a function of wavelength (61).

Another method is circular dichroism (CD)(8), which measures the different molar extinction coefficients of left and right circularly polarized beams. Both optical rotation and ellipticity depend on the environment of the absorbing chromophore. ORD and CD depend, therefore, on the secondary structure of the protein molecule. By these methods, it is also possible to distinguish among the four possible secondary structures of proteins.

The second possible conformation of a protein molecule is the pleated sheet. In this arrangement, protein molecules form interchain hydrogen bonding running parallel or antiparallel to each other. In the former, all chains have their nitrogen terminals at the same end and, in the latter, every other chain points in the opposite direction. Transition from the α-helix to the pleated sheet is possible. For example, if α-keratin is steam heated, the intrachain hydrogen bonds of the α-helix break and, after stretching and cooling of the protein, a pleated-sheet conformation develops.

Some proteins cannot form an α-helix or pleated sheet because of side-chain interactions due to steric hindrance or to repelling electric charges. These proteins exist as random coils, the third possible secondary structure of a protein. Transition to random-coil conformation from α-helical or pleated-sheet conformation can be caused by pH change, heating of the solution, or the formation of hydrogen bonds between the protein and strongly interacting solvents. Thermodynamic aspects of these transitions were described by Giacometti (19).

The fourth type of secondary structure is encountered in collagens. One-third of this protein consists of proline and hydroxyproline, amino acids that have no hydrogen attached to the nitrogen bonds necessary for the α-helical or pleated-sheet configurations. Instead, collagen forms a superhelix made up of three protein chains twisted around one another. This conformation is stabilized by hydrogen bondings of amino acids other than proline or hydroxyproline.

Enzymes are specifically arranged protein molecules. In their arrangement, the α-helix and the pleated-sheet configurations may play some role, subordinated to the overall influence of tertiary structure, to be discussed later. The relative proportions of different secondary structural arrangements in an enzyme can be estimated by ORD and CD. More precise information can be obtained, however, by crystallography (44). For example, it was demonstrated by this method that about 25% of the enzyme lysozyme is in the α-helical form (9). It was also shown that lysozyme is compactly arranged, with little space for water within its structure. Chymotrypsinogen also contains residues in α-helical conformation. In contrast to these, cytochrome c contains only pleated sheets and no α-helices. The secondary structure of enzymes is associated with a certain degree of rigidity that contributes to the well-defined, specific tertiary arrangement of enzymes.

Tertiary Structure of Enzymes

In general, enzymatic activity depends on a specific, three-dimensional conformation of the tertiary structure of its molecule. It is known that loss of this three-dimensional conformation, for example by denaturation, causes loss of activity. As already discussed, the primary structure of an enzyme determines, to a great extent, its secondary and tertiary structures. This statement, however, needs some clarification. Some enzymes do not lose their activity if some of the amino acids are removed. Enzymatic catalysis depends on the fact that certain amino acids, sometimes far removed from each other, acquire unusual positions in the three-dimensional arrangement of the molecule. This phenomenon may occur without continuity in the primary structure. Ribonucleases, which contain 124 amino acids, may be cleaved between the 20th and 21st amino acids by the enzyme subtilisin (35). If the two resulting peptides are separated, neither has enzymatic activity under any condition. If, however, the two peptides are mixed at pH 7.0, enzymatic activity is restored. This finding indicates that physical proximity of certain groups is necessary for catalytic activity, and that covalent binding is not an absolute prerequisite to such activity. Furthermore, it is evident that enzyme proteins are synthesized on the ribosome with an amino acid sequence that, aided by a well-defined tertiary structure, guarantees the proximity of the amino acids necessary for catalysis. This tertiary structure is "fixed" by different binding forces, such as hydrogen bonding, apolar interactions, and ionic forces (50).

More detailed investigations have also shown that such "fixed" structures may be flexible and that the location of certain amino acids is not the same in the presence or absence of the substrate. These findings are contrary to Emil Fischer's lock-and-key theory. According to his hypothesis, the active site has a rigid three-dimensional arrangement that, due to steric hindrance, repulses molecules structurally different from the normal substrate. A more recent theory (29) explains the catalytic activity by an "induced-fix" mechanism. According to this latter hypothesis, the substrate induces alterations in the protein structure that orient the groups involved toward configurations favorable for enzymatic activity. The changed conformation is unfavorable thermodynamically, but favorable with respect to catalytic activity. A poor substrate may interact with the enzyme, but its binding forces are too small to change the enzyme into a catalytically active form. A good substrate can induce these changes. The energetically unfavorable conformation may facilitate the removal of the reaction products from the active site, which then returns to its initial conformation.

Recent X-ray diffraction studies on pure enzymes and enzyme–substrate complexes have shown that the conformation may be different in these two states. These studies were made with carboxypeptidase (35), lysozyme (45), and α-chymotrypsin (10). The three-dimensional model of α-chymotrypsin, based on X-ray analysis, is shown in Figure 1-3.

Similar conclusions were drawn about metalloenzymes (54). In these studies, the spectra of metalloenzymes reflected the thermodynamic conditions in the vicinity of the metal, and these spectra did not differ in the presence and absence of substrates.

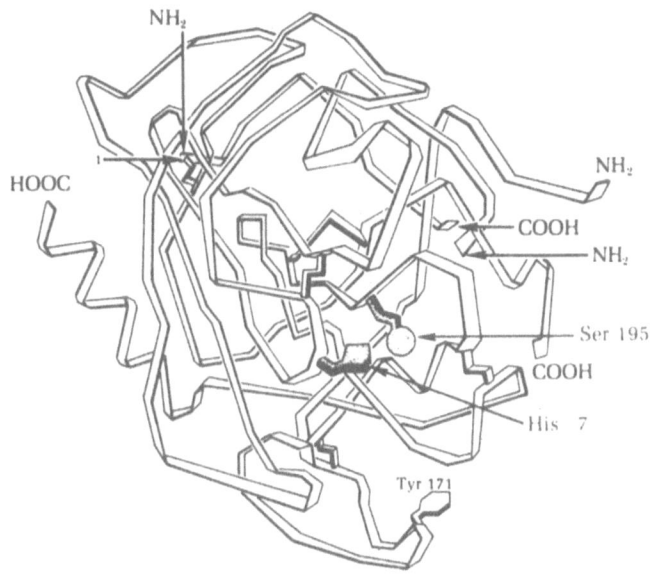

Figure 1-3.   Three-dimensional model of α-chymotrypsin.

The specific interaction forces (48) in the enzyme protein that are responsible for its conformation can be broken up by placing the enzyme into certain environments. The transition from the active conformation to a "disordered" random conformation is referred to as denaturation. Denaturation changes the orientation and location of the amino acid residues and thereby disrupts the steric requirements for catalytic activity. Denaturation of enzymes can be achieved in several ways. Strong hydrogen bond-forming solvents, such as concentrated guanidine hydrochloride or urea, disrupt hydrogen bonding and break up hydrophobic interactions. Similar effects can be achieved by raising the temperature of the enzyme solution. Detergents with hydrophobic side chains can combine with hydrophobic side chains of the enzyme and pull the enzyme into solution by the polar head of the detergent molecule. Also, environments with very low ionic concentrations may cause similar polar orientation on the surface of enzymes, leading to internal repulsion of the group involved and the development of a more extended random enzyme structure. Changes in pH of the solution, by altering dissociation of ionic groups, may cause structural variations similar to those caused by low ionic concentration of the environment. Since different conformations have different optical activities (24), the changes induced by denaturation can be determined by measurements of optical rotation. Ultraviolet spectroscopy can also be used for the measurement of these changes. The chromophore groups of trypsin and tryptophan within the protein molecule interact with the polar groups of their internal environment and are inaccessible for interaction with solvents. After denaturation, these groups can interact with the solvent and a spectral change occurs in the 270 to 290-nm region (21, 57). Recently, electron paramagnetic resonance measurements (EPR) were used for the study of conformational changes. Proteins labeled with paramagnetic radicals cause dramatic changes in the paramagnetic electron resonance spectrum of the label during the conformational change. The changes of the label in the EPR are apparently due to the addition to or the removal of steric restrictions on the free rotational tumbling of the paramagnetic label within the macro-molecule.

It is interesting that substrates often increase the stability of enzymes. For example, ribonuclease, trypsin, and chymotrypsin are stabilized against de-naturation in the presence of their substrates. Certain coenzymes, like nico-tine adenine dinucleotide (NAD) and pyridoxal phosphate, can, similarly, exert stabilizing effects. In other cases, mostly with metabolic regulatory enzymes, binding of the molecule loosens the protein structure and, by inducing conformational changes, facilitates denaturation (7).

Changes in the tertiary structure can alter the electrophoretic behavior of enzymes. Exposure of differently charged amino acids is believed to be the reason for this phenomenon. Enzyme variants with different electrophoretic characteristics are called conformers (27) or subforms (4). Methods for the investigation of these conformational changes were reviewed by Nazaki and Tanford (40) and by Timasheff (52).

Quaternary Structure of Enzymes

Enzymes whose kinetics can be altered by effector molecules are called regulatory enzymes, and usually contain more than one discrete polypeptide chain. These chains are not connected by covalent bonds. In addition, each of these chains is characterized by its own primary, secondary, and tertiary structures. The characteristic manner in which the individual polypeptide chains, the protomers, are held together in the native oligomeric enzyme is called the quaternary structure. These oligomeric enzymes have great stability due to a very tight fit, which is guaranteed by the combined action of many weak binding forces operative in each contact area between the protomers. These forces include hydrophilic binding, hydrogen binding, dipole–dipole interactions, and ionic bindings. These binding forces can be broken up with agents such as urea or guanidine hydrochloride. Detergents (e.g., sodium dodecylsulfate) that interact with hydrophobic binding sites can also be used to dissociate enzymes to protomers or to active subunits. Some enzymes dissociate when treated with dilute acid or concentrated salt solutions, or when frozen. If the original solution conditions are reestablished, the proto-mers reassociate to their oligomeric form. These findings indicate that the primary structure of an enzyme codes not only its tertiary form, but also the structure of its protomer-binding contact sites.

The number of protomers in each enzyme can be determined by molecular-weight measurements of the native enzyme and its protomers. From the number of terminal amino acids per active enzyme and from their peptide maps, it may be determined whether the enzyme has identical or different types of protomers.

The protomer and subunit structure of enzymes is of interest because it appears to be the cause of the regulating effect of small molecules on enzyme activity. These small molecules, by inducing changes in the tertiary structure, alter the conformation of the active site and those of the contact areas of the protomers. These changes are referred to as allosteric or regulatory effects.

The changes in the quaternary structure of many enzymes have been investigated. Thus, for example, it was found that the tetrameric rabbit muscle glyceraldehyde-3-phosphate dehydrogenases is dissociated by its effector adenosine triphosphate (ATP) or by KCl into dimers or monomers (11, 12). Dissociation is partial at 12°C, but complete to monomers at 0°C. If the ATP or KCl is removed and the solution is warmed, the enzyme regains its activity and its tetrameric structure.

L-Arabinose isomerase has a molecular weight of 362,000 daltons (43). It was shown by ultracentrifugation and peptide mapping that protonation at pH 2.0 or treatment with 8 M urea resulted in dissociation of the enzyme into six identical protomers.

Yeast enolase, which has a molecular weight of 57,000, dissociates to two protomers when treated with KCl (18). Measurement of the dissociation rates has revealed a large change in the heat capacity of the activation process, indicating the exposure of hydrophilic groups, like valyl residues, to solvent in the monomeric state.

As already mentioned, changes in the quaternary structure are related to alterations in tertiary structure, especially at the points of attachment of the protomers. Recently, the relation of the tertiary and quaternary structural changes attracted the attention of many investigators. Two major theories were advanced regarding the relationship of the tertiary and quaternary structural changes. The first, by Monod et al. (36, 37), assumed that two states differing in the distribution of energy of interprotomer bonds are reversibly accessible to the oligomers, that these states differ in their affinity for ligands, and that the transition between them occurs with conservation of molecular symmetry. The binding of individual ligands is considered independent of the binding of others. The transition between the two states is characterized by an equilibrium constant, L. When L is large, the allosteric effect exerted by ligand binding is large. The characteristic kinetic curve is sigmoidal and is suggestive of cooperative substrate binding. This curve fits the experimental oxygen-uptake curve of hemoglobin and is in agreement with the kinetics of several allosteric enzymes. Cooperativity means that the interprotomer binding energy accompanying the attachment of a ligand is not uniform, but depends on the extent to which the protein is bound to the ligand. It was suggested (42) that change in the energy of protomer binding may be the parameter of cooperativity. This parameter can be estimated from kinetic data (48).

An alternate theory of allosteric effects implies (13, 14, 31) that there are many transitions possible between the tense conformation brought about by interprotomeric bonds and the relaxed conformation of the isolated protomers. This theory is based on the assumption that the energy involved in binding of protomers can be broken up into three components: (a) standard free-energy change of conformational transformation apart from interactions of protomers; (b) standard free-energy change due to interaction of protomers in one conformation; and (c) a component similar to (b) in the other conformation. At its limits, this theory is identical with the symmetry theory of Monod et al. (36, 37). In between, however, protomers can behave noncooperatively, that is, the binding of one ligand does not change the conformation of all protomers.

Several other theories of conformational changes are summarized by both Whitehead (59) and Pohl (47). The validity of the two principal models has been tested by physicochemical methods. Investigations (22) with a spin-labeling technic indicated that the allosteric interaction model of Koshland et al. (31) is more realistic than the Monod model. No suitable model, however, has been found for the investigation of allosterism. Because of the great interest in the evolutionary adjustments of species mitigated by the oligomeric regulatory enzymes, the problems related to allosterism are being investigated extensively.

References

1. Ackers, G. K., Analytical gel chromatography of proteins. *Adv. Protein Chem.* **24**, 343 (1970).
2. Aszalos, A., Kirschbaum, J., Ratych, O. T., Kraemer, N., Kocy, O., Frost, D., and Casey, J. P., Reversible dissociation of *L* asparaginase of Escherichia coli B. *J. Pharm. Sci.* **61**, 791 (1972).
3. Bailey, J. L., "Techniques in Protein Chemistry," 2nd ed. Am. Elsevier, New York, 1967.
4. Banks, B. E. C., Doonan, S., Lawrence, A. J., and Vernon, C. A., The molecular weight and other properties of aspartate amino transferase from pig heart muscle. *Eur. J. Biochem.* **5**; 528 (1968).
5. Bender, M. L., and Kaiser, E. T., The mechanism of trypsin-catalyzed hydrolyses. The cinnamoyl-trypsin intermediate. *J. Am. Chem. Soc.* **84**; 2556 (1962).
6. Bernhard, S. A., "The Structure and Function of Enzymes," p. 170. Benjamin, New York, 1968.
7. Bernhard, S. A., and Rossi, G. L., On the substrate-induced stabilization of native enzyme protein conformation. *In* , "Structural Chemistry and Molecular Biology" (A. Rich and N. Davidson, eds.), p. 98. Freeman, San Francisco, California, 1968.
8. Beyehok, S., Circular dichroism of poly-α-amino acids and proteins. *In* "Poly-α-Aminoacids" (G. D. Fasman, ed.), p. 293. Dekker, New York, 1967.
9. Blake, C. C. F., Johnson, L. N., Mair, G. A., North, A. C. T., Phillips, D. C., and Sasma, V. R., Crystallographic studies of the activity of hen egg-white lysozyme. *Proc. R. Soc. London, Ser. B* **167**; 378 (1967).

10. Blow, D. M., Birktoft, J. J., and Hartley, B. S., Role of a buried acid group in the mechanism of action of chymotrypsin. *Nature (London)* **221**, 337 (1969).
11. Constantinides, S. M., and Deal, W. C., Jr., Reversible dissociation of tetrameric rabbit muscle glyceraldehyde 3-phosphate dehydrogenase into dimers or monomers by adenosine triphosphate. *J. Biol. Chem.* **244**, 5695 (1969).
12. Constantinides, S. M., and Deal, W. C., Jr., Reversible dissociation of tetrameric 7.4 S rabbit muscle glyceraldehyde 3-phosphate dehydrogenase into 4.4 S dimers by ammonium sulfate and into 3.2 S monomers by KCl. *J. Biol. Chem.* **245**, 246 (1970).
13. Cornish-Bowden, A. J., and Koshland, D. E., Jr., The influence of binding domains on the nature of subunit interactions in oligomeric proteins. Application to unusual kinetic and binding patterns. *J. Biol. Chem.* **245**, 6241 (1970).
14 Cornish-Bowden, A. J., and Koshland, D. E., Jr., The quaternary structure of proteins composed of identical subunits. *J. Biol. Chem.* **246**; 3092 (1971).
15. Creeth, J. M., and Pain, R. H, The determination of molecular weights of biological macromolecules by ultracentrifuge methods. *Prog. Biophys. Mol. Biol.* **17**, 217 (1967).
16. Edman, P., and Begg, G.; A protein sequenator. *Eur. J. Biochem.* **1**, 80 (1967).
17. Frieden, C., Glutamic dehydrogenase. I. The effect of co-enzyme on the sedimentation velocity and kinetic behavior. *J. Biol. Chem.* **234**, 809 (1959).
18. Gawronski, T. H., and Westhead, E. W., Equilibrium and kinetic studies on the reversible dissociation of yeast enolase by neutral salts. *Biochemistry* **8**, 4261 (1969).
19. Giacometti, G., Recent experimental approaches to the thermodynamics of coupsmational transitions in polypeptides. *In* "Structural Chemistry and Molecular Biology" (A. Rich and N. Davidson, eds.), p. 67. Freeman, San Francisco, California 1968.
20. Grazi, E., Meloche, H., Martinez, G., Wood, W. A., and Horecker, B. L., Evidence for Schiff base formation in enzymatic aldol condensations. *Biochem. Biophys. Res. Commun.* **10**, 4 (1963).
21. Greatzer, W. B., Ultraviolet absorption spectra of polypeptides. *In* "Poly α-Aminoacids" (G. D. Fasman, ed.), p. 177. Dekker, New York, 1967.
22. Hamilton, C. L., and McConnell, H. M., Spin labels. *In* "Structural Chemistry and Molecular Biology" (A. Rich and N. Davidson, eds.), p. 115. Freeman, San Francisco, California, 1968.
23. Haurowitz, F., ed., "The Chemistry and Function of Proteins," p. 65. Academic Press, New York, 1963.
24. Haurowitz, F., ed., "The Chemistry and Function of Proteins," p. 143. Academic Press, New York, 1963.
25. Hirs, C. H. W., and Timasheff, S. N., eds., "Methods in Enzymology," Vol. 25, Academic Press, New York, 1972.
26. Ho, P. P. K., and Milikin, E. B., Multiple forms of L-asparaginase. *Biochim. Biophys. Acta* **206**, 196 (1970).
27. Kaplan, N. O., Nature of multiple molecular forms of enzymes. *Ann. N. Y. Acad. Sci.* **151**, 382 (1968).
28. Kirschbaum, J., Wriston, J. C., and Ratych, O. T., Subunit structure of L-asparaginase from Escherichia coli B. *Biochim. Biophys. Acta* **194**, 161 (1967).
29. Koshland, D. E., Jr., Mechanism of transfer enzymes: *In* "The Enzymes" (P. D. Boyer, H. Lardy and K. Myrbäck, eds.), 2nd ed., Vol. 1, p. 305. Academic Press, New York, 1959.
30. Koshland, D. E., Jr., The active site and enzyme action. *Adv. Enzymol.* **22**, 45 (1960).
31. Koshland, D. E., Jr., Neméthy, G., and Filmer, D., Comparison of experimental binding data and theoretical models in proteins containing subunits. *Biochemistry* **5**, 365 (1966).
32. Kumar, S., Dorsey, J. K., and Porter, J. W., Mechanism of dissociation of pigeon liver fatty acid synthetase complex into half-molecular weight subunits and their reassociation to enzymatically active complex. *Biochem. Biophys. Res. Commun.* **40**, 825 (1970).
33. Leach, S. L., ed., "Physical Principles and Techniques of Protein Chemistry," Part B. Academic Press, New York, 1970.

34. LeJohn, H. B., McCrea, B. E., Suzuki, I., and Jackson, S., Association-dissociation reactions of mitochondrial isocitric dehydrogenase induced by protons and various ligands. *J. Biol. Chem.* **244**, 2484 (1969).

35. Lipscomb, W. N., Hartsuck, J. A., Reeke, G. N., Jr., Guicho, F. A., Bethe, P. H., Ludwig, M. C., Steitz, T. A., Muirhead, H., and Coppola, J. C., The structure of carboxypeptidase.A. VII. *Brookhaven Symp. Biol.* **21**, 24 (1968).

36. Monod, J., Changeux, J. P., and Jacob, F., Allosteric proteins and cellular control systems. *J. Mol. Biol.* **6**, 306 (1963).

37. Monod, J., Wyman, J., and Changeux, J. P., On the nature of allosteric transitions: A plausible model. *J. Mol. Biol.* **12**, 88 (1965).

38. Morse, D. E., and Horecker, B. L., The mechanism of action of aldolases. *Adv. Enzymol.* **31**, 125 (1968).

39. Narasinga Rao, M. S., and Kegeles, G., An ultracentrifuge study of the polymerization of α-chymotrypsin. *J. Am. Chem. Soc.* **80**, 5724 (1958).

40. Nazaki, Y., and Tanford, C., Investigation of conformational changes, examination of titration behavior. *Methods Enzymol.* **11**, 715 (1967).

41. Needleman, S. B., ed., "Protein Sequence Determination." Springer-Verlag, New York, 1970.

42. Noble, R. W., Relation between allosteric effects and changes in the energy of bonding between molecular subunits. *J. Mol. Biol.* **39**, 479 (1969).

43. Patrick, J. W., and Lee, N., Subunit structure of L-arabinose isomerase from Escherichia coli. *J. Biol. Chem.* **244**, 4277 (1969).

44. Perutz, M. F., The first Sir Hans Krebs lecture. X-ray analysis, structure and function of enzymes. *Eur. J. Biochem.* **8**, 445 (1969).

45. Phillips, D. C., The three-dimensional structure of an enzyme molecule. *Sci. Am.* **215**, 78 (1966).

46. Pizer, L. I., Studies of the phosphoglyceric acid mutase reaction with radioactive substrates. *J. Am. Chem. Soc.* **80**, 4431 (1958).

47. Pohl, F. M., Cooperative conformational changes in globular proteins. *Angew. Chem.* **84**, 931 (1972).

48. Reich, J. G., Wangerman, G., Falch, N., and Rhode, K., A general strategy for parameter estimation from isosteric and allosteric-kinetic data and binding measurements. *Eur. J. Biochem.* **26**, 368 (1972).

49. Schachman, H. K., "Ultracentrifugation in Biochemistry." Academic Press, New York, 1959.

50. Steiner, R. F., "The Chemical Foundation of Molecular Biology: Forces Involved in the Stabilization of Protein Structure," p. 133. Van Nostrand-Reinhold, Princeton, New Jersey, 1965.

51. Storke, G. R., Recent developments in chemical modification and segmental degradation of proteins. *Adv. Protein Chem.* **24**, 2906 (1969).

52. Timasheff, S. N., Some physical probes of enzyme structure in solution. *In* "The Enzymes" (P. D. Boyer, ed.), 3rd ed., Vol. 2, p. 371. Academic Press, New York, 1970.

53. Vallee, B. L., and Riordan, J. F., Chemical approaches to the properties of active sites of enzymes. *Annu. Rev. Biochem.* **38**, 733 (1969).

54. Vallee, B. L., and Williams, R. J., Metalloenzymes: The entatic nature of their active sites. *Proc. Natl. Acad. Sci. U.S.A.* **59**, 498 (1968).

55. Vessel, E. S., ed., "Multiple Molecular Forms of Enzymes," Ann. N.Y. Acad. Sci. No. 157. N.Y. Acad. Sci., New York, 1968.

56. Vinograd, J., and Brunes, P., Band centrifugation of macromolecules in self-generating density gradients. II. *Biopolymers* **4**, 131 (1966).

57. Wetlaufer, D. B., Ultraviolet spectra of proteins and amino acids. *Adv. Protein Chem.* **17**, 304 (1962).

58. White, J. G., and Heagen, B., The fine structure of cell-free sickled hemoglobin. *Am. J. Pathol.* **58**, 1 (1970).

59. Whitehead, E., The regulation of enzyme activity and allosteric transition. *Prog. Biophys. Mol. Biol.* **21**, 321 (1970).
60. Wofsy, L., Metzger, H., and Singer, S. J., Affinity labelling. *Biochemistry* **1**, 1031 (1962).
61. Yang, J. T., Optical rotatory dispersion. *In* "Poly α-Aminoacids" (G. D. Fasman, ed.), p. 239. Dekker, New York, 1967.

Chapter 2

Isolation of Enzymes

A. A. Aszalos

In living organisms, enzymes are components of mixtures that contain numerous organic and inorganic substances. These substances may interact with the enzymes, coenzymes, substances, or products of the enzymatic reaction. Study of the properties of an enzyme requires that it be purified. At present, several hundred enzymes are available in purified form. It should be remembered that the characteristics of purified enzymes may not be the same as those they possess in their natural environment. Such variable factors as the stabilizing effect of mitochondria, diffusion rates, and inhibitory feedback mechanisms may alter the behavior of enzymes *in vivo*. Competition of other substances (e. g., nonspecific proteins) for substrates and inhibitors may also change enzyme activity *in vivo*. Nevertheless, the investigation of enzyme activity *in vivo* and the possibility of increasing or inhibiting this activity for experimental or therapeutic purposes is greatly facilitated by studies *in vitro* of the properties of purified enzymes.

Investigation of the properties of purified enzymes requires the availability of: (a) a test system; (b) extraction methods; and (c) fractionation methods. These concepts are considered in detail in various texts (15, 22, 49).

Test Systems

The first requirement for the purification of an enzyme is the development of a quantitative test of its activity. The progress of purification is followed by the appropriate test until maximum activity per weight has been reached. It is recommended by the International Commission on Enzymes that enzyme assays should, whenever possible, be based on the measurement of the initial rates of activity; in this way, complications due to reversibility of reactions or to formation of inhibitory products are avoided. The concentration of the substrate should be large enough to saturate the enzyme, so that the reaction rate approximates a zero-order reaction. Under these test conditions, one unit of enzyme is defined as the amount that will catalyze the transformation of 1 μmole of substrate per minute. Use of this absolute unit permits a comparison of different enzymes. Specific activity is expressed as units of enzyme per milligram of protein. When the enzyme has a catalytic center, the concentration of which can be determined, the catalytic power can be expressed in terms of catalytic center activity, that is the number of molecules of substrate acted upon per minute per catalytic center.

Enzyme reactions may be measured by a variety of methods. Many substrates or products absorb in the visible or ultraviolet range, and their formation or disappearance can be followed by spectrophotometric methods (73). Instruments are available that record optical densities continuously at any desired wavelength. These instruments are equipped with thermostatically controlled cuvette holders for maintaining constant temperature during the assay. Enzyme reactions that involve the production or consumption of gases or the production of acids can be followed by manometric methods (19). In case acid is formed, the reaction is carried out in the presence of a salt such as bicarbonate, which, on interaction with acid, releases gas. Oxidases, decarboxylases, hydrogenases, or carbonic anhydrases can be studied by this method. Another method requires measurement of the time necessary for the decoloration (reduction) (determined visually) of a certain amount of dye by the enzyme. This method is used extensively in the study of dehydrogenases. Enzyme reactions that involve pH changes can be followed by continuous titration. The pH is kept approximately constant by the addition of alkali and the reaction velocity is determined from the amount of alkali added per minute, as measured by an automatic apparatus like the Radiometer-pH Stat (37). Polarimetric measurements can also be used to follow enzyme reactions when an optically active substrate is converted to an inactive product. Sometimes it is necessary to follow the formation of products by chromatography (65). Many enzyme reactions can be followed by withdrawing samples from the reactor at intervals and measuring the product by chemical methods.

Extraction

The choice of an extraction method, the first step in the purification of an enzyme, is determined by the source of the enzyme. It is easier to extract enzymes from fluids (e.g., plasma) or from animal or plant cells than from mitochondria or from bacteria.

Enzymes can usually be extracted from animal tissues with water or a salt solution, after disruption of the tissue with a homogenizer or a Waring Blendor. Extraction of enzymes from membranes requires special techniques.

Enzymes that are not bound, but are present in a space that is surrounded by a matrix, are water soluble and can be extracted by the addition to the homogenate of a salt solution or an acidic water solution. Centrifugation at 100,000g leaves the soluble enzymes in the supernatant. Enzymes that are bound to lipids, to membrane metals, such as Ca^{2+} or Mg^{2+}, or to other proteins are more difficult to extract. Sonication (35), prolonged dialysis against alkaline water or buffer (33), and extraction with concentrated urea solution (39), alcohol (47), or acetone (62) are useful techniques for the separation of enzymes in such cases. If the enzyme is bound to cationic groups of membranes, the use of metal-chelating agents, like ethylenediamine tetraacetic acid (EDTA), is recommended. Surfactant reagents have been widely used to extract lipoproteins or water-soluble proteins from membranes. The most frequently used surfactants are salts such as Triton X-100, sodium dodecyl sulfate, and Tween (11). Changes in the net electrical charge of membrane-bound enzymes can result in change of water solubility. Chemical modification, such as succinylation of the ε-amino group of lysine, has been shown to be highly effective in increasing the water solubility of some enzymes (43). Treatment of membranes with enzymes that degrade lipids or proteins may result in release of the enzyme to be purified. For the digestion of lipids and proteins, lipases (66) and proteolytic enzymes like trypsin (36), respectively, are used.

The disruption of bacteria, necessary for the liberation of their enzymes, requires special techniques. In the case of intracellular enzymes, mechanical methods, such as extrusion on a Manton-Gaulin homogenizer, are used. The disruption of micrococci, streptococci, and yeast is more difficult and may require freezing with liquid nitrogen, followed by thawing (42). Other methods of cell disruption utilize glass beads in different mill devices, or the x-press, which extrudes frozen cells. Enzymes attached to the surface of *Escherichia coli* or *Salmonella* may be released by osmotic-shock treatment. In this method, cells are treated for 10 min with 20% sucrose that contains 30 mM Tris at pH 7.2–7.5 and 0.1–1.0 mM EDTA. After centrifugation, the cells are rapidly dispersed in cold water. Enzymes are released by the sudden shift of medium (13) from a high to a low osmotic strength. Some microorganisms can be broken by autolysis. This technique is used in large-scale operations, mostly for the extraction of yeast enzymes (63). Digestion of the cell wall may also be carried out by other enzymes (e.g., lysozymes).

Enzymes extracted by any of the above methods are fractionally separated from any inactive material extracted together with them.

Fractionation

Crude enzyme extracts contain, in addition to the enzyme, numerous other materials of both small and large molecular weight. Molecules much smaller or larger than the enzyme to be isolated may be separated from each other by dialysis, by pH changes, or by heat precipitation. The separation of molecules with similar molecular weights requires chromatography, electrophoresis, or molecular sieving.

A useful fractionation method, especially for enzymes of animal origin, is pH precipitation. It is particularly useful for the separation of colloid-size materials (e.g., nucleoproteins). The enzyme solution is adjusted to a pH at which the enzyme remains in solution and the other substances coagulate and are then removed by centrifugation (40). In another method (7), the solution is heated for 10–15 min to a temperature just below that at which the enzyme would be denatured. The resulting precipitate is separated after the solution has cooled. By varying the salt concentration, the pH, or both, or by adding organic solvents during heat precipitation, conditions optimal for the purification of a particular enzyme can be achieved.

Organic solvents like alcohol (34) can be used to precipitate nonsensitive enzymes from solutions of relatively low salt concentrations at low temperature (3, 45). More sensitive enzymes can be fractionally precipitated by the use of salts. Optimum conditions for this technique are produced by regulating the pH, the temperature, the nature of the salt, and the concentration of the enzyme solution. Because of its high water solubility, ammonium sulfate (14, 25) is used in most cases. The methods for fractionating enzymes by salting out have been discussed in several publications (20, 21, 70).

For some enzymes, specific precipitating agents, like heavy metals (5) and tannic acid (16), have to be used. Water-soluble, high-molecular-weight polymers are useful protein-precipitating agents. Some of the most useful polymers are polyethylene glycol, polyvinylpyrrolidone, and dextran (37a, 54). Fractional adsorption is another method that is used extensively in large-scale procedures. In most cases, the enzyme is absorbed on the adsorbent and is subsequently eluted with slightly alkaline buffer or salt solution. In other cases, impurities are adsorbed and the purified enzyme remains in solution. Adsorbents like calcium phosphate (38), bentonite (41), aluminum hydroxide (31), charcoal, and polysaccharides can also be used for this purpose.

Column chromatography is frequently used for enzyme purification. The separation may depend on adsorption, ion exchange, or molecular-sieve effects. Usually, the enzyme is diluted in the same buffer with which the column has been equilibrated. Elution from the column is accomplished by the use of salt (2, 64) or pH gradients (58), that is, with gradual changes in salt concentration or in pH. The eluted fractions are collected by a fraction collector, and the activities of the individual fractions are measured. Adsorbents like hydroxyapatite (71), cellulose (46), and glass beads (10) are used for column chromatography. For ion-exchange chromatography, resins like IR C-50 (69), Amberlite CG-50 (67), Duolite A-2 (48), CM-cellulose (44), and DEAE-cellulose (6, 28) are most suitable. Gel filtration is also used frequently for enzyme purification. Gels with predetermined pore sizes permit molecules smaller than the pore size to enter the pores and be held there, while molecules larger than pore size are eluted. Molecules smaller than pore size are separated on the column according to their sizes. Gels used most frequently are Sephadex (50, 59), polyacrylamide (9), and agarose (27). Recently preparations of glass bead (32) have also been used for the same purpose. The combined effects of molecular sieving and ion exchange can be achieved with CM-Sephadex (30) or DEAE-Sephadex (51, 57).

Affinity chromatography is a very selective separation method. The principle of this method is that ligands bound covalently to a support can withhold the ligand-specific enzyme selectively on the column. The technique has been reviewed recently by Fernstein (26). Porous glass can be used as the support in this method. Immunoadsorption is a specific type of affinity chromatography. By this technique the antiserum to the enzyme to be purified is attached to activated adsorbent, like of Sepharose 2B and the crude enzyme preparation is passed through. Only the enzyme to be purified is retained on the column and is eluted then by suitable solvent. Carboxypeptidase G_1 was isolated recently this way on a large scale (12a).

After partial purity of an enzyme has been achieved, crystallization can be attempted. Crystallinity, however, is not an absolute proof of purity. Several crystallization steps may be necessary before the specific activity reaches a constant maximum value (17). Crystallization from salt (60) or alcohol (4) solutions may be accomplished by gradually increasing the salt or alcohol concentration, respectively. A crystallization step may take several weeks. Sometimes the metal salt of the enzyme has to be crystallized, because the native enzyme does not crystallize (1, 12). Human lysozyme was recently crystallized by this technique. It is believed that this compound, because of its ability to dissolve bacteria, will play an important role in the chemotherapy of infectious diseases.

Electrophoresis separates proteins according to their electrophoretic mobility. The technique is used primarily for the preparation of small quantities of enzyme. It can also be used for following the progress of the purification. The process can be carried out with the standard Tiselius apparatus or with a paper-electrophoresis apparatus. The apparatus designed by Grassman and Hannig (29) separates proteins continuously by running a buffer solution down a sheet of filter paper and collecting the eluted proteins into a series of tubes. An electric current is passed at right angles to the flow from the left edge of the paper to the right, and the enzyme solution is applied at a point near the top. Various modifications of electrophoretic techniques have been described (8). One of these is electrodecantation (53), a method that depends on horizontal stratification of colloids. Other variants, such as cellulose- (55), agarose- (23), polyacrylamide- (18, 52), and starch-gel (56, 61) electrophoresis, are used for the separation of small quantities of enzymes. These methods usually give very sharp separations.

Isoelectric focusing is another method widely used for small-scale separation of proteins. This method is based on the maintenance of a stable pH gradient between the anode and cathode in an electrolysis cell. This gradient is achieved by the electrolysis of a water solution of a mixture of suitable ampholytes called the "carrier ampholytes." The best ampholytes are poly-amine–polycarboxylic acid mixtures (68, 72). If proteins are added to this system, each protein will migrate to and focus at its isoelectric point. Proteins differing in isoelectric point by as little as 0.01 of a pH unit may be separated by this method. The method can also be used to determine the isoelectric point of an enzyme. Protein mixtures fractionated by this method can be separated from the carrier ampholytes by dialysis or molecular sieving.

Concentration of an enzyme solution may be necessary during or after purification of the enzyme. A commonly used method of concentration is freeze-drying (lyophilization), in which the frozen water is evaporated *in vacuo*. Concentration can also be accomplished by ultrafiltration, in the course of which water and salts pass through a special filter that retains the enzyme.

In another technique used for the isolation of very sensitive enzymes, the enzyme is reversibly deactivated with a ligand (24) or by the dissociation of its protomers during some of the purification steps. After completion of the purification process, the enzyme is reactivated.

Finally, some consideration should be given to the sequence of steps in the fractionation process. Usually, heat precipitation is attempted first and crystallization at the end. If heat precipitation causes a large loss of enzyme activity, salt precipitation, chromatography or, in special cases, affinity chromatography may be used instead. The optimal sequence of steps must be determined by experimentation for each enzyme. The capacity of the available refrigerated centrifuges limits the volumes that can be handled at any one time. Dialysis is a relatively slow process and should be used only in later stages, when the volumes have already been reduced.

References

1. Allan, B. J., Keller, P. J., and Neurath, H., Procedures for the isolation of crystalline bovine pancreatic carboxypeptidase A.I. Isolation from acetone powders of pancreas glands. *Biochemistry* **3**, 40 (1964).
2. Arima, K., Yu, J., Iwasaki, S., and Tamura, G., Milk-clotting enzymes from microorganisms. V. Purification and crystallization of mucor rennin from *Mucor pusillus* var. Lindt. *Appl. Microbiol.* **16**, 1727 (1968).
3. Askonas, B. A., The use of organic solvents at low temperature for the separation of enzymes. Application to aqueous rabbit muscle extract. *Biochem. J.* **48**, 42 (1951).
4. Balls, A. K., Concerning trypsinogen. *Proc. Natl. Acad. Sci. U.S.A.* **53**, 392 (1965).
5. Barker, S. A., Bourne, E. J., and Peat, S., The enzymic synthesis and degradation of starch. Part IV. The purification and storage of the Q-enzyme of the potato. *J. Chem. Soc., Part III* p. 1705 (1949).

6. Baughman, D. J., and Waugh, D. F., Bovine thrombin. *J. Biol. Chem.* **242**, 5252 (1967).

7. Bergeret, B., Chatagner, F., and Fromageot, C., Etude des decarboxylations de l'acide L-cystéinesulfinique de l'acide L-cystéique et de l'acide L-glutamique pardivers organes du lapin. Influence du phosphate depyridoxal et des groupements thiols. *Biochim. Biophys. Acta* **22**, 329 (1956).

8. Bier, M., Preparative electrophoresis. *Methods Enzymol.* **5**, 33 (1962).

9. Boman, H. G., and Hjertén, S., Molecular sieving of bacterial RNA. *Arch. Biochem. Biophys.*. **276**, Suppl. I (1962).

10. Bowie, E. J. W., and Owen, C. A., Jr., Some factors influencing platelet retention in glass bead columns including the influence of plastics. *Am. J. Clin. Pathol.* **56**, 479 (1971).

11. Bromstein, R., Goldberger, R., and Tisdale, H., Studies on the electron transport system. XXXIV. Isolation and properties of mammalian cytochrome. *Biochim. Biophys. Acta* **50**, 527 (1961).

12. Brubacher, L. J., and Bender, M. L., The preparation and properties of trans-cinnamoyl-papain. *J. Am. Chem. Soc.* **88**, 5871 (1966).

12a. Carnell, R., and Charm, S. E., Purification of carboxypeptidase G_1 by immunoadsorption. *Biotechnol. Bioeng.* **18**, 1171 (1976).

13. Cedar, H., and Schwartz, J. H., Localization of the two-L-asparaginase in anaerobically grown *Escherichia coli*. *J. Biol. Chem.* **242**, 3753 (1967).

14. Chow, R. B., and Kassel, B., Bovine pepsinogen and pepsin. I. Isolation, purification, and some properties of the pepsinogen. *J. Biol. Chem.* **243**, 1718 (1968).

15. Colowick, S. P., and Kaplan, N. O., eds., "Methods in Enzymology," Vols. 1-23. Academic Press, New York, 1955-1971.

16. Coulthard, C. A., Michaelis, R., Short, W. F., Sykes, G., Skrimshire, G. E., Standfast, A. F. B., Birkinshaw, J. H., and Raistrick, H., Notatin: An antibacterial glucose-aero-dehydrogenase from *Pencillium notatum* Westling and *Penicillium testiculosum* sp. nov. *Biochem. J.* **39**, 24 (1945).

17. Czok, R., and Bucher, T., Crystallized enzymes from the myogen of rabbit skeletal muscle. *Adv. Protein Chem.* **15**, 315 (1960).

18. Davis, B. J., Disc electrophoresis. II. Method and application to human serum proteins. *Ann. N.Y. Acad. Sci.* **121**, 404 (1964).

19. Dixon, M., "Manometric Methods." Cambridge Univ. Press, London and New York, 1951.

20. Dixon, M., A nomogram for ammonium sulphate solutions. *Biochem. J.* **54**, 457 (1953).

21. Dixon, M., and Webb, E. C., Enzyme fractionation by salting-out: A theoretical note. *Adv. Protein Chem.* **16**, 197 (1961).

22. Dixon, M., and Webb, E. C., "Enzymes." 2nd ed. Academic Press, New York, 1964.

23. Dymling, J. F., Separation of serum and placental alkaline phosphatase by agarose gel electrophoresis and Sephadex chromatography. *Scand. J. Clin. Lab. Invest.* **18**, 129 (1966).

24. Englund, D. T., King, T. P., Craig, L. C., and Walti, A., Studies on ficin. I. Its isolation and characterization. *Biochemistry* **7**, 163 (1968).

25. Farago, A., and Denes, G., Mechanism of arginine biosynthesis in Chlamydomonas reinhardti. II. Purification and properties of N-acetylglutamase-5-phosphotransferase, the allosteric enzyme of the pathway. *Biochim. Biophys. Acta* **136**, 6 (1967).

26. Feinstein, G., Affinity chromatography of biological macromolecules. *Naturwissenschaften* **58**, 389 (1971).

27. Fish, W. W., Reynolds, J. A., and Tanford, C., Gel chromatography of proteins in denaturing solvents. *J. Biol. Chem.* **245**, 5166 (1970).

28. Folk, J. E., and Schirmer, E. W., Chymotrypsin C. I. Isolation of the zymogen and the active enzyme; preliminary structure and specificity studies. *J. Biol. Chem.* **240**, 181 (1965).

29. Grassman, W., and Hannig, K., Ein einfaches Verfahren zur kontinuierlichen Trennung von Stoffgemischen auf Filterpapier durch Elektrophorese. *Naturwissenschaften* **37**, 397 (1950).

30. Guy, O., Gratecos, D., Rovery, M., and Desnuelle, D., Contribution à l'étude du chymotrypsinogene B de boeuf. *Biochim. Biophy. Acta* **115**, 404 (1966).

31. Hall, D. A., and Czerkawski, J. W., The purification of the proteolytic component of elastase. *Biochem. J.* **73**, 356 (1959).
32. Haller, W., Correlation between chromatographic and diffusional behaviour of substances in beds of pore controlled glass. *J. Chromatogr.* **32**, 676 (1968).
33. Harris, J. R., Release of a macromolecular protein component from human erythrocyte ghosts. *Biochim. Biophys. Acta* **150**, 534 (1968).
34. Ho, P. P., Frank, B. H., and Bruck, D. J., Crystalline L-asparaginase from *Escherichia coli* B. *Science* **165**, 510 (1969).
35. Horstman, L. L., and Racker, E., Partial resolution of the enzymes catalyzing oxidative phosphorylation. *J. Biol. Chem.* **245**, 1336 (1970).
36. Ito, A., and Sato, R., Purification by means of detergents and properties of cytochrome B5 from liver microsomes. *J. Biol. Chem.* **243**, 4922 (1968).
37. Jacobsen, C. F., Leonis, J., Linderstrom-Lang, K., and Ottesen, M., The pH-stat and its use in biochemistry. *Methods Biochem. Anal.* **4**, 171 (1957).
37a. Jakoby, W. B., ed., "Methods in Enzymology," Vol. 22, p. 238. Academic Press, New York, 1971.
38. Jayaram, H. N., Ramakrishnan, T., and Vaidyanathan, C. S., L-Asparaginases from Mycobacterium tuberculosis strains $H_{37}R_v$ and $H_{37}R_a$. *Arch. Biochem. Biophys.* **126**, 165 (1968).
39. Kagawa, Y., and Racker, E., Partial resolution of the enzymes catalyzing oxidative phosphorylation. *J. Biol. Chem.* **241**, 2461 (1966).
40. Kimmel, J. R., and Smith, E. L., Crystalline papain. I. Preparation, specificity and activation. *J. Biol. Chem.* **207**, 515 (1954).
41. Kunitz, M., Crystalline soybean trypsin inhibitor. *J. Gen. Physiol.* **29**, 149 (1946).
42. Lilly, D. M., and Dunnill, P., Isolation of intracellular enzymes from microorganism. *Ferment. Adv., Pap. Int. Ferment. Symp., 3rd, 1968* p. 225 (1969).
43. MacLennan, D. H., Tzagoloff, A., and Rieske, J. S., Studies on the electron transfer system. 63. Solubilization and fractionation of mitochondrial proteins by succinylation. *Arch. Biochem. Biophys.* **109**, 383 (1965).
44. McConn, J. D., Tsurn, D., and Yasunobu, K. T., *Bacillus subtilis* neutral proteinase. I. A zinc enzyme of high specific activity. *J. Biol. Chem.* **239**, 3706 (1964).
45. Malacinski, G. M., and Rutter, W. J., Multiple molecular forms of α-amylase from the rabbit. *Biochemistry* **8**, 4382 (1969).
46. Mitchell, H. K., Gordon, M., and Haskins, F. A., Separation of enzymes on the filter-paper chromatopile. *J. Biol. Chem.* **180**, 1071 (1949).
47. Morton, R. K., Separation and purification of enzymes associated with insoluble particles. *Nature (London)* **166**, 1092 (1950).
48. Murachi, T., Yasui, M., and Yasuda, Y., Purification and physical characterization of stem bromelain. *Biochemistry* **3**, 48 (1964).
49. Meister, A., ed., "Advances in Enzymology," Vol. 1. Wiley Interscience, New York, 1941.
50. O'Dor, R. K., Parker, C. D., and Copp, D. H., Biological activities and molecular weights of ultimobranchial and thyroid calcitonins. *Comp. Biochem. Biophys.* **29**, 295 (1969).
51. Peterson, E. A., and Sober, H. A., Column chromatography of proteins. *Methods Enzymol.* **5**, 3 (1962).
52. Peterson, R. F., The applicability of acrylamide gel electrophoresis to determination of protein purity. *Methods Enzymol.* **25**, 178 (1972).
53. Polson, A., Multi-membrane electrodecantation and its application to isolation and purification of proteins and viruses. *Biochim. Biophys. Acta* **11**, 315 (1953).
54. Polson, A., Potgieter, G. M., Largier, J. F., Mears, G. E. F., and Joubert, F. J., The fractionation of protein mixtures by linear polymers of high molecular weight. *Biochim. Biophys. Acta* **82**, 463 (1964).
55. Poráth, J., and Hjertén, S., Some recent developments in column electrophoresis in granular media. *Methods Biochem. Anal.* **9**, 193 (1962).

56. Poulik, M. D., Heterogeneity and structure of ceruplasmin. *Ann. N.Y. Acad. Sci.* **151**, 476 (1968).

57. Prahl, J. W., and Neurath, H., Pancreatic enzymes of the spiny pacific dogfish I. Catonic chymotrypsinogen and chymotrypsin. *Biochemistry* **5**, 2131 (1966).

58. Ryan, C. A., Chicken chymotrypsin and turkey trypsin. I. Purification. *Arch. Biochem. Biophys.* **110**, 169 (1965).

59. Schroeder, D. D., and Shaw, E., Chromatography of trypsin and its derivatives. Characterization of a new active form of bovine trypsin. *J. Biol. Chem.* **243**, 2943 (1968).

60. Shotton, D. M., Hartley, B. S., Camerman, N., Hofmann, T., Nyburg, S. C., and Rao, L., Crystalline porcine pancreatic elastase. *J. Mol. Biol.* **32**, 155 (1968).

61. Show, C. R., and Koen, A. L., Starch gel zone electrophoresis of enzymes. *In* "Chromatographic and Electrophoretic Techniques" (I. Smith, ed.), p. 325. Wiley, New York, 1968.

62. Singer, S. J., The properties of proteins in nonaqueous solvents. *Adv. Protein Chem.* **17**, 1 (1962).

63. Snoke, J. E., Isolation and properties of yeast glutathione synthetase. *J. Biol. Chem.* **213**, 813 (1955).

64. Sonneborn, H. H., and Pfleiderer, G., Zur Evolution der Endopeptidasen. VII. Eine Protease vom Molekulargewicht 12500 aus Larven von Vespa orientalis F. mit cymotrypschen Eigenschaften. *Hoppe-Seyler's Z. Physiol. Chem.* **350**, 389 (1969).

65. Stahl, E., "Dünnschicht = Chromatographie" 1st ed. Springer Verlag, Berlin, 1962.

66. Strittmatter, P., and Verlick, S. F., The isolation and properties of microsomal cytochrome. *J. Biol. Chem.* **221**, 253 (1956).

67. Subramanian, A. R., and Kalnitsky, G., The major alkaline proteinase of Aspergillus oryzae, aspergillopeptidase B. I. Isolation in homogeneous form. *Biochemistry* **3**, 1861 (1964).

68. Svensson, H., Isoelectric fractionation, analysis, and characterization of ampholytes in natural pH gradients. I. The differential equation of solute concentration at a steady state and its solution for simple cases *Acta Chem. Scand.* **15**, 325 (1961).

69. Tang, J., Wolf, S., Caputto, R., and Trucco, R. E., Isolation and crystallization of gastricsin from human gastric juice. *J. Biol. Chem.* **234**, 1174 (1959).

70. Thuma, E., Schirmer, R. H., and Schirmer, I., Preparation and characterization of a crystalline human ATP:AMP phosphotransferase. *Biochim. Biophys. Acta* **268**, 81 (1972).

71. Tiselius, A., Hjertén, S., and Levin, O., Protein chromatography on calcium phosphate columns. *Arch. Biochem. Biophys.* **65**, 132 (1956).

72. Vesterberg, O., Synthesis and isoelectric fractionation of carrier ampholytes. *Acta Chem. Scand.* **23**, 2653 (1969).

73. Webb, E. C., and Morrow, P. F. W., The activation of an arylsulphatase from ox liver by chloride and other anions. *Biochem. J.* **73**, 7 (1959).

Chapter 3

Enzyme Kinetics

A. A. Aszalos

The function of an enzyme is to catalyze a chemical reaction. The progress of the catalysis can be followed by quantitative measurements of the rate of formation or disappearance of the molecules participating in the reaction. Investigation of the rate of catalysis under various conditions contributes to a clarification of the mechanism of action of an enzyme, to an understanding of its biologic role *in vivo*, and to a definition of the unit of the particular enzyme under study. All these possible applications of enzyme kinetics involve the mathematical formulation and analysis of the behavior of the system under study. The mechanism of action has been elucidated for only a few enzyme reactions. Therefore, the kinetics of most enzymes are based on postulated pathways and mechanisms of action.

In the following pages, various aspects of enzyme kinetics will be discussed. More detailed treatment of the subject, particularly more extensive mathematical analysis utilizing computer techniques, has been discussed in several recent textbooks and publications (4, 6, 16, 18, 21).

General Consideration of Enzyme Kinetics

Numerous factors influence the kinetics of enzyme reactions. Enzyme kinetics should be studied under the simplest conditions possible. For example, most reaction velocities at zero time are influenced only by: (a) concentration of the enzyme; (b) concentration of the substrate; (c) activators and/or inhibitors; (d) temperature; and (e) pH.

To appreciate the influence of these factors on enzyme kinetics, it is advisable to consider first the reaction system. Distinction has to be made between open and closed systems. In open systems, matter is exchanged with the surroundings. Reactants are supplied to the system from the environment and products are removed from the system to the environment. When both the concentration of substances in the surroundings and the space available for the uptake of products are constant and the exchange of substrates and products is determined by their prevailing concentrations, the system will reach a time-independent, or stationary, state. This state is described by the formula $S \rightleftharpoons ES \rightleftharpoons P$ when these parameters are set to be independent of time. In this formula, $S =$ the substrate concentration, $ES =$ the enzyme-substrate complex, and $P =$ the concentration of the product. In a closed system, the outside factors have no influence and the stationary state is the equilibrium state, at which the forward and reverse rates of each step in the entire process become equilibrated. It may be assumed that, at such a time, there is an equilibrium not only with regard to substrates and products, but also with regard to all intermediates. Such a system describes adequately enzyme reactions with linear equations, and will be the basic system used in the ensuing discussion. It should be remembered, however, that some enzyme reactions can only be described by nonlinear equations that describe oscillation (28), and are applicable when the substrate is added unevenly, in an oscillating fashion. Under these conditions, the rate of the enzymatic reaction is also variable (oscillating). Even in closed systems in which the substrate concentration is much larger than the enzyme concentration, oscillation may occur. This concentration difference makes possible different time scales for the substrate and the enzyme intermediary complexes, and results in a sustained stationary state. In a stationary state, oscillation is possible, but in a closed system it is *damped*. However, most enzyme reactions are in an open stationary state, as described above. In the case of coupled reactions, i.e., when the product of the first reaction serves as the substrate for the subsequent reaction, it is possible to have oscillations in the concentrations of the intermediate products. The first substrate, from an external source, and the last product, which leaves the system, may exist on a time scale that is very much larger than that of the intermediate substrates and products. It has been shown that oscillation around the stationary-state level of these intermediate substrates and products is thermodynamically possible. Although oscillating reactions may describe some enzyme systems in living organisms, there is no intention here to describe this phenomenon in detail.

Influence of Enzyme Concentration

The enzyme molecule, E, and the substrate molecule, S, in solution approach each other under the influence of intermolecular forces and form a complex, ES. This complex undergoes a reaction that eventually yields the product, P, and regenerates the enzyme. One may assume, on the basis of this mechanistic interpretation, that more molecules of enzymes will produce more product during a given time. The velocity, V, is proportional to enzyme concentration, E, therefore:

$$V = K[E] \tag{1}$$

Equation (1) applies to most, but not to all, enzymatic reactions. Plotting the data obtained from this equation yields a straight line. Deviations from this linear relationship may occur under various circumstances discussed in the following paragraphs.

When one of the products of an enzyme reaction is a gas, or the product liberates gas from the test system, it is possible to follow the course of the reaction by measuring manometrically the amount of gas formed. For example, manometric measurement of the quantity of carbon dioxide liberated from bicarbonate by acetic acid, one of the products of the enzymatic hydrolysis of acetylcholine, is the basis of Ammon's (2) method for determining the activity of cholinesterases. If some of the gas formed is dissolved in the water of the test system, manometric measurements will not reflect accurately the course of the reaction, and the plot of the reaction will deviate from a straight line. This phenomenon occurs in the course of the enzymatic hydrolysis of succinylcholine: Succinylcholine is first broken down relatively rapidly to succinylmonocholine and choline; subsequently, succinylmonocholine is hydrolyzed much more slowly to succinic acid and choline (11, 29) (see p. 81).

Finally, impurities not associated with the enzyme, but present in the test system, may result in a plot that is a straight line that does not pass through the zero point of the coordinate system, but intersects the abscissa some distance from it.

Influence of Substrate Concentration

The substrate concentration is an important factor from the point of view of enzyme catalysis. If the results of the measurements of velocity (V) are plotted against the substrate concentration (S), a hyperbolic curve is obtained. Mathematically, the curve can be expressed as:

$$(a - v) \cdot (b + S) = \text{constant} \tag{2}$$

Mechanically, the phenomenon was explained by Michaelis and Menten (24). According to them, first the enzyme quickly forms a complex with its substrate. This complex, ES, breaks down comparatively slowly and remains in equilibrium with the enzyme, E, and with the substrate S, as the enzyme reaction is proceeding. The reactions involved are:

$$E + S \rightleftharpoons ES \tag{3}$$

$$ES \rightarrow E + \text{product} \tag{4}$$

The equilibrium constant for this reaction can be written as:

$$K_m = \frac{(e-p)s}{p} \tag{5}$$

where e is the total enzyme concentration, p is the concentration of the enzyme–substrate complex, s is the free substrate concentration, and K_m is the Michaelis constant.

For enzymes with very high catalytic activity, the assumptions of Michaelis and Menton may not be true, and p may differ from its equilibrium value. Mathematical treatment (17) applicable to such so-called steady-state systems are based on the assumption that the change in p is negligible in comparison to the amount of substrate already formed. In fact, p can be regarded as constant over the short period necessary for a velocity measurement. In a steady-state system, the maximum velocity, V_{max}, is usually obtained under conditions at which the enzyme is saturated with the substrate (30). The concentration of the enzyme–substrate complex can be expressed as

$$p = \frac{es}{K_m + s} \tag{6}$$

According to the Michaelis–Menten theory, the velocity v of the reaction is determined by the breakdown of the complex

$$v = kp \tag{7}$$

Substituting Equation (6) in Equation (7) we obtain

$$v = \frac{ek}{1 \pm \dfrac{K_m}{s}} \tag{8}$$

If s is large compared with K_m, when the enzyme is saturated with the substrate, ek will equal v. The maximum velocity, V_{max}, obtained for this case can be incorporated in Equation (8).

$$v = \frac{V_{max}}{1 + \dfrac{K_m}{s}} \tag{9}$$

This equation is the Michaelis–Menten equation. Rewritten:

$$(V_{max} - v)(K_m + s) = V_{max} \cdot K_m \tag{10}$$

Like Equation (2), (10) is the equation of a hyperbole. Therefore, the Michaelis–Menten equation fits the experimental curve, which is a hyperbole (see Fig. 3-1).

Equation (9) indicates that when s is equal to K_m, then v is equal to $V_{max}/2$. K_m can be determined from different types of graphs prepared from experimental data. The most common method is to plot $1/v$ against $1/s$, according to Lineweaver and Burk (20). The intersection of the line obtained

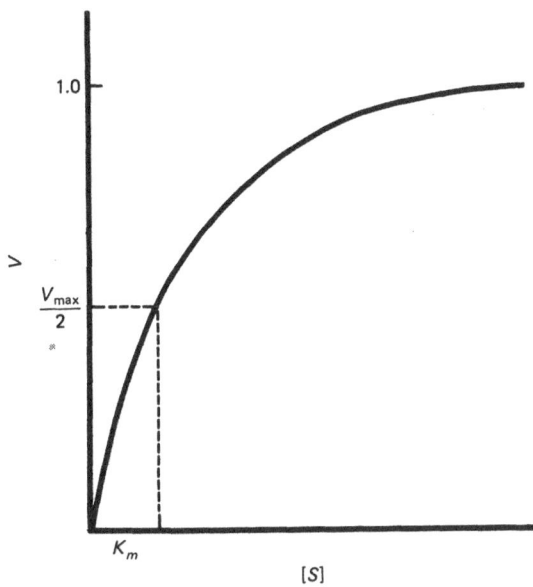

Figure 3-1. The characteristic substrate concentration–velocity curve of heterotropic enzymes.

Figure 3-2. Effect of modulators on velocities of heterotropic regulatory enzymes. (a) Change in affinity for the substrate (K_m) without change in V_{max}; (b) Change in V_{max} without effect on K_m.

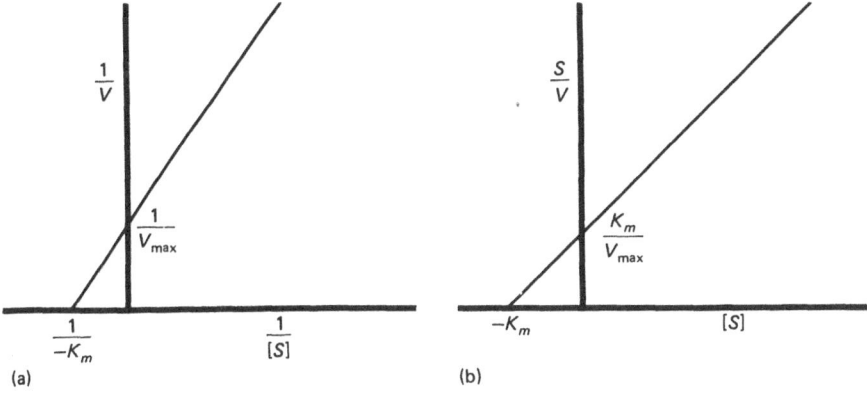

with the abscissa gives the value of $-1/K_m$. If s is plotted against S/v, the intersection of the line with the abscissa gives the value of $-K_m$. V_{max} can be obtained from the same plots. In the first type of plot, $1/V_{max}$ is determined by the intersection of the line and the ordinate. In the second type of plot, the same point represents the value of K_m/V_{max} (see Fig. 3-2).

The determination of K_m and V_{max} is important. K_m is the equilibrium constant of the reaction (3) and is the reciprocal of the affinity of the enzyme for the substrate. V_{max} is the measure of the velocity of the breakdown of the complex.

Determination of K_m and V_{max} is more difficult in some cases, as for example when two substrate molecules react with the same enzyme. Mathematical handling of such cases has been discussed by Haldane (14) and Florini and Vestling (10).

The determination of K_m and V_{max} is also complicated when two enzymes act simultaneously (7) on one substrate, or when nonspecific enzymes act on more than one substrate.

Systems in which two substrates compete for the same enzyme can be used to determine whether a system contains a single nonspecific enzyme or two specific enzymes. In the first case, the overall velocity will be less than it would be if only higher-velocity substrate were present. With two specific enzymes in the system, the observed velocity will be greater in the presence of both substrates than in the presence of either substrate alone.

Influence of Activators or Inhibitors

The velocity of an enzyme reaction may decrease at high substrate concentrations, even if, at low concentrations, the reaction follows the Michaelis law. High substrate concentration may inhibit enzyme activity by various mechanisms. The substrate may be strongly hydrated and, if water is a reactant, the velocity of the reaction will be reduced at very high substrate concentrations because of the decreased availability of water. Or, if a cofactor is involved in the reaction, substrate may bind the cofactor to such an extent that the enzymatic reaction is hindered. It is also possible that the substrate is capable of binding not only to the active site, but also to another, nonspecific, site of an enzyme. If these sites are located in close proximity to one another, then the binding of the substrate to the inactive sites may inhibit binding to the active sites, and thereby decrease the rate of its own enzymatic breakdown.

The opposite of the above situation may also occur, and the substrate may act as an activator. In these cases, however, some of the substrate molecules combine with the enzyme at points distant from the active center. The overall result of such interaction is a nonlinear, sigmoidal kinetic curve. The reaction equations for such cases are:

$$E + S \rightleftharpoons ES \tag{3}$$

$$E + S \rightleftharpoons ESa \tag{11}$$

$$ESa + S \rightleftharpoons ESSa \tag{12}$$

where a denotes the substrate that binds at the activator site. The altered kinetic behavior may also be explained for these allosteric enzymes by a different mechanism; the binding of a substrate or modular molecule to an enzyme may change the binding capacity of the active sites (7, 12). The binding of a second ligand may further change the binding constant of the remaining sites. The changes in the binding constants may be equal or different, as described by Frieden (12) and Monod *et al.* (25). It is also possible that enzymes that have subunits may alter their conformation from subunit to subunit. The influence of different subunits on each other and on the kinetics involved is discussed by Koshland *et al.* (19). If the enzyme dissociates to a form containing fewer subunits, its binding constant may change in parallel with the change in its conformation. Such a situation was discussed in general by Frieden (12) and for the enzyme L-asparaginase by Aszalos *et al.* (2). The change in kinetics caused by the substrate itself is called homotropic interaction. The rate increase caused by homotropic interaction may be due to an increased affinity (lower K_m) of the substrate for the enzyme or to an increase in the maximal reaction rate (V_{max}).

Figure 3-3. Two methods of plotting the substrate concentration against the velocity for the determination of K_m and V_{max}.

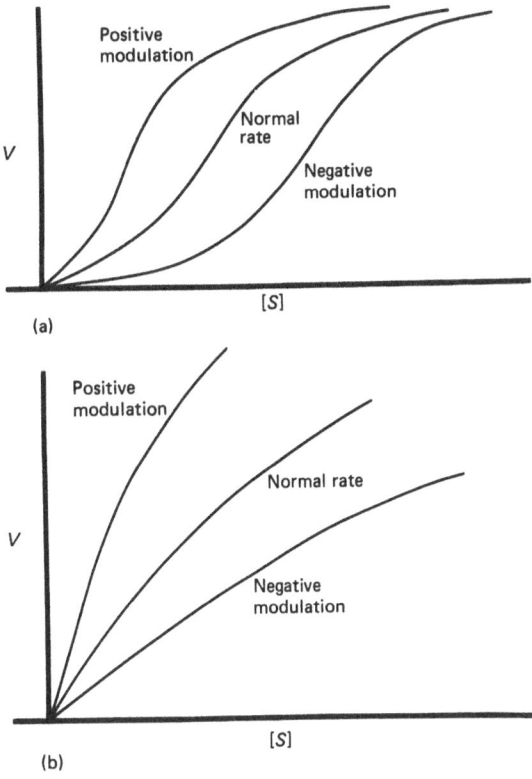

With certain enzymes, the modulator or effector molecule that alters the kinetics is not the substrate itself. The kinetics of many enzymes may be altered by several different modulator molecules. These enzymes are usually referred to as heterotropic enzymes. Heterotropic enzymes play an important role in many biosynthetic processes. Thus, for example, glutamine synthetase, the regulatory enzyme with transaminating activity, is inhibited by adenosine monophosphate, cytosine triphosphate, histidine, tryptophan, and carbamyl phosphate, all metabolites of the synthesis of glutamine.

The modulator molecules may either increase (lower K_m or higher V_{max}) or decrease (higher K_m or lower V_{max}) the activity of heterotropic enzymes. Figure 3-3 illustrates typical substrate-concentration curves of heterotropic enzymes.

The measurement and interpretation of K_m is complicated in systems in which the enzyme combines with the substrate first and with the cofactor later. For example, peroxidase combines rapidly with H_2O_2, and the equilibrium constant can be measured for this reaction if no hydrogen donor molecule is added to the system. The overall reaction rate depends directly on the concentration of the complex of enzyme, substrate, and cofactor. Graphical illustration of the kinetics of such reactions shows a rise in velocity during the formation of the complex, followed by a gradual decline to zero. Mathematical treatment of this and similar enzyme systems has been described by Chance (3), Gutfreund (13), and Slater (27). Complicated kinetics are found with *E. coli.* alkaline phosphatase, which forms intermediates before breaking down to product and enzyme, as shown in Equation 13.

$$E + RO\!-\!\overset{\displaystyle O}{\overset{\|}{P}}\!-\!OH \underset{}{\overset{k_1}{\rightleftharpoons}} [E.RO\!-\!\overset{\displaystyle O}{\overset{\|}{P}}\!-\!OH] \underset{}{\overset{k_2}{\rightleftharpoons}} E\!-\!\overset{\displaystyle O}{\overset{\|}{P}}\!-\!OH + ROH \tag{13}$$

$$\underset{\underset{OH}{|}}{} \qquad \underset{\underset{OH \ A}{|}}{} \qquad \underset{\underset{OH \ B}{|}}{}$$

$$\Big\downarrow k_3$$
$$E + P_i$$

In this equation, k_1 and k_{-1} are the forward and reverse reaction rate constants of the formation of the enzyme (E)–substrate (phosphate) complex (A); k_2 is the reaction rate constant of the hydrolysis of the enzyme–phosphate complex (B); k_3 is the reaction rate constant of the hydrolysis of intermediate to enzyme and inorganic phosphate (P_i). Study of the hydrolysis of different phosphate esters showed that the formation of the A and B complexes is rapid. Subsequently, B undergoes a slow conformational change, which is the rate-determining step. The decomposition of the complex to E and P_i is again rapid (15). Some of these intermediates have been isolated in crystalline form.

Even more complicated kinetics than the ones described exist for enzymes acting in heterogeneous systems. Soluble enzymes acting on insoluble substrates, or enzymes acting at interfaces or in crystalline form, have such complicated kinetics. These systems have been reviewed by McLaren and Packer (22).

Influence of Temperature

Temperature has a significant effect on the velocity of most enzyme reactions. The effect of temperature may be due to denaturation, to changed velocity of the breakdown of the enzyme–substrate complex, to changes in the ionization of groups at the catalytic site, and to alterations in the conformation of the enzymes. The temperature effects can be measured experimentally if care is taken to keep all other variables constant. It should be remembered, however, that temperature changes may affect simultaneously several determinants of enzyme kinetics, and that the overall effect, depending on the circumstances, may be either an increased or a decreased reaction rate.

Enzymes, being proteins, can be denatured by heat. The denaturation temperature, time, and entropy vary from enzyme to enzyme and from condition to condition. Usually, the initial reaction velocity increases with increasing temperatures. However, a trend toward a decrease in velocity can be detected as denaturing temperatures are approached. The time necessary for the inactivation of an enzyme decreases with increasing temperatures. Most enzymes are instantaneously deactivated at temperatures greater than 100°C. The rate of inactivation of enzymes depends, in most cases, on the protein concentration of the system. In the presence of large amounts of inactive proteins, denaturation may occur more slowly than otherwise. Substrates or other small molecules may also protect enzymes against denaturation. It is interesting that structural mutations involving the replacement of a single amino acid in the enzyme can cause considerable changes in stability (26).

Inactivation rates are, in most cases, also influenced by the pH of the solution. Generally, there is a pH zone of maximum stability that usually, but not always, coincides with the isoelectric point of the enzyme.

Many denatured enzymes can be reactivated by cooling. The reactivation can be explained by the re-formation of hydrogen bonds that were broken during denaturation. This re-formation restores the original conformation of the enzyme and reestablishes its activity.

As already mentioned, temperature may affect the ionization of the groups of the catalytic site. Determination of the change due to altered ionization is complicated by the fact that the pH function of these groups changes simultaneously with the changing temperature.

Temperature also has a significant effect on the overall enzyme reaction that is independent of denaturation and the ionization of the active sites. Thus, for example, decreasing temperature decreases V_{max}. Decreased activity of the involved groups, as well as an effect on the reaction itself, are cited as reasons for these phenomena (9).

Influence of pH

The pH has a strong influence on the catalytic activity of enzymes (5). In most cases, enzymes have a definite pH optimum of activity. There are several reasons for this limited range of pH-activity relationship. First of all, the pH influences the state of ionization of the active groups of the enzyme, of the substrate, and of the enzyme–substrate complex. Changes in the ionization of groups at the active site have the most significance; changes at other sites may not be so important. Second, the stability of an enzyme may be different at various pH values. Third, the affinity of the substrate for the active center may vary on both sides of the pH optimum. The effect of pH on the stability of an enzyme can be determined experimentally by exposing the enzyme to various pH values before the test. Decreased activity of an enzyme after exposure to a certain pH indicates instability. The effect of pH on the affinity of the substrate for the active center can be eliminated in the above measurements by using high concentrations. A more detailed discussion of this complex topic is presented in the publications of Michaelis (23), Alberthy and Massey (1), and Dixon and Webb (8).

References

1. Alberthy, R. A., and Massey, V., Study of protein-ion interaction by the moving-boundary method. *J. Am. Chem. Soc.* **73**, 3220 (1951).
2. Aszalos, A., Kirschbaum, J., Ratych, O. T., Kraemer, N., Kocy, O., Frost, D., and Casey, J. P., Reversible dissociation of L-asparaginase of *Escherichia coli* B. *J. Pharm. Sci.* **61**, 791 (1972).
3. Chance, B., The kinetics of the enzyme-substrate compound of peroxidase. *J. Biol. Chem.* **151**, 553 (1943).
4. Cleland, W. W., The statistical analysis of enzyme kinetic data. *Adv. Enzymol.* **29**, 1 (1967).
5. Dixon, M., The effect of pH on the affinities of enzymes for substrates and inhibitors. *Biochem. J.* **55**, 161 (1953).
6. Dixon, M., and Webb, E. C., "Enzymes," p. 54. Academic Press, New York, 1964.
7. Dixon, M., and Webb, E. C., "Enzymes," p. 87. Academic Press, New York, 1964.
8. Dixon, M., and Webb, E. C., "Enzymes," p. 123. Academic Press, New York, 1964.
9. Dixon, M., and Webb, E. C., "Enzymes," p. 150. Academic Press, New York, 1964.
10. Florini, J. R., and Vestling, C. S., Graphical determination of the dissociation constants for two-substrate enzyme systems. *Biochim. Biophys. Acta* **25**, 575 (1957).

11. Foldes, F. F., Succinylcholine iodide: Its enzymatic hydrolysis and neuromuscular activity. *Proc. Soc. Exp. Biol. Med.* **83**, 187 (1953).
12. Frieden, C., Treatment of enzyme kinetic data. II. The multiside case: Comparison of allosteric models and a possible new mechanism. *J. Biol. Chem.* **242**, 4045 (1967).
13. Gutfreund, H., Catalytic site of trypsin. *Discuss. Faraday Soc.* **20**, 167 (1955).
14. Haldane, J. B. S., "Enzymes." Longmans, Green, New York, 1930.
15. Halford, S. E., Bennett, N. G., and Trentham, D. R., A substrate-induced conformation change in the reaction of alkaline phosphatase from *Escherichia coli. Biochem. J.* **114**, 243 (1969).
16. Hearon, J. Z., Bernhard, S. A., Friess, S. L., Botts, D. J., and Morales, M. F., Enzyme kinetics. *In* "The Enzymes" (P. D. Boyer, H. Lardy, and K. Myrbäck, eds.), 2nd rev. ed., Vol. 1, p. 49. Academic Press, New York, 1959.
17. King, E. L., and Altman, C., A schematic method of deriving the rate lows for enzyme-catalyzed reactions. *J. Phys. Chem.* **60**, 1375 (1956).
18. Koshland, D. E., Jr., and Neet, K. E., The catalytic and regulatory properties of enzymes. *Annu. Rev. Biochem.* **37**, 359 (1968).
19. Koshland, D. E., Jr., Neméthy, G., and Filmer, D., Comparison of experimental binding data and theoretical models in proteins containing subunits. *Biochemistry* **5**, 365 (1966).
20. Lineweaver, H., and Burk, D., The determination of enzyme dissociation constants. *J. Am. Chem. Soc.* **56**, 658 (1934).
21. Mahler, H. R., and Cordes, E. H., "Biological Chemistry," 2nd ed., p. 267. Harper, New York, 1966.
22. McLaren, A. D., and Packer, L., Some aspects of enzyme reactions in heterogeneous systems. *Adv. Enzymol.* **33**, 245 (1970).
23. Michaelis, L., "Die Wasserstoffionenkonzentration." Springer-Verlag, Berlin, 1922.
24. Michaelis, L., and Menten, M. L, Die Kinetick der Invertinwirkung. *Biochem. Z.* **49**, 333 (1913).
25. Monad, J., Wyman, J., and Changeux, J. P., On the nature of allosteric transitions: A plausible model. *J. Mol. Biol.* **12**, 88 (1965).
26. Pollack, M. R., The range and significance of variations amongst bacterial penicillinases. *Ann. N.Y. Acad. Sci.* **151**, 502 (1968).
27. Slater, E. C., Calculation of the rate constants of the reaction between an enzyme and its substrate from the overall kinetics of the reaction catalyzed by the enzyme. *Discuss. Faraday Soc.* **20**, 308 (1955).
28. Walter, C., "Enzyme Kinetics, Open and Closed Systems," p. 71. Ronald, New York, 1966.
29. Whittaker, V. P., and Wijesundera, S., The separation of esters of choline by paper chromatography. *Biochem. J.* **49**, xlv (1951).
30. Zerner, B., Bond, R. P. M., and Bender, M. L., Kinetic evidence for the formation of acyl-enzyme intermediates in the α-chymotrypsin-catalyzed hydrolysis of specific substrates. *J. Am. Chem. Soc.* **86**, 3674 (1964).

Chapter 4

Enzymatic Catalysis

A. A. Aszalos

Chemical reactions take place because molecules acquire an "activated state" in which enough energy is present for formation or breakage of a chemical bond. This amount of energy, referred to as the "activation energy," is required to bring 1 mole of substrate to the activated state, at a given temperature. The activation energy of a reaction is lower if the reaction is catalyzed. Catalysts combine with the reactants and produce an activated state of the substances. These activated substances have less free energy than the inactivated substances in noncatalyzed reactions.

These considerations are also valid for biologic reactions catalyzed by enzymes. In addition, enzymatic reactions usually proceed through several intermediate steps and, because of this, the energy requirements are lower than if the reaction were to occur in one step. Enzymatic reactions require less activation energy than do the same reactions catalyzed chemically. A prerequisite of this enzymatic mechanism is a thermodynamically stable enzyme–substrate complex. The formation of such complexes is not possible with chemical catalysts.

After these brief comments on thermodynamic considerations, the meaning of catalytic mechanisms in enzymatic reactions should be discussed. The first event in the course of enzymatic catalysis is the formation of the enzyme–substrate complex by stoichiometric amounts of substrate and enzyme. This complex, usually referred to as the Michaelis complex, forms without the development of new covalent bonds. In the course of a simple enzyme-catalyzed process, this complex, in which the substrate is present in an activated form, breaks down to products and free enzyme. The breakdown of the complex is considerably slower than the formation of the complex. In most enzymatic reactions, however, the situation is more complex. In addition to the Michaelis complexes, other complexes or intermediates are formed during the enzymatic process. Some of these intermediates involve covalent bonding between substrate and enzyme. There may be as many as five or even more intermediates involved. A knowledge of the number and structure of these intermediates is essential to an understanding of the mechanism of enzyme-catalyzed reactions. In reality, for an understanding of the mechanisms involved, it is not necessary to know the structure of the entire enzyme. It is enough to know the relative positions of the groups that interact with the enzyme at the active sites, and the nature of this interaction. In fact, the structure of the whole enzyme has been determined only in about 20 cases, by X-ray crystallography. However, there is considerable information available on the chemical processes involved in enzymatic catalysis. In contrast, the physical aspects of these processes are still mostly speculative.

A distinction between the chemical and the physical aspects of enzymatic mechanisms is significant. Although each chemical step in the reaction pathway can be studied and understood, at least in principle, theories to explain the 10^9-order rate enhancement in enzyme-catalyzed reactions are still very elementary. In the following pages, enzymatic catalysis are discussed under three headings: (a) methods of investigation of enzyme-catalyzed reactions; (b) factors responsible for enzymatic catalysis; (c) reaction pathways of enzyme-catalyzed reactions. These discussions are nonquantitative, but descriptive, to fit the requirements of this book.

Methods of Investigation of Enzyme-Catalyzed Reactions

This subject is dealt with selectively and only methods of current interest, (e.g., chemical, crystallographic, and kinetic) are considered.

Chemical Methods

The selective reactivity of amino acid residues in proteins is the basis of the biologic activity of enzymes. This selective reactivity, however, cannot be explained simply in terms of organic chemistry, because it is restricted to specifically organized steric structures and disappears when the steric configuration is destroyed.

Such a specific reactivity was shown to exist (3), for example, between α-chymotrypsin and diisopropylfluorophosphate (DFP). DFP reacts stoichiometrically with α-chymotrypsin to yield an inactive enzyme. It was found that the covalent bond formed resists hydrolysis, and that the reaction involves the binding of DFP to a unique serine residue. Subsequent investigations have shown that the active site of α-chymotrypsin, trypsin, and of many other esterases contains a serine residue. Many other reagents can be used to pinpoint, by covalent labeling, such serine residues at active sites of enzymes (35).

Another example of such methods is the labeling of the carboxyl group of the active site of pepsin with 1, 2-epoxy-3-(p-nitrophenoxy)=propane (21). In this case, the carboxyl group was shown to be in a side chain of aspartic acid. This information, together with rate-constant measurements, led to a clarification of the mechanism of action of pepsin.

Active-site labeling techniques, together with affinity labeling, have already been discussed under the heading "primary structure of enzyme." Useful information on this subject can also be obtained from studies of active-site-directed irreversible inhibitors (1).

Another interesting method was used to characterize the active-site residue in α-chymotrypsin. This method was derived from kinetic studies of the hydrolysis of p-nitrophenyl-acetate by this enzyme (2). It was shown that this reaction is, in the beginning, a first-order and, later, a zero-order reaction. It was postulated, therefore, that at first a Michaelis complex forms, followed by the acylation of the active residue and simultaneous liberation of the p-nitrophenyl group. Subsequently, the acyl-enzyme is slowly hydrolyzed, and this hydrolysis becomes the rate-determining step. The slow hydrolysis of the acyl-enzyme makes possible its isolation under suitable conditions. It was shown subsequently that the serine residue acetylated was identical with the one that interacted with DFP, as described above. These experiments also throw light on the mechanism of this enzymatic reaction by demonstrating the formation of an acyl-enzyme intermediate. Similar results were obtained when the methyl or p-nitrophenyl esters of N-acetyl-L-tryptophan were used as substrates (5).

In connection with the active-site studies of α-chymotrypsin, another line of investigation deserves mention. In these studies, the catalytic rates of an analogous series of poor substrates were measured. From the variation in hydrolysis rates of these substrates, it was possible to draw conclusions about the fit of the various substrates to the binding sites. It was also possible to identify groups in the substrates that represent steric hindrance to the binding. For example, it was shown (32) that the aromatic binding site of α-chymotrypsin is planar, elongated, and curved. Information obtained from such investigations, together with the knowledge of the participating side chains of the active sites, may yield pictures that approximate those obtained from crystallographic studies.

As mentioned above, the differential reactivity of identical amino acids in enzymes is due to their specific locations. A method based on this fact was used to determine accessible and "buried" tyrosine residues in glyceralde-hyde-3-phosphate dehydrogenases (16). The determination was based on the reaction of accessible tyrosine with iodine and showed that, of nine tyrosine moieties per protomer, three reacted rapidly with iodine without much conformational change, three other tyrosine moieties reacted after small conformational change, and the last three reacted only after denaturation of the enzyme. Individual tyrosine moieties could then be localized in the sequence after peptide mapping. The active site of alcohol dehydrogenase was also elucidated by chemical methods (33).

In general, the studies of active sites by chemical methods are still in their infancy but, as illustrated above, they have considerable potential.

Crystallographic Methods

Precise information on tertiary structures and on the changes of these structures in the course of catalysis can be obtained by crystallographic methods. These methods require the preparation of enzyme derivatives that are isostructural with the native enzyme, but contain heavy atoms. If these derivatives can be crystallized, their X-ray diffraction can be recorded and their electron-density maps can be prepared. By this method, the tertiary structures of α-chymotrypsin (28), ribonuclease (25), papain (14), and several other enzymes were elucidated. It was shown that a trypsin side chain in carboxypeptidase-A moves 14 Å on combining with the substrate, glycyl-L-tyrosine (34). This "movement" can be taken as a specific example of Koshland's induced-fit hypothesis, which was discussed earlier under the heading of "tertiary structures" (Chapter 1).

Knowledge of the precise tertiary structure of an enzyme and of the enzyme–substrate complex, together with information obtained by chemical and kinetic methods, permits the elucidation of the mechanism of its catalytic activity. The conclusions arrived at from such studies are, however, valid only if the enzyme possesses the same characteristics in solution as in crystalline form, a point that can be resolved only for enzymes that are also active in crystalline form (36). Yet another problem exists in the case of enzymes that are active only if they are membrane bound. It will be necessary to investigate more enzymes in the crystalline state before these complicated issues can be elucidated.

Kinetic Methods

The kinetic analysis of most enzyme reaction mechanisms is based on the classical Michaelis–Menten equation. It is assumed in these analyses that a steady state exists and that the reverse reactions can be ignored. Although this equation can account for the dependence of reaction velocity on substrate concentration, it does not explain on a molecular basis the changes that occur during catalysis.

From the study of mathematical models based on kinetic data (19) obtained in enzyme reactions with one to four intermediates, it became apparent that neither the number of intermediates nor the value of any individual rate coefficient could be determined by this method. It was observed that, when an enzyme was studied with several substrates, in many cases the values of the kinetic parameters were independent of the characteristics of the substrate. This finding indicates that a "common" intermediate exists for all these substrates. In the case of phosphomonoesterase, this intermediate was thought to be the phosphorylated enzyme. If, however, a catalytic reaction has four intermediates, then the common kinetic parameter in mathematical models may arise from a rate-determining conformational change of the Michaelis complex rather than from the slow hydrolysis of an enzyme intermediate, usually one containing a covalent bond (e.g., phosphorylated enzyme).

When an enzyme-catalyzed reaction involves two or more substrates, and their concentrations can be varied independently, it is possible to obtain information regarding the mechanism of action of the enzyme from kinetic experiments. For two substrates, two possibilities exist: (a) both substrates can associate and dissociate freely with the enzyme prior to the formation of the triple complex; and (b) one of the substrates must associate with the enzyme before the second substrate can associate with it. By studying rate constants for the catalysis of one of the substrates at various concentrations of the other substrate, a distinction can be made between the two possible pathways.

Information on the sequence of events in the catalytic pathway and on the existence of covalent enzyme derivatives in the pathway can be obtained by studying the kinetics of product-inhibited reactions. In such experiments, products are added to the reaction mixture and the resulting slow initial velocities can be determined easily. By this method, it can be established mathematically which product is formed first (11). In case the two substrate fragments are dissociated from the enzyme in random order, the possibility that the existence of a covalent enzyme derivative is involved in the rate-determining step can be eliminated.

The velocity of enzyme-catalyzed reactions changes with pH, presumably because of the different ionization states of enzymes. Of these ionized states, only one is catalytically active (13). Therefore, it is possible to determine experimentally, from the measured relationship of maximum velocity and pH, the dissociation constants involved in the reaction. From the dissociation constants obtained, the nature of the ionizing group present at the active site can be identified. Such a technique was used successfully with α-chymotrypsin (22) and ribonuclease (17) to identify an imidazole group in the active site of each enzyme.

There are several other kinetic methods, such as "transient kinetics" and "relaxation" methods, that are useful for the elucidation of catalytic mechanisms (6, 15).

Factors Responsible for Enzymatic Catalysis

The overall rate of an enzyme-catalyzed reaction, as compared with a noncatalyzed reaction, is increased by a factor of 10^9. This great increase indicates that enzymes, in general, lower the free energies of activation required for chemical reactions by about 0.5 electron volts (eV). There is little information available on the rates of the various steps involved in enzymatic reactions. Various theories proposed for the explanation of this enormous rate increase are summarized below.

One of these theories, the proximity-effect theory (9), implies that the concentration of substrate in the vicinity of the active site gives "juxtaposition" and thereby creates a situation favorable for the chemical reaction. This juxtaposition increases the probability for the participation of neighboring groups in the reaction, which accounts for the rate enhancement. It has been shown (10) that this "neighboring-group effect" can play a role in noncatalyzed organic reactions. However, Koshland and Neet (27) pointed out that this effect cannot be working in enzymatic reactions because of the relatively low concentrations of the active sites.

A more accepted theory is based on acid–base catalysis. Since many enzymatic reactions involve proton movement, and since this proton movement is pH dependent, acid or base catalysis may be an acceptable mechanism for enzymatic reactions. However, side-chain groups of enzymes involved in proton-transfer reactions do not enhance reaction rates as much as do the intact enzymes. A possible explanation of these findings is that these groups are much stronger acids or bases in enzymes than in free solution. But this is unlikely, since the pK_a values of the conjugate acids in side chains of enzymes and those of corresponding functional groups in free solution are nearly identical. The suggestion that proton donation and acceptance occur synchronously in these reactions and that the formation of fully charged groups is thereby avoided also seems an unlikely explanation for rate enhancement, since the occurrence of such synchronous reactions has never been convincingly demonstrated in free solution (12). Therefore, although "acid–base" processes do occur in enzyme-catalyzed reactions, it is likely that they describe mechanisms, rather than account for the large rate enhancement (7). Nucleophilic- or electrophilic-catalysis theories also fail to explain the large rate enhancement in enzyme-catalyzed reactions.

Another theory proposes to explain the rate increase of enzyme-catalyzed reactions by the formation of covalent enzyme–substrate intermediates. Again, the demonstration of the existence of these intermediates clarifies the reaction sequence, but does not explain the physical processes of enzymatic catalysis.

A more plausible explanation of the rate enhancement of enzymatic catalysis is offered by the "strain theory." According to this hypothesis, after binding of the substrate, a conformational change occurs in the enzyme and this, in turn, causes distortion of the substrate molecule. This distortion raises the energy level of the substrate molecule to a transitional state. This assumption was corroborated by X-ray crystallographic studies on the interaction of the enzyme lysosyme and a mucopolysaccharide substrate (31). It was suggested (23) that strain can be induced in a substrate even by a rigid enzyme, and that a conformational change of the enzyme protein molecule is not absolutely necessary. The conformational changes of the substrate are probably caused by strongly electrostatic forces.

Further development of the "strain theory" leads to the following picture: The binding forces at the active site that can induce a conformational change in the substrate can be electrostatic forces; this electrostatically induced conformational change can be the reason for the catalytic rate enhancement, e.g., in the case of lysosyme, the formation of the carbonium ion can be electrostatically induced by the aspartate-52 ion, which is about 3 Å distant from the glycosidic oxygen that forms the carbonium ion. Which of the two mechanisms, the conformational change or "the ion-pair effect," is more important for the rate enhancement could only be determined if the rate coefficient of the carbonium-ion formation were measurable.

According to yet another theory, forces operating at the active sites are orientation forces. These forces orient the substrates by steric effects into positions favorable for their interaction with the active sites of enzymes. For example, in the case of hexokinase, the 6-hydroxyl group of glucose is oriented toward the terminal phosphate of ATP (20). This theory is essentially an extension of the proximity-effect theory for more than one substrate.

In enzymatic reactions, substrates are translocated from aqueous media to a low-dielectric region of the protein. Although reactions involving charged groups can be accelerated in media with low dielectric constants, this circumstance cannot explain the acceleration of reactions involving addition or displacement of groups. Therefore, solvent effects can explain enzymatic catalysis only where charged or partially charged groups are involved, as in the case of α-chymotrypsin.

Recent theories of enzymatic catalysis are based on polarization–depolarization cycles of both the enzyme (polarizer) and the substrate (polarizant) that result from transduction of external heat energy into conformational changes of the enzyme and, ultimately, into electrical potential energy. These theories are summarized well by Green (18) and Ji (24).

Reaction Pathways of Enzyme-Catalyzed Reactions

The term "reaction pathway" denotes the step-by-step description of the sequence of chemical events involved in enzyme-catalyzed reactions. The reaction pathway of the enzyme α-chymotrypsin (4, 8) will be used as an example.

Two different pathways were proposed for the catalytic activity of this enzyme. In the first of these, represented by Equation (1), it is assumed that no covalent intermediate is formed between the substrate and the enzymes. In the second, represented by Equations (2) and (3), it is assumed that one of the intermediates is an acyl-enzyme.

$$E + RCONR' = EHRCONR' = RCODH + H_2NR' \tag{1}$$

$$EH + RCONR' = EHRCONR' = ECOR + H_2NR' \tag{2}$$

$$ECOR + H_2O = EH + RCOOH \tag{3}$$

For some substrates, like p-nitrophenyl acetate, the second pathway was demonstrated conclusively by isolation of the acyl-enzyme intermediate. For some substrates, like esters of N-substituted aromatic acids, only indirect, kinetic evidence exists.

For further examples, the reader is referred to the pertinent literature (26, 29, 30, 33, 37).

References

1. Baker, B. R., Factors in the design of active-site-directed irreversible inhibitors. *J. Pharm. Sci.* **53**, 347 (1964).
2. Balls, A. K., and Aldrich, F. L., Acetyl-chymotrypsin. *Proc. Natl. Acad. Sci. U.S.A.* **41**, 190 (1955).
3. Balls, A. K., and Jansen, E. F., Stoichiometric inhibitor of chymotrypsin. *Adv. Enzymol.* **13**, 321 (1952).
4. Bender, M. L., and Kezdy, F. J., Mechanism of action of proteolytic enzymes. *Annu. Rev. Biochem.* **34**, 49 (1965).
5. Bender, M. L., and Kezdy, F. J., The current status of the α-chymotrypsin mechanism. *J. Am. Chem. Soc.* **86**, 3704 (1969).
6. Bernhard, S. A., "The Structure and Function of Enzymes," p. 117. Benjamin, New York, 1968.
7. Bernhard, S. A., "The Structure and Function of Enzymes," p. 175. Benjamin, New York, 1968.
8. Blow, D. M., Birktoft, J. J., and Hartley, B. S., Role of a buried acid group in the mechanism of action of chymotrypsin. *Nature (London)* **221**, 337 (1969).
9. Bruice, T. C., Proximity effects and enzyme catalysis. *In* "The Enzymes" (P. D. Boyer, ed.), 3rd ed., Vol. 2, p. 217. Academic Press, New York, 1970.
10. Bruice, T. C., and Benkovic, S. J., "Bioorganic Mechanisms," Vol. 1, p. 1. Benjamin, New York, 1966.
11. Cleland, W. W., The kinetics of enzyme-catalyzed reactions with two or more substrates or products. I. Nomenclature and rate equations. *Biochim. Biophys. Acta* **67**, 104 and 188 (1963).
12. Cram, D. J., and Guthrie, R. D., Electrophilic substitution at saturated carbon. XXVII. Carbanions as intermediates in the base-catalyzed methylene-azomethine rearrangement. *J. Am. Chem. Soc.* **88**, 5761 (1966).
13. Dixon, M., and Webb, E. C., "Enzymes," 2nd ed., p. 118. Academic Press, New York, 1964.
14. Drenth, J., Jansonius, J. N., Koekoek, R., Swen, H. M., and Wolthers, B. G., Structure of papain. *Nature (London)* **218**, 929 (1968).
15. Eigen, M., and de Maeyes, L., Relaxation methods. *In* "Investigations of Rates and Mechanisms of Reactions" (S. L. Friess, E. S. Lewin, and A. Weissberger, eds.), Vol. 8, Part II, p. 896. Wiley (Interscience), New York, 1963.

16. Elodi, P., Libor, S., and Mora, S., Localization of functional groups in dehydrogenases. *FEBS Symp.* **18**, 17 (1970).
17. Findlay, D., Mathias, A. P., and Rabin, B. R., The active site and mechanism of action of bovine pancreatic ribonuclease. *Biochem. J.* **85**, 139 (1962).
18. Green, D. E., A framework of principles for the unification of bioenergetics. *Ann. N.Y. Acad. Sci.* **227**, 6 (1974).
19. Gutfreund, H., "Introduction to the Study of Enzymes." Wiley, New York, 1965.
20. Hamilton, C. L., Niemann, C., and Hammond, G. S., A quantitative analysis of the binding of N-acyl derivatives of alpha-amino acids by alpha chymotrypsin. *Proc. Natl. Acad. Sci. U.S.A.* **55**, 664 (1966).
21. Hartsuck, J. A., and Tang, J., The carboxylate ion in the active center of pepsin. *J. Biol. Chem.* **247**, 2575 (1972).
22. Inagami, T., and Sturtevant, J. M., Nonspecific catalyses by α-chymotrypsin and trypsin. *J. Biol. Chem.* **235**, 1019 (1960).
23. Jenks, W. P., Strain and conformation change in enzymatic catalysis. *In* "Current Aspects of Biochemical Energetics" (N. O. Kaplan and E. P. Kennedy, eds.), p. 273. Academic Press, New York, 1966.
24. Ji, S., Energy and negentropy in enzymatic catalysis. *Ann. N.Y. Acad. Sci.* **227**, 419 (1974).
25. Kartha, G., Bello, J., and Harker, D., Tertiary structure of ribonuclease. *Nature (London)* **213**, 862 (1967).
26. Keleti, T., Foldi, J., Erdei, S., and Tro, T. Q., Some thermodynamic data on d-glyceraldehyde-3-phosphate dehydrogenase action under optimal conditions. *Biochim. Biophys. Acta* **268**, 285 (1972).
27. Koshland, D. E., Jr., and Neet, K. E., The catalytic and regulatory properties of enzymes. *Annu. Rev. Biochem.* **37**, 359 (1968).
28. Matthews, W., Sigler, P. B., Henderson, R., and Blow, D. M., Three-dimensional structure of tosyl-α-chymotrypsin. *Nature (London)* **214**, 652 (1967).
29. Nordlie, R. C., and Johns, P. T., The inhibition of microsomal glucose-6-phosphatase by metal-binding agents. *Biochemistry* **7**, 1473 (1968).
30. Nordlie, R. C., and Lygre, D. G., The inhibition by citrate of inorganic pyrophosphate-glucose phosphotransferase and glucose 6-phosphatase. *J. Biol. Chem.* **241**, 3136 (1966).
31. North, A. T. C., and Phillips, D. C., X-ray studies of crystalline proteins. *Prog. Biophys. Mol. Biol.* **19**, 84 (1969).
32. Pattabiraman, T. N., and Lawson, W. B., Stereochemistry of the active site of α-chymotrypsin. *J. Biol. Chem.* **247**, 3029 (1972).
33. Rabin, B. R., Evans, N., and Rashed, N., The active site sulfhydryl group of alcohol dehydrogenases: A mechanism for the enzymatic catalysis. *FEBS Symp.* **18**, 27 (1970).
34. Reeke, G. N., Hartsuck, J. A., Ludwig, M. L., Quiocho, F. A., Steitz, T. A., and Lipscomb, W. N., The structure of carboxypeptidase A. VI. Some results at 2.0-Å resolution and the complex with glycyltyrosine at 2.8-Å resolution. *Proc. Natl. Acad. Sci. U.S.A.* **58**, 2220 (1967).
35. Singer, S. J., Covalent labeling of active sites. *Adv. Protein Chem.* **22**, 1 (1967).
36. Stryer, L. Implications of X-ray crystallographic studies of protein structure. *Annu. Rev. Biochem.* **37**, 25 (1968).
37. Vallee, B. L., and Williams, R. J. P., Metalloenzymes: The entatic nature of their active sites. *Proc. Natl. Acad. Sci. U.S.A.* **59**, 498 (1968).

Chapter 5

Enzyme Cofactors

A. A. Aszalos

For their catalytic activity, certain enzymes depend on the availability of a protein structure alone, while others require, in addition, nonproteinous structures. These structures are referred to as cofactors. The enzyme cofactor complex is called a holoenzyme, and the protein structure remaining after removal of the cofactor is called an apoenzyme. Cofactors are classified into two groups: (a) metals or metalloorganic compounds; and (b) organic molecules, or coenzymes. Coenzymes can be further divided into two subgroups. In the first of these, the coenzyme is attached to the active site and can be separated, usually reversibly, from it. Thiamine pyrophosphate and pyridoxal phosphate are good examples of such coenzymes. The coenzymes of the second group are not parts of the active site, but are specific and necessary reagents of the catalyzed reactions. An example of such a coenzyme is nicotinamide adenine dinucleotide (NAD^+), which serves as an electron acceptor or donor in oxidation–reduction reactions.

On the following pages, the structures and mechanisms of action of the most common cofactors are discussed. Because of limitations of space, the role of cofactors in many important metabolic pathways can not be considered here. These aspects of enzymology have been discussed in numerous textbooks and monographs of biochemistry (1–4).

Coenzyme I and II

Coenzyme I is composed of nicotinamide, 5-phosphoribose, and adenosine monophosphate (AMP) (see Figure 5-1). In its oxidized form, it is usually referred to as nicotinamide adenine dinucleotide (NAD^+), and in its reduced form as NADH. In coenzyme II, AMP is phosphorylated at the 2′-hydroxyl

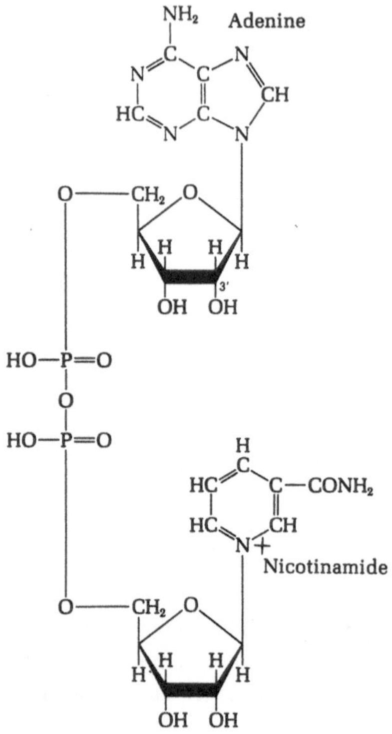

Figure 5-1.   Structures of coenzyme I.

group and is called nicotinamide adenine dinucleotide phosphate, $NADP^+$ and NADPH in its oxidized and reduced forms, respectively.

The oxidation of glyceraldehyde-3-phosphate to 1,3-diphosphoglycerate is catalyzed by glyceraldehyde-3-phosphate dehydrogenase, and requires NAD^+ as coenzyme. NAD^+ or $NADP^+$ is essential for many other enzymatic reactions, some of which are listed in Table 5-1.

The nicotinamide part of nicotinic acid is an essential vitamin in both animal and human nutrition. However, other parts of coenzymes I and II are biosynthesized.

Enzymatic reactions involving coenzyme I or II can be followed by ultraviolet spectroscopy, because the reduced nicotinamide has an absorption peak at 340 nm, whereas the oxidized form does not.

The Flavin Coenzymes

Flavin coenzymes contain riboflavin (vitamin B_2), which is an essential growth factor. Riboflavin is composed of a 6,7-dimethylisoalloxazine ring and D-ribitol, a sugar alcohol. The flavin coenzymes contain, in addition to

Table 5-1.   Some enzymatic reactions that require the presence of
coenzyme I (NAD$^+$ or NADH) or coenzyme II (NADP$^+$ or NADPH)

Enzyme	Substrate	Product	Coenzyme
Glucose-6-phosphate dehydrogenase	Glucose-6-phosphate	6-Phospho-gluconic acid	NADP$^+$
Glutamic dehydrogenase	L-Glutamic acid	α-Ketoglutarate +NH$_3$	NAD$^+$
Glutathione reductase	Oxidized glutathione	Reduced glutathione	NADPH
Quinone reductase	p-Benzoquinone	Hydroquinone	NADPH
Nitrate reductase	Nitrate	Nitrite	NADPH
Alcohol dehydrogenase	Ethanol	Acetaldehyde	NAD$^+$
Isocitric dehydrogenase	Isocitrate	α-Ketoglutarate +CO$_2$	NAD$^+$
α-Glycerolphosphate dehydrogenase	L-α-Glycerol-phosphate	Dihydroxyacetone phosphate	NAD$^+$
Lactic dehydrogenase	Lactate	Pyruvate	NAD$^+$
Malic enzyme	L-Malate	Pyruvate	NAD$^+$
Glyceraldehyde-3-phosphate	Glyceraldehyde-3-phosphate + H$_3$PO$_4$	1,3-Diphospho-glyceric acid	NAD$^+$

riboflavin, a phosphate group, flavin mononucleotide (FMN), or an adenosine diphosphate group, flavin adenine dinucleotide (FAD) (see Figure 5-2).

In contrast to NAD and NADP, the flavin coenzymes are firmly bound to the protein structure and are carried along during purification of the enzymes. They can be reversibly separated, however, with acidified water. Some of the enzyme reactions in which flavin coenzymes participate are listed in Table 5-2.

Pyridoxal Phosphate

The coenzyme pyridoxal phosphate is composed of a substituted pyridine ring and a phosphate group (see Figure 5-3). The substituted pyridine ring, 2-methyl-3-hydroxy-4-methylal-5-hydroxymethyl pyridine, is an essential vitamin (vitamin B$_6$) for most microorganisms and for higher species.

Table 5-2.   Some reactions catalyzed by flavin-containing enzymes

Enzyme	Electron donor	Product	Coenzyme	Electron acceptor
α-Amino acid oxidase	α-Amino acids	α-Ketoacids $+NH_3$	FAD	O_2
NADH dehydrogenase	NADH	NAD^+	FAD	O_2
Fatty acyl-CoA dehydrogenase	Fatty acyl-CoA	Δ_{23}Fatty acyl-CoA	FAD	CoQ
Glycolic acid oxidase	Glycolate	Glyoxylate	FMN	O_2
Succinic dehydrogenase	Succinate	Fumarate	FAD	CoQ
Nitrate reductase	NADPH	$NADP^+$	FAD	Nitrate
Lipoyl dehydrogenase	Reduced lipoic acid	Oxidized lipoic acid	FAD	NAD^+

It has been shown that pyridoxal phosphate is bound to the ε-amino group of lysine residues of enzymes by a Schiff-base type linkage. During catalytic activity, the amino group of the incoming substrate displaces the enzyme ε-amino group to form the enzyme–coenzyme–substrate complex. The formation of this new Schiff-base is the first step of the catalytic sequence mediated by this coenzyme. Incidentally, many of the same enzymatic reactions are catalyzed by pyridoxal phosphate, in the absence of the specific enzyme. In such cases, however, the reaction rates are ten times slower, presumably because of the lack of steric specificity provided by the enzyme protein (proximity effect).

Thiamine Pyrophosphate

This coenzyme is the pyrophosphate ester of thiamine, or vitamin B_1 (see Figure 5-4). Vitamin B_1 is an essential growth factor for many microorganisms and for most vertebrate species.

Thiamine pyrophosphate may be associated with α-keto acid decarboxylases, α-keto acid oxidases, transketolases, and phosphoketolases. It is essential for the function of decarboxylating enzymes, such as pyruvate decarboxylase or pyruvate dehydrogenase.

As mentioned before, thiamine pyrophosphate also participates in reactions with transketolase and transaldolase enzymes. Bound to enzymes that participate in the phosphogluconate pathway, thiamine pyrophosphate carries

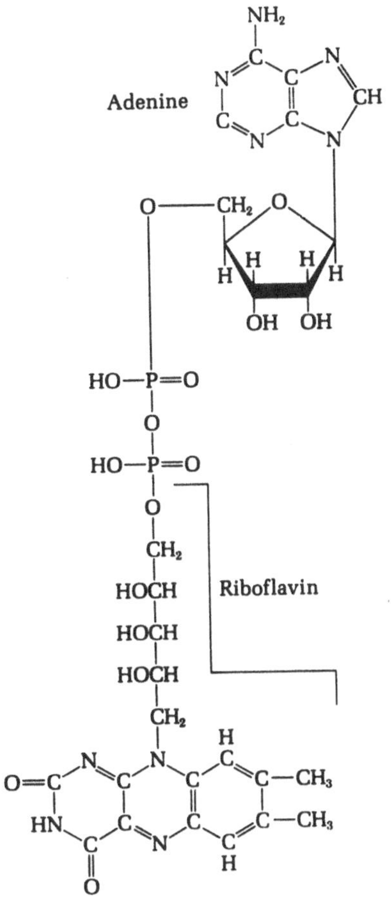

Figure 5-2. Structure of flavin adenine dinucleotide.

Figure 5-3. Structure of pyridoxal phosphate.

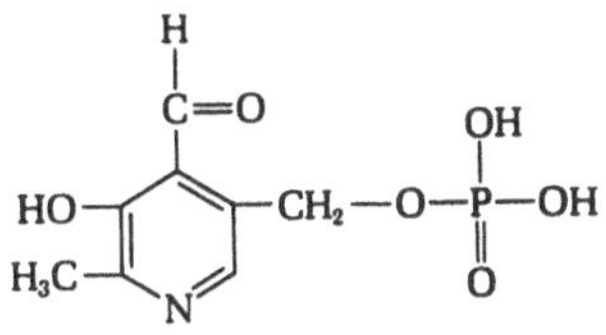

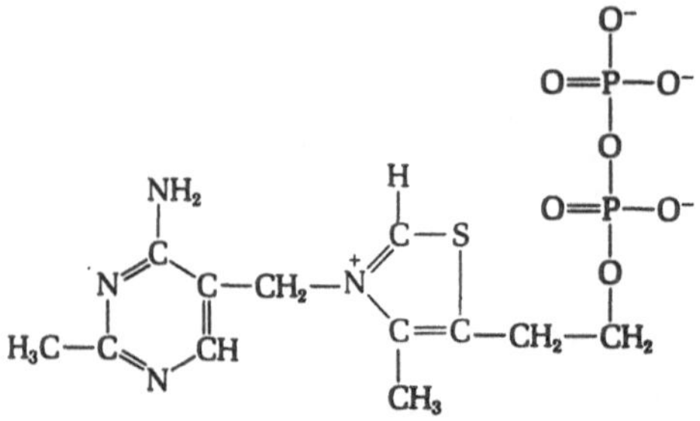

Figure 5-4. Structure of thiamine (vitamin B$_1$) pyrophosphate.

glycolaldehyde or dihydroxyacetone. These pathways are important for the interconversion of three-, four-, five-, six-, and seven-carbon sugars.

Biotin

The coenzyme biotin, like pyridoxal phosphate, is bound to ε-N-lysine residues of enzymes. In this case, however, the linkage is a permanent amide bond. Therefore, it is difficult to separate biotin from the apoenzyme; it can be done only with acid treatment. The structure of this vitamin–coenzyme is shown in Figure 5-5.

The biochemical function of biotin is the mediation of carboxylation reactions. ATP, and in certain cases Mg^{2+} ion, is also necessary for the reaction. In the activated form of biotin, the proton of the ring nitrogen is replaced by a —COO$^-$ radical. The activated biotin can transfer a —COO$^-$ radical to an α-carbon of a CoA-bound carboxyl or to an α-carbon of a CoA-bound crotonyl group. These α-carbons are nucleophilic. The positively charged carbon of the active —COO$^-$ radical bound to the nitrogen of biotin readily accepts these nucleophiles, thereby completing the transfer of —COO$^-$.

Figure 5-5. Structure of the coenzyme biotin.

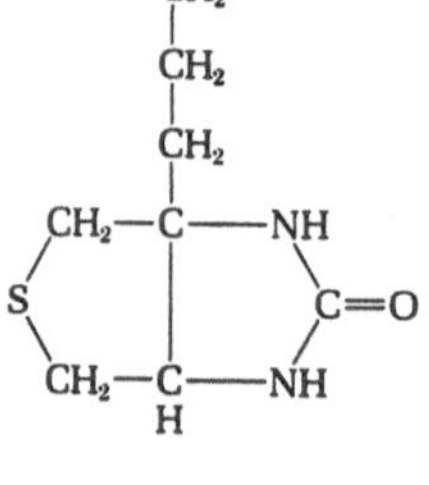

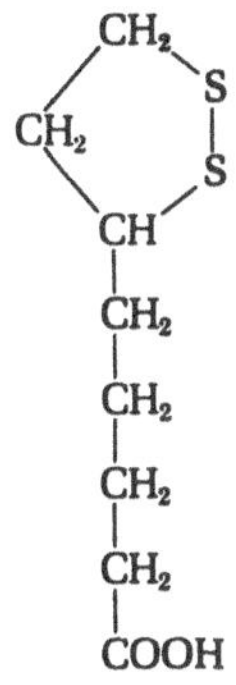

$$\begin{array}{c}
\quad CH_2 \\
\quad\quad \diagdown S \\
CH_2 \quad\quad | \\
\quad\quad\quad S \\
\quad CH \diagup \\
\quad | \\
\quad CH_2 \\
\quad | \\
\quad CH_2 \\
\quad | \\
\quad CH_2 \\
\quad | \\
\quad CH_2 \\
\quad | \\
\quad COOH
\end{array}$$

Figure 5-6. Structure of lipoic acid.

Biotin is involved in intermediary metabolic reactions, like the ones that are catalyzed by pyruvate carboxylase, acetyl-CoA carboxylase, methyl-crotonyl-CoA carboxylase, and carboxyl transferases.

Lipoic Acid

Lipoic acid (see Figure 5-6) belongs to the thiol coenzyme group. Essentially, it is an eight-carbon saturated fatty acid, in which the 6- and 8- carbons are joined by a disulfide group. It is bound by a peptide linkage to the ε-amino group of a lysine residue of enzymes, such as lipoyl reductase-transacetylase. This enzyme is part of the pyruvate dehydrogenase system, which mediates the oxidative decarboxylation of pyruvate to acetyl-CoA and CO_2.

Lipoid acid-containing enzymes are important in the generation of acyl groups, in acyl transfer, and in electron-transport systems. The mechanism of action of lipoic acid is based on its ability to function as a redox system and to form thermodynamically unstable thiol esters. The redox system works on the principle that the disulfide linkage of lipoic acid is reduced, while the substrate is oxidized and gets covalently bound to lipoic acid. The thiolester linkage formed is thermodynamically less stable than the corresponding oxygen compound.

Coenzyme A

Coenzyme A (CoA, CoASH) also belongs to the thiol coenzyme group. It consists of β-amino ethanethiol, adenosine diphosphate, and pantothenic acid. Pantothenic acid is an essential vitamin for microorganisms, as well as for higher species. The structure of coenzyme A is shown in Figure 5-7.

The mode of action of CoA is the same as that of lipoic acid. Reactions with acyl-CoA can proceed via nucleophilic displacement at the carbonyl carbon or by electrophilic attack at the α-carbon. CoA, however, does not

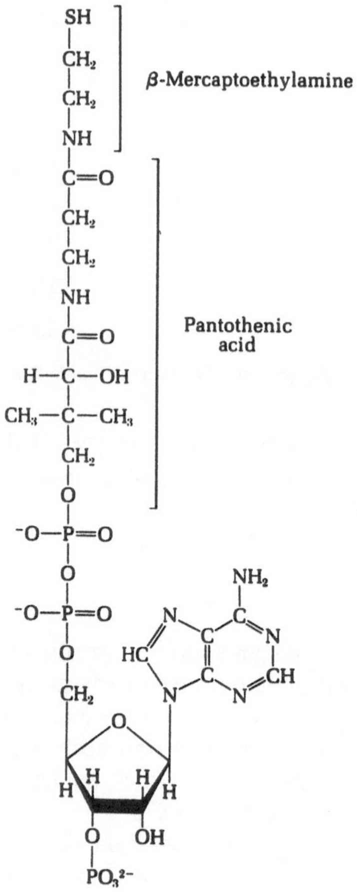

Figure 5-7.   Structure of coenzyme A (CoA, CoASH).

possess the oxidation–reduction ability of lipoic acid. CoA is primarily involved in acyl transfers, condensation reactions, α–β elimination, or β = hydroxylation. These reactions are involved in a variety of biochemical events in most living organisms.

Glutathione

The third member of the thiol coenzyme group is glutathione. Structurally, it is a γ-glutamylcysteinylglycine. The sulfhydryl group of its cysteine residue is responsible for its coenzyme function. The mechanism of action of glutathi-

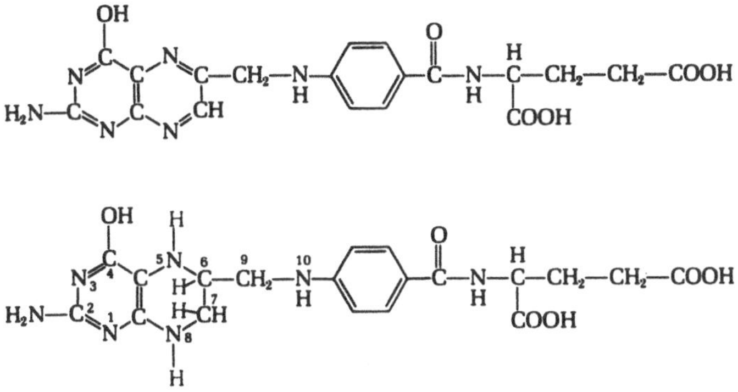

Figure 5-8. Structure of folic acid (F) and tetrahydrofolic acid (FH$_4$).

one is similar to that of CoA. Glutathione is involved in the enzymatic oxidation of aldehydes to acids by NAD$^+$ and in isomerization reactions.

Acyl Carrier Protein (ACP)

The fourth member of the thiol coenzyme group is the acyl carrier protein. It serves as an acyl carrier during the biosynthesis of fatty acids. Basically, it is a polypeptide with a molecular weight of about 10,000 and contains a 4'-phosphopantothiene moiety, like CoA. This moiety is linked to a serine residue in the polypeptide chain via a phosphoric acid–ester linkage. The function of ACP in fatty acid synthesis is to hold the acyl intermediates in thiolester form during the reaction by which the aliphatic chain is built up.

Tetrahydrofolic Acid

Folic acid is the precursor of this coenzyme. Both compounds contain glutamic acid, p-aminobenzoic acid, and 6-methyl-2-amino-4-hydroxypterin moieties. One of these components, p-aminobenzoic acid, is an essential vitamin. In the presence of p-aminobenzoic acid, folic acid (F) is biosynthesized in most organisms. F is then further reduced enzymatically to dihydro- (FH$_2$) and tetrahydrofolic acid (FH$_4$). The structures of F and FH$_4$ are shown in Figure 5-8.

The biochemical function of FH$_4$ is based on its ability to mediate the transfer of single-carbon moieties. If these reactions do not occur, the amino acid and nucleic acid metabolism is blocked. This blocking can be achieved intentionally by using folic acid analogs that are inhibitors of the enzyme dihydrofolic acid reductase, which catalyzes the reaction FH$_2 \rightarrow$ FH$_4$. Such folic acid analogs are used in the chemotherapy of certain leukemias.

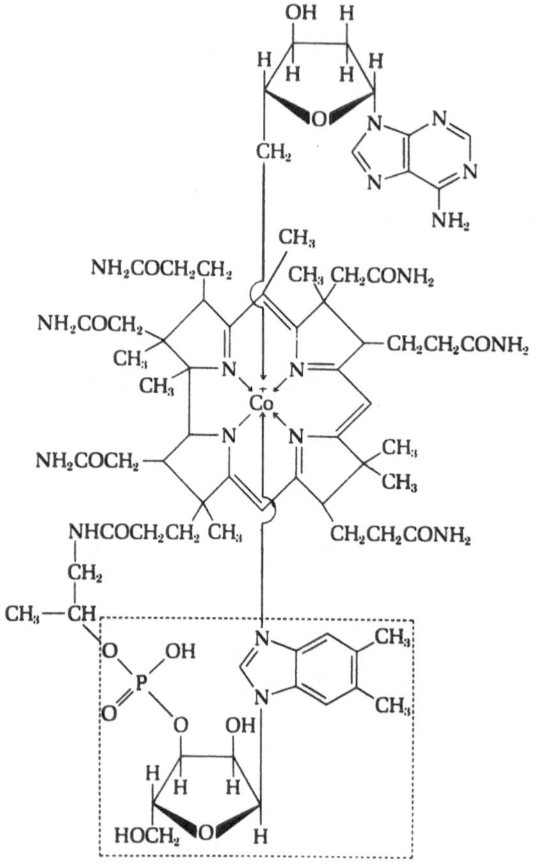

Figure 5-9. Structure of the coenzyme cobamide.

Cobamide

The coenzyme cobamide (see Figure 5-9) is a derivative of vitamin B_{12}, which is an essential vitamin not only for man, but probably also for most animal and plant species. Vitamin B_{12}, cobalamine, is probably present in nature as the coenzyme, but is degraded to vitamin B_{12} during isolation.

The central feature of cobamide is a cobalt atom with six coordinational valencies. One of these is bound to a 5′-deoxyadenosyl group in cobamide, or to cyanide, chloride, or hydroxyl groups in different forms of vitamin B_{12}. The Co^{2+} atom can undergo reduction to Co^+, which is an extremely strong nucleophile. Co^+ can serve as a carrier of alkyl groups in biosynthetic reactions. This ability of cobamide is the basis of its coenzyme activity. For example, in the formation of methane gas from methylated compounds by anaerobic bacteria, cobamide mediates the transfer of the methyl group.

Nonheme Iron Proteins, Cytochromes, and Chlorophylls

Nonheme iron proteins, cytochromes, and chlorophylls participate in electron-transfer processes.

Nonheme iron proteins have a molecular weight of 6000 to 12,000. They bind two to eight iron atoms through sulfur linkages of their cysteine residues. Their mode of action is still unclear; it may be associated with oxidation–reduction processes in a one-electron transfer step. Nonheme iron proteins participate in reactions involving NADH dehydrogenase and succinate dehydrogenase and are also important for photosynthesis and nitrogen fixation.

Cytochromes are metal ion-containing proteins in which the metal ion is part of a prosthetic group. They particpate in the sequential electron transfer from flavoproteins to moleculr oxygen in aerobic systems. The prosthetic group containing the metal ion is a porphyrin moiety. Porphyrins are tetrapyrrole compounds. The various prophyrins differ from one another with regard to the composition of their side chains and the relative order of the substitution of these side chains on the tetrapyrrole frame. Porphyrins form chelate complexes with metals such as iron (Fe), magnesium (Mg), nickel (Ni), cobalt (Co), and copper (Cu). The Fe^{2+} complex is called "heme" and the Fe^{3+} complex is called "hemin." Figure 5-10 shows the structure of heme A. Four coordination valencies of Fe^{2+} in heme A are bound to the pyrrole nitrogens of the porphyrin structure, the fifth to an imidazole ring of a histidine, and the sixth to oxygen, carbon monoxide, or cyanide in myoglobin and hemoglobin. In cytochromes b, c, and c_1, both the fifth and sixth coordination valencies are occupied by amino acids of the proteins.

The Fe^{2+} in hemoglobin and myoglobin does not change reversibly to Fe^{3+} during oxygen uptake. In cytochromes, however, Fe^{2+} changes to Fe^{3+} during electron transport. Cytochromes a and a_3 contain (Cu). Cu^+ changes reversibly to Cu^{2+} during electron transport to oxygen.

Figure 5-10.   Structure of heme A.

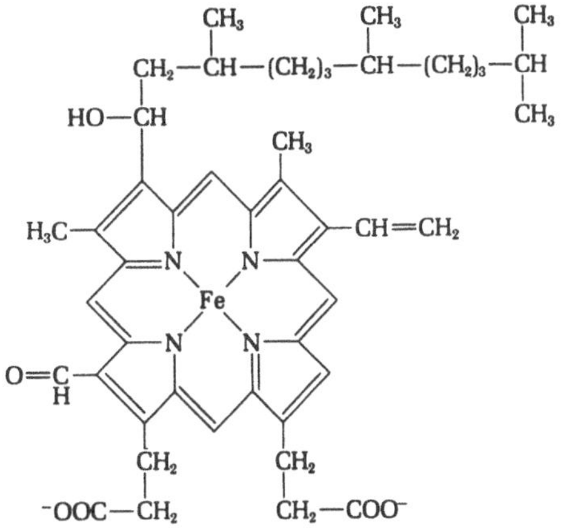

The catalases and peroxidases also contain heme. The modes of action of these hemes are not fully understood.

Chlorophylls also contain pyrrole-type porphyrins. In chlorophylls, the chelating metal is Mg. They are classified, according to the side-chain substitutions on the porphyrin ring, as chlorophylls a, b, c, and d.

All chlorophylls absorb visible light efficiently, because of their many conjugated double bonds. The light energy (photons) absorbed as the double bonds spreads through the electronic structure of the chlorophyll molecule. This energy is further transmitted to a compound called P700, which emits the energized electrons required for photosynthetic electron transport.

Metal Ions

Metal ions are necessary cofactors of about one-fourth of the known enzymes. The specific metal requirements of some enzymes are listed in Table 5-3.

Metal ions, like protons, are "Lewis acids." This means that they can form an ε-bond with a donor of an electron pair. Unlike protons, however, they can accept electrons into their low-lying vacant orbitals to form π bonds.

Table 5-3. The metal-ion requirements of some enzymes

Enzyme	Metal ion
Phosphohydrolases Phosphotransferases	Mg^{2+}
Arginase Phosphotransferases	Mn^{2+}
Cytochromes Peroxidase Catalase Ferredoxin	Fe^{2+} or Fe^{3+}
Tyrosinase Cytochrome oxidase	Cu^{2+}
Alcohol dehydrogenase Carbonic anhydrase Carboxypeptidase	Zn^{2+}
Pyruvate phosphokinase (also requires Mg^{24})	K^+
Plasma-membrane ATPase (also requires K^+ and Mg^{24})	Na^+

They can function as three-dimensional templates (molds) for the binding of bases. Their ability to form ε and π bonds is based on their polarization ability and electron configuration. The mode of action of metal ion-mediated enzymes is very complex and could not be detailed here.

References

1. Bernhard, S., "The Structure and Function of Enzymes." Benjamin, New York, 1968.
2. Boyer, P. D., Lardy, H., and Myrbäck, K., eds., "The Enzymes," 2nd rev. ed., Vols. 2 and 3. Academic Press, New York, 1960.
3. Hutchison, D. W., "Nucleotides and Coenzymes." Wiley, New York, 1965.
4. Wagner, A. F., and Folkers, K., "Vitamins and Coenzymes." Wiley (Inter-Science), New York, 1964.

Chapter 6

Immobilized Enzymes

A. A. Aszalos

Because of their increasing importance in industry and medicine, a brief discussion of immobilized enzymes is warranted.

Nonimmobilized enzymes are usually water soluble. After their administration *in vivo*, their duration of action is limited because they are excreted and are broken down by proteolytic enzymes. In contrast, enzymes that are immobilized by adsorption, covalent bonding, or encapsulation are insoluble in water, not readily excreted, and not attacked by proteolytic enzymes.

The least expensive method for the large-scale immobilization of enzymes is adsorption, e.g., to bentonite (8), cellulose (18), collagen (21), or surface-active glass (15). The disadvantage of this method is that the adsorbed enzyme is slowly washed away from the adsorbent.

Enzymes bound covalently to water-insoluble polymeric carriers are more stable and can be easily removed from solution. These preparations can also be used for the removal from solution of enzyme inhibitors or substrates.

The immobilization not only of enzymes, but also of other proteins, by covalent bonding to polymers has recently attracted considerable interest. Two types of materials are frequently used for this purpose: (a) polyanionic substances with which the enzyme is bound to a carboxylic polymer by an amide linkage; and (b) modified starch or cellulose derivatives to which the enzyme is bound by a diazo linkage.

Amino groups of enzymes and of other proteins may be coupled to insoluble polymers by azid (12) and carbodiimid (26) reagents. For binding to cellulose or Sephadex, sym-trichlorotriazine (20), dichloro-sym-triazinyl dye (28), or cyanogen bromide (1) can be used. Proteins containing relatively large amounts of aromatic amino acids can be coupled to the polydiazonium salts of *p*-aminobenzylcellulose (10) or to the bismethylenedianiline derivative of oxidized starch (19).

Certain enzymes can be bound covalently to thin films or to membranes (14). Such films can be made from a mixture of albumin and pepsin that undergoes self digestion at low pH (16).

Recently, urease (25), trypsin (23), papain (22), and acetylcholinesterase (2) have been bound to porous glass by diazo group coupling.

Other inorganic carriers, such as NiO (7) and aluminum (24), have been used for the covalent bonding of enzymes.

The third method of enzyme immobilization is encapsulation (6). The method involves the mixing of the buffered aqueous enzyme solution in an organic liquid with a detergent. Monomers of the polymer film to be formed are also dissolved in the system, and polymerization takes place almost entirely at the interface of the organic solvent and the emulsified protein solution. With suitable techniques, 60- to 100-μm diameter semipermeable capsules of nylon, polyacrylamide, and other materials can be produced. Enzymes remain encapsulated, but the substrates can diffuse through the semipermeable membranes and can interact with the enzyme.

The stability of encapsulated enzymes is good and can be improved by crosslinking the enzyme to the membrane, for example, with glutaraldehyde. Incorporation of large concentrations of a neutral protein, such as hemoglobin (3), also increases enzyme stability.

Encapsulated enzymes had already been used to a limited extent in biology and medicine. Encapsulated catalase was used for the treatment of acatalasaemic mice (5). Encapsulated urease decreased the concentration of blood urea *in vivo* (9), and encapsulated L-asparaginase was used for the treatment of asparagine-dependent lymphosarcoma (4).

L-asparaginase is the first enzyme used with greater success in cancer chemotherapy. Therefore, besides regular encapsulation methods, L-asparaginase was encapsulated in red blood cells (20a) to avoid serious allergic reactions. Also, extension in the *in vivo* half-life of L-asparaginase and the possibility of targeting this agent into the reticuloendothelial system could be achieved by this encapsulation method.

Not only the solubility and stability, but also the kinetic properties, of immobilized enzymes are different from those of free enzymes. The factors capable of affecting the kinetics of immobilized enzymes include: (a) the chemical composition of the carrier; (b) steric restrictions imposed by the carrier; and (c) the rate of diffusion of the substrate to the enzyme buried in the carrier.

The influence of pH changes on the activity of enzymes bound to a polyionic support is different from that on the activity of the free enzyme. This difference is due to the influence of repulsive ionic forces, and can be counteracted by using substrate solutions of high ionic strength (27) that overcome the uneven distribution of ionic species among the charged solid phase, the enzyme, and the surrounding solution. In the case of small uncharged substrates, enzyme kinetics are influenced only by diffusion rates.

In the case of large substrates, in addition to diffusion rates, steric hindrance by the supporting network may also influence their access to the encapsulated or otherwise entrapped enzyme. For example, the water-insoluble polytyrosyl-trypsin hydrolyzes casein at 15 to 30% of the rate of the free enzyme. Interestingly, the Michaelis constant (K_m) of an immobilized enzyme may be lower than that of the free enzyme (17). This increase in affinity of a substrate for an immobilized enzyme can occur if favorable conformational changes develop upon immobilization. In other cases, the electrical charge of the support to which the enzyme is bound may be opposite to that of the substrate. Under these circumstances, the maximum velocity of the enzymatic reaction (V_{max}) is reached with lower substrate concentrations than are needed for the free enzyme.

For a detailed discussion of the kinetic behavior, pH-activity profile, and electrostatic potentials in the neighborhood of an enzyme bound to a solid support, the reader is referred to the literature (13, 29).

References

1. Axén, R., Poráth, J., and Ernback, S., Chemical coupling of peptides and proteins to polysaccharides by means of cyanogen halides. *Nature (London)* **214**, 1302 (1967).
2. Baum, G., Ward, F. B., and Weetall, H. H., Stability, inhibition and reactivation of acetylcholinesterase covalently coupled to glass. *Biochim. Biophys. Acta* **268**, 411 (1972).
3. Chang, T. M., Stabilization of enzymes by microencapsulation with a concentrated protein solution or by microencapsulation followed by cross-linking with glutaraldehyde. *Biochem. Biophys. Res. Commun.* **44**, 1531 (1971).
4. Chang, T. M., The in vivo effects of semipermeable microcapsules containing L-asparaginase on 6C3HED lymphosarcoma. *Nature (London)* **229**, 117 (1971).
5. Chang, T. M., and Poznansky, M. J., Semipermeable microcapsules containing catalase for enzyme replacement in acatalasaemic mice. *Nature (London)* **218**, 243 (1968).
6. Chang, T. M., MacIntosh, F. C., and Mason, S. G., Semipermeable aqueous microcapsules. I. Preparation and properties. *Can. J. Physiol. Pharmacol.* **44**, 115 (1966).
7. Corning Glass Works, German Patent 1,944,418.
8. Durand, G., Enzymologie des sols. Modifications de l'activité de l'uréase en présence de bentonite. *C.R. Hebd. Seances Acad. Sci.* **259**, 3397 (1964).
9. Gardner, D. L., Falb, R. C., Kim, B. C., and Emmerling, D. C., Possible uremic detoxification via oral-injected microcapsules. *Trans. Am. Soc. Artif. Intern. Organs* **17**, 239 (1971).
10. Goldstein, L., Pecht, M., Blumberg, S., Atlas, D., and Levin, Y., Water-insoluble enzymes. Synthesis of a new carrier and its utilization for preparation of insoluble derivatives of papain, trypsin, and subtilopeptidase A. *Biochemistry* **9**, 2322 (1970).
11. Gurvich, A. E., Quantitative determination of antibody content by means of protein antigens on paper. *Biokhimiya* **22**, 977 (1957).
12. Hornby, W. E., Lilly, M. D., and Crook, E. M., The preparation and properties of ficin chemically attached to carboxymethylcellulose. *Biochem. J.* **98**, 420 (1966).
13. Katchalski, E., Silman, I., and Goldman, K., Effect of the microenvironment on the mode of action of immobilized enzymes. *Adv. Enzymol.* **34**, 445 (1971).
14. Kay, G., Lilly, M. D., Sharp, A. K., and Wilson, R. J. H., Preparation and use of porous sheets with enzyme action. *Nature (London)* **217**, 741 (1968).

15. Kobamoto, N., Lofroth, G., Camp, P., Van Amburg, G., and Augenstein, L., Specificity of trypsin adsorption onto cellulose, glass and quartz. *Biochem. Biophys. Res. Commun.* **24**, 622 (1966).

16. Mazia, D., and Hayashi, T., The activity of pepsin-albumin films. *Arch. Biochem. Biophys.* **43**, 424 (1953).

17. McLaren, A. D., and Packer, L., Some aspects of enzyme reactions in heterogeneous systems. *Adv. Enzymol.* **33**, 245 (1971).

18. Siegel, B. Z., and Siegel, S. M., Enhancement of peroxidase action by polysaccharides. *Nature (London)* **186**, 391 (1960).

19. Silman, I. H., Albu-Weissenberg, M., and Katchalski, E., Some water-insoluble papain derivatives. *Biopolymers* **4**, 441 (1966).

20. Surinov, B. P., and Manoilov, S. E., Production and properties of insoluble compounds of certain enzymes with cellulose. *Biokhimiya* **31**, 337 (1966).

20a. Updike, S. J., Wakamiya, R. T., and Lightfoot, E. N., Jr., Asparaginase entrapped in red blood cells: Action and survival. *Science* **193**, 681 (1976).

21. Vieth, W. R., Gilberg, S. G., and Wang, S. S., Performance of collagen-invertase complex membrane in a biocatalytic module. *Trans. N.Y. Acad. Sci.* **34**, 454 (1972).

22. Weetall, H. H., Alkaline phosphatase insolubilized by covalent linkage to porous glass. *Nature (London)* **223**, 959 (1968).

23. Weetall, H. H, Trypsin and papain covalently coupled to porous glass: Preparation and characterization. *Science* **166**, 615 (1969).

24. Weetall, H. H., Storage stability of water-soluble enzymes, enzymes covalently coupled to organic and inorganic carriers. *Biochim. Biophys. Acta* **212**, 1 (1970).

25. Weetall, H. H., and Hersh, L. S., Urease covalently coupled to porous glass. *Biochim. Biophys. Acta* **185**, 464 (1969).

26. Weliky, N., and Weetall, H. H., The chemistry and use of cellulose derivatives for the study of biological systems. *Immunochemistry* **2**, 293 (1965).

27. Wharton, C. W., Crook, E. M., and Brocklehurst, K., The nature of the perturbation of the Michaelis constant of the bromelain-catalyzed hydrolysis of α-N-benzoyl-L-arginine ethyl ester consequent upon attachment of bromelain to carboxymethyl-cellulose. *Eur. J. Biochem.* **6**, 572 (1968).

28. Wilson, R. J. H., Kay, G., and Lilly, M. D., The preparation and kinetics of lactate dehydrogenase attached to water-insoluble particles and sheets. *Biochem. J.* **108**, 845 (1968).

29. Zaborsky, O., ed., "Immobilized Enzymes." CRC Press, Cleveland, Ohio, 1972.

Chapter 7

Classification of Enzymes

A. A. Aszalos

In the past, several systems have been used for the classification of enzymes. At present, the classification system introduced by the International Commission on Enzymes is used almost exclusively. This commission, founded in 1955, considered "the classification and nomenclature of enzymes and coenzymes, their units of activity and standard methods of assay, together with the symbols used in the description of enzyme kinetics." More than 1300 known enzymes were classified according to the type of reaction catalyzed by them. The size, structure, source, subunit numbers, or cofactor requirements used in earlier classifications do not play any role in this classification system.

Each enzyme is assigned a number made up of four parts. The first part indicates the main group to which the enzyme belongs. The main groups are:

Group 1 Oxidoreductases
Group 2 Transferases
Group 3 Hydrolases
Group 4 Lyases
Group 5 Isomerases
Group 6 Ligases

The second part indicates the subgroup and the third part the sub-subgroup to which the enzyme belongs. The fourth part is the number of the enzyme in that particular sub-subgroup (1).

Oxidoreductases

A large number of enzymes belong to this group. Some of these were discovered in the past century. Oxidoreductases transfer electrons from the substrate and, in some cases, transfer oxygen directly to the substrate.

Enzymes that oxidize the primary substrate have been designated dehydro-genases. Those enzymes that utilize molecular oxygen as an immediate electron acceptor, forming H_2O_2 or H_2O, are called oxidases. In many biologic reactions, where the end result is the addition of molecular oxygen, the catalyzing enzymes are called oxygenases.

Oxidoreductases act on substrates like alcohol, aldehyde, ketone, CH—CH groups, amines, imines, sulfur-containing groups, and diphenols. The acceptor molecule can be nicotinamide-adenine dinucleotide, cytochrome, O_2, quinone, disulfide, etc.

1.1 Oxidoreductases of this subgroup act on the —CH—OH type of donor. Many biologically important reactions are catalyzed by these enzymes. The study of some oxidoreductases dates back as far as the end of the nineteenth century.

1.1.1. Oxidoreductases of this sub-subgroup utilize nicotinamide-adenine dinucleotide (NAD) or nicotinamide-adenine dinucleotide phosphate (NADP) as acceptors. A great variety of reactions is catalyzed by these enzymes. For example, alcohol dehydrogenase catalyzes the reaction:

Alcohol + NAD = aldehyde or ketone + reduced NAD

Alcohol dehydrogenase (NADP) utilizes NADP as acceptor exclusively in the same general type of reaction. Glycerol dehydrogenase catalyzes the reaction:

Glycerol + NAD = dihydroxyacetone + reduced NAD

Mannitol phosphate dehydrogenase catalyzes the reaction:

D-Mannitol 1-phosphate + NAD = D-fructose 6-phosphate + reduced NAD

An important reaction common to all living organisms, the oxidation of uridine diphosphate (UDP) D-glucuronate, is catalyzed by UDP glucose dehydrogenase. Equally important is the enzyme glucose-6-phosphate dehy-drogenase, which produces D-glucono-γ-lactone 6-phosphate from D-glucose 6-phosphate. Lactate dehydrogenase acts on L-lactate:

L-Lactate + NAD = pyruvate + reduced NAD

Cortisone reductase catalyzes the reaction:

20-Dehydrocortisone + NAD = cortisone + reduced NAD

1.1.2. This sub-subgroup of oxidoreductases utilizes cytochrome as accep-tor. For example, lactate dehydrogenase catalyzes the reaction:

L-Lactate + ferricytochrome c = pyruvate + 2 ferrocytochrome c

1.1.3. Enzymes of this sub-subgroup are oxidoreductases acting on —CH—OH groups and utilizing O_2 as the acceptor molecule. For example, lactate oxidase catalyzes the reaction:

L-Lactate + O_2 = acetate + CO_2 + H_2O_2

Other members of this subgroup are glycolate oxidase and glucose oxidase.

1.1.9. Enzymes of this subgroup carry out oxidation–reduction reactions of —CH—OH groups, utilizing acceptors other than O_2, cytochrome, NAD, or NADP. An interesting member of this subgroup is choline dehydrogenase, which catalyzes the reaction:

Choline + acceptor = betaine aldehyde + reduced acceptor

1.2. Oxidoreductases of this subgroup catalyze reactions using an aldehyde or keto group as donor. This subgroup is divided again according to the acceptor group.

1.2.1. Enzymes of this sub-subgroup utilize NAD or NADP as acceptors. Formaldehyde dehydrogenase, for example, catalyzes the reaction:

Formaldehyde + NAD + H_2O = formate + reduced NAD

Different aldehyde dehydrogenases, like benzaldehyde-, betainealdehyde-, and aspartate semialdehyde-dehydrogenase, belong to this sub-subgroup.

1.2.2. Members of this sub-subgroup utilize cytochrome as acceptor. An example of this type of reaction, catalyzed by pyruvate dehydrogenase, is:

Pyruvate + ferricytochrome b_1 = acetate + CO_2 + ferrocytochrome b

1.2.3. Enzymes of this sub-subgroup act on aldehydes or ketones as donors and utilize O_2 as acceptor. Aldehyde oxidase, for example, catalyzes the reaction:

Aldehyde + H_2O + O_2 = acid + H_2O_2

1.2.4. Enzymes of this sub-subgroup utilize oxidized lipoate as acceptor. For example, pyruvate dehydrogenase catalyzes the reaction:

Pyruvate + oxidized lipoate = 6-S-acetylhydrolipoate + CO_2

1.2.99. This sub-subgroup of enzymes utilizes other acceptors than do the enzymes in subgroups 1.2.1.—1.2.4.

1.3. Oxidoreductases of this subgroup dehydrogenate —CH—CH groups. Further classification again depends on the nature of the acceptor.

1.3.1. Enzymes of this sub-subgroup utilize NAD or NADP as an acceptor. An example of this type of reaction, catalyzed by cortisone β-reductase, is

4,5,β-Dihydrocortisone + NADP = cortisone + reduced NADP

1.3.2. Cytochrome is used as acceptor by this sub-subgroup of oxidoreductases. For example, galactonolactone dehydrogenase catalyzes the reaction:

L-Galactone-γ-lactone + 2 ferricytochrome c = L-ascorbate
+ 2 ferrocytochrome c

1.3.3. This sub-subgroup of enzymes utilizes O_2 as acceptor and —CH—CH— groups as donors.

1.3.99. This sub-subgroup of enzyme utilizes acceptors other than O_2, cytochrome, NAD, or NADP. For example, butyryl-CoA dehydrogenase, acyl-CoA dehydrogenase, and 3-ketosteroid Δ^1-dehydrogenase belong to this sub-subgroup.

1.4. Oxidoreductases of this subgroup act on donors containing —CH—NH_2 groups and utilize O_2, NAD, or NADP as acceptors.

1.4.1. NAD or NADP is utilized as acceptor by this sub-subgroup of enzymes acting on CH—NH_2 groups. Alanine dehydrogenase, for example, catalyzes the reaction:

L-Alanine + H_2O + NAD = pyruvate + NH_3 + reduced NAD

Glycine dehydrogenase, D-proline reductase, and glutamate dehydrogenase are other members of this subgroup.

The deamination of naturally occurring amines and many pharmacologically interesting substances are catalyzed by amine oxidases. These enzymes act by oxidative removal of the amino group. Examples of substrates are 5-hydroxytryptamine, epinephrine, tyramine, and histamine.

1.4.3. Oxygen is used as acceptor by this sub-subgroup of enzymes. L-Aspartate oxidase, L-aminoacid oxidase, arginine oxidase, and ethanolamine oxidase belong to this sub-subgroup.

1.5 A C—NH group is the donor for this subgroup of oxidoreductases.

1.5.1. This sub-subgroup of enzymes utilizes NAD or NADP as acceptor. For example, pyrroline-2-carboxylate reductase catalyzes the reaction:

L-Proline + NAD = Δ^1-pyrroline-2-carboxylate + reduced NAD

An important member of this sub-subgroup is tetrahydrofolate dehydrogenase.

1.5.3. Members of this enzyme sub-subgroup utilize O_2 as acceptor. Sarcosine oxidase and spermine oxidase belong to this sub-subgroup.

1.6. Oxidoreductases of this subgroup oxidize reduced NAD or NADP. The acceptor can be NAD, NADP, cytochrome, disulfides, quinones, or other molecules.

1.6.1. NAD as acceptor. NAD or NADP transhydrogenase catalyzes the reaction:

Reduced NADP + NAD = reduced NAD + NADP

1.6.2. Cytochrome as acceptor. Cytochrome b_5 reductase catalyzes the reaction:

Reduced NAD + 2 ferricytochrome b_5 = NAD + 2 ferrocytochrome b_5.

1.6.4. Glutathione as acceptor. Glutathione reductase catalyzes the reaction:

Reduced NADP + oxidized glutathione = NADP + glutathione

1.6.6. Nitrate as acceptor. Nitrate reductase catalyzes the reaction:

Reduced $NAD + nitrate = NAD + nitrate + H_2O$

1.6.99. Nonspecific acceptors. Reduced NADP dehydrogenase is not specific in its acceptor requirements and can use ferricyanide, methylene blue, or benzoquinone as acceptor.

1.7. Oxidoreductases of this group act on nitrogenous compounds other than those with $-CH-NH_2$ or $-C-NH$ groups.

1.7.3. This sub-subgroup utilizes O_2 as acceptor; the 1.7.99 sub-subgroup is not specific in its acceptor requirements.

1.8. Oxidoreductases of this subgroup utilize sulfur-containing compounds as donors. Further classification is based on the nature of the acceptor molecule.

1.8.1. Enzymes of this sub-subgroup utilize NADP as acceptor. Sulfate reductase catalyzes the reaction:

$H_2S + 3\ H_2O = sulfite + 3$ reduced NADP

1.8.3. Enzymes of this sub-subgroup utilize O_2 as acceptor. Sulfite reductase catalyzes the reaction:

$Sulfite + O_2 + H_2O = sulfate + H_2O_2$

1.9. Oxidoreductases of this subgroup act on heme-type donors. Enzymes of the sub-subgroup 1.9.3, like cytochrome oxidase, utilize O_2 as acceptor, and enzymes in the sub-subgroup 1.9.6., like nitrate reductase, utilize nitrate as acceptor.

1.10. Oxidoreductases of this subgroup act on diphenol and related donors and use O_2 as acceptor. For example, ascorbate oxidase (1.10.3) catalyzes the reaction:

2 L-Ascorbate + O_2 = 2 dehydroascorbate + 2 H_2O

1.11. Oxidoreductases of this subgroup utilize H_2O_2 as the acceptor molecule. Cytochrome oxidase, for example, catalyzes the reaction:

2 Ferrocytochrome $c + H_2O_2 + 2\ H^+ = 2$ ferricytochrome $c + 2\ H_2O$

Catalase, an iron-containing enzyme, catalyzes the biologically important reaction:

2 $H_2O_2 = O_2 + 2\ H_2O$

Peroxidase and NAD peroxidase also belong to this subgroup.

1.12. Oxidoreductases of this subgroup utilize hydrogen as donor molecule.

1.13. Oxidoreductases of this subgroup are called oxygenases and incorporate O_2 into a single acceptor molecule. For example, meso-inositol oxygenase catalyzes the reaction:

meso-Inositol + O_2 = D-glucuronate

Other enzymes of this subgroup are tryptophan oxygenase, catechol oxygenase, 3-hydroxyanthranilate oxygenase, and lipoxygenase. These enzymes play an important role in the food industry by oxidizing unsaturated fatty acids and carotenes.

1.14. Oxidoreductases of this subgroup act on paired donors, but incorporate oxygen into only one acceptor.

1.14.1. Enzymes of this sub-subgroup utilize either reduced NAD or NADP as one of the two donors. For example, aryl-4-hydroxylase catalyzes the reaction:

Aniline + reduced $NADP + O_2 = $ 4-hydroxyaniline + $NADP + H_2O$

Other members of this subgroup are steroid 11-β-hydroxylase, steroid 17-α-hydroxylase, and salicylate hydroxylase.

1.14.2. Enzymes of this sub-subgroup use ascorbate as one of the two donors. Dopamine hydroxylase and p-hydroxyphenyl pyruvate hydroxylase are members of this sub-subgroup.

1.14.3. Pteridine is one of the two donors utilized by these enzymes. For example, tyrosine-3-hydroxylase catalyzes the reaction:

L-Tyrosine + tetrahydropteridine + $O_2 = $ 3,4-dihydroxy-L-phenylalanine + dihydropteridine + H_2O

Transferases

Enzymes of the second main group catalyze transfer reactions of chemical groups such as methyl, carboxyl, aldehyde, glycosyl, phosphorus, sulfur, and nitrogen radicals.

2.1. This subgroup is capable of transferring methyl, hydroxymethyl, carboxyl, and amidino groups.

2.1.1. Enzymes in this sub-subgroup catalyze the transfer of a methyl group. For example, betaine homocysteine methyltransferase catalyzes the reaction:

Betaine + L-homocysteine = dimethylglycine + L-methionine

The enzyme catechol methyltransferase catalyzes the reaction:

S-Adenosylmethionine + nicotinate = S-adenosylhomocysteine + N-methylnicotinate

2.1.2. Enzymes of this sub-subgroup catalyze the transfer of hydroxymethyl, formyl, and related groups. For example, serine hydroxymethyltransferase catalyzes the reaction:

L-Serine + tetrahydrofolate = glycine + 5,10,-methylenetetrahydrofolate

N-Formiminolglycine is deformylated by glycine formiminotransferase, and the acceptor group is tetrahydrofolate. Tetrahydrofolate is also the acceptor group in the reaction in which *N*-formyl-L-glutamate is deformylated by formulglutamate formyltransferase.

2.1.3. Carboxyl groups are transferred by enzymes of this sub-subgroup. For example, carbamoylphosphate and L-aspartate are transformed by phosphate, and *N*-carbamoyl-L-aspartate is transformed by aspartate carbamoyltransferase. Ornithine and oxamate carbamoyltransferases also belong to this subgroup.

2.1.4. Glycine amidinotransferase catalyzes the reaction:

L-Arginine + glycine = L-ornithine + guanidinoacetate

2.2. Enzymes of this subgroup, such as transaldolase, transfer aldehyde or ketone residues.

2.3.1. These acyltransferases are enzymes that transfer acyl groups from CoA, or formyl groups from tetrahydrofolic acid, to an acceptor. Choline acetyltransferase transfers acetyl groups from acetyl-CoA to choline. Phosphate-, carnitine-, glucosamine phosphate-, and AcCoA-acetyltransferases transfer acetyl groups to phosphate, carnitine, glucosaminephosphate, and AcCoA, respectively, from AcCoA. Carnitine palmitoyltransferase transfers palmitoyl groups to carnitine.

2.3.2. This sub-subgroup of enzyme transfers amino acids.

2.4. Glycosyltransferases are enzymes capable of transferring glycosyl groups. These enzymes occupy a key position in the biologic activity of living cells, because they synthesize many complex polysaccharides that serve as energy stores or as structural materials.

2.4.1. Hexosyltransferases. Enzymes of this sub-subgroup transfer hexosyl residues. For example, glucose from α-glucan, maltose, or cellobiose is transferred to orthophosphate by α-glucan phosphorylase, maltose phosphorylase, or cellobiose phosphorylase, respectively. Many enzymes catalyze the general reaction:

UDPsugar + sugar acceptor = UDP + sugar—sugar acceptor.

Among these enzymes are UDPglucose–glycogen glucosyltransferase, UDPglucose–fructose glucosyltransferase, UDPglucose–fructose phosphate glucosyltransferase, and UDP glucuronyltransferase.

2.4.2. Pentosyltransferases. Many of the enzymes of this sub-subgroup catalyze the general reaction:

Nucleoside + phosphate = nucleic acid + sugarphosphate

Purine nucleoside phosphorylase, uridinephosphorylase, thymidine phosphorylase, and guanosine phosphorylase belong to this group.

Another general type of reaction catalyzed by these subgroups of enzymes is

Nucleotide + phosphate = nucleic acid + phosphosugar—phosphate

For example, adenine phosphoribosyltransferase, hypoxanthine phosphoribosyltransferase, and uracil phosphoribosyltransferase catalyze such reactions.

2.5. Enzymes of this subgroup catalyze the transfer of allyl or related groups. For example, dimethylallytransferase (2.5.1) transfers a dimethylallyl group from dimethylallyl pyrophosphate to isopentenyl pyrophosphate, forming geranyl pyrophosphate. Thiamine monophosphate is produced by thiamine phosphate pyrophosphorylase from 2-methyl-4-amino-5-hydroxymethylpyrimidine pyrophosphate as donor and 4-methyl-5-(2'-phosphoethyl)-thiazole as acceptor.

2.6. Enzymes of this subgroup transfer nitrogeneous groups.

2.6.1. An amino group is transferred, usually from an amino acid to a keto acid, by the enzymes of this sub-subgroup. The mechanism of the enzyme reaction involves reversible transformation of enzyme-attached vitamin B_6 coenzyme. No free ammonia forms during this transfer. Examples of some of the enzymes and the reactions catalyzed by them are

Aspartate aminotransferase:

L-Aspartate + 2-oxoglutarate = oxaloacetate + L-glutamate

Alanine aminotransferase:

L-Alanine + 2-oxoglutarate = pyruvate + L-glutamate

Glycine + 2-oxoglutarate = glyoxylate + L-glutamate

Pyridoxamine pyruvate transaminase:

Pyridoxamine + pyruvate = pyridoxal + L-alanine

A very important enzyme of the heart is glutamic oxalacetate transaminase, which catalyzes the reaction:

Glutamate + oxalacetate = ketoglutarate + aspartate

2.7. Enzymes of this subgroup transfer phosphorous-containing groups, notably, phosphoryl, phosphoanhydride, or phosphate ester groups, from donor to acceptor molecules. These reactions are generally reversible.

2.7.1. Enzymes of this sub-subgroup transfer phosphorous-containing groups to an alcohol acceptor. For example, hexokinase forms ADP and D-hexose- 6-phosphate from ATP and a D-hexose or some related compound (e.g., acetylated hexose, a 2-deoxyhexose, hexoacid). Enzymes involved in these reactions are present in all types of cells that are capable of metabolizing hexoses. Some specific enzymes of this subgroup are glucokinase, fructokinase, galactokinase, phosphofructokinase, gluconokinase, and ribulokinase. All these enzymes catalyze the transfer of phosphoryl groups to the corresponding sugars.

Other members of this subgroup utilize acceptors other than sugars. For example, adenosine kinase catalyzes the reaction:

$$ATP + adenosine = ADP + AMP$$

Enzymes like riboflavin kinase, pyridoxal kinase, protein kinase, pyruvate kinase, and polyphosphate glucokinase also belong to this sub-subgroup.

2.7.2. Enzymes of this sub-subgroup transfer phosphorous-containing groups to carboxylic groups. An interesting reaction catalyzed by carbamate kinase is

$$ATP + H_3N + CO_2 = carbamoylphosphate$$

Other members of this sub-subgroup are acetate kinase, formate kinase, asparate kinase, and butyrate kinase.

2.7.3. Enzymes of this sub-subgroup transfer phosphorous-containing groups to a nitrogen-containing group.

Guanidinoacetate kinase, for example, catalyzes the reaction:

$$ATP + guanidinoacetate = ADP + phosphoguanidinoacetate$$

Creatine kinase, arginine kinase, and taurocyanine kinase also belong to this sub-subgroup.

2.7.4. Enzymes of this sub-subgroup transfer phosphorous-containing groups to a phospho-group. Polyphosphate kinase catalyzes the reaction:

$$ATP + (phosphate)_n = ADP + (phosphate)_{n+1}$$

Enzymes like adenylate kinase nucleosidemonophosphate kinase, nucleosidediphosphate kinase, and guanylate kinase, all utilize ATP as phosphate donor and transfer it to a nucleoside monophosphate or to a nucleosidediphosphate.

2.7.5. Enzymes of this sub-subgroup transfer a phosphorous-containing group from a diphosphate molecule to the analog of the same molecule. Members of this subgroup are phosphoglucomutase and phosphoglyceromutase, both widespread in all types of tissues.

2.7.6. Pyrophosphotransferases. This sub-subgroup transfers pyrophosphate, mostly from ATP. For example, ribophosphate pyrophosphokinase produces 5-phospho-α-ribosyl-pyrophosphate from D-ribose- 5-phosphate.

2.7.7. Nucleotidyltransferases. Some members of this sub-subgroup participate in the transfer of genetic information. For example, RNA-nucleotidyltransferase catalyzes the reaction:

$$m\text{-Nucleosidetriphosphate} + RNA_n = m\text{-pyrophosphate} + RNA_{n+m}$$

DNA nucleotidyltransferase acts the same way on deoxynucleoside triphosphates. A reversible reaction is catalyzed by polynucleotidephosphorylase:

$$RNA_{n+1} + orthophosphate = RNA_n + a\ nucleosidediphosphate$$

Two groups are transferred by GDP (guanosinediphosphate)-mannose-pyrophosphorylase in the reaction:

GTP + α-D-mannose 1-phosphate = pyrophosphate + GDP mannose

An important cyclic phosphate is formed by ribonuclease, which transfers the 3'-phosphate of a pyrimidine nucleotide residue of polynucleotides from the 5'-position of the adjoining nucleotide to the 2'-position of the particular pyrimidine nucleotide. sRNA-adenylyltransferase and sRNA-cytidylyl-transferase add an adenyl or cytidyl residue, respectively, to sRNA. A specific member of this sub-subgroup is CMP (cytidinemonophosphate)-sialate-synthase, which catalyzes the reaction:

CTP + N-acetylneuraminate = pyrophosphate + CMP-N-acetylneuraminate

2.7.8. This sub-subgroup includes the transferases of substituted phosphorous-containing groups. Cholinophosphotransferase is a member of this sub-subgroup.

2.8. Members of this subgroup transfer sulfur-containing groups. They are isolated mostly from plant extracts, but animal tissues, like liver and heart, also contain them.

2.8.1. Sulfurtransferases. An example of this type of enzyme is thiosulfate sulfurtransferase, which catalyzes the reaction:

Thiosulfate + cyanide = sulfite + thiocyanate

2.8.2. Sulfotransferases. This sub-subgroup of enzymes is capable of transferring a sulfate group from a sulfate ester.

2.8.3. CoA-transferases. Enzymes of this sub-subgroup catalyze the transfer of CoA. 3-Ketoacid CoA-transferase and 3-oxoadipate CoA transferase belong to this sub-subgroup.

Hydrolases

Hydrolytic enzymes cleave covalent bonds, such as the peptide and other N—N, the glycosidic C—O, the ester C—O, the anhydride O—O, the phosphoric acid ester P—O, the C—C, and the thioester S—C bonds. Different subclasses of enzymes catalyze the cleavage of these various bonds. In the following, we will discuss briefly the function of the most important enzyme subgroups of this main group.

3.1. Enzymes acting on ester groups.

3.1.1. Carboxylic esterhydrolases. Enzymes of this sub-subgroup hydrolyze carboxylic esters by reacting first with the ester to produce an acyl-enzyme intermediate, which, in turn, can react with acyl acceptors. Under suitable conditions, the reverse reaction, that is, ester formation from free carboxylic acids, is catalyzed by the same enzymes. The pancreatic and plasma esterases of this group hydrolyze a wide variety of esters of physiologic and pharmacologic importance. The acetylesterases preferentially attack acetyl esters. The most important member of this group is acetylcholinesterase, which hydrolyzes acetylcholine:

$$(CH_3)_3-N-C_2H_4-O-\overset{\overset{\displaystyle O}{\|}}{C}-CH_3 + H_2O \rightleftharpoons$$
$$(CH_3)_3-N-C_2H_4-OH + CH_3-COOH$$

Other esterases hydrolyze other esters of choline (e.g., butyryl-or propionylcholine). Brain and liver extracts contain triacetinase, which hydrolyzes triacetin. Cholesterolesterase catalyzes the hydrolysis and formation of cholesterol esters.

The phospholipases break down phospholipids, like lecithin and cephalin. These enzymes can be extracted from muscle, heart, liver, kidney, and pancreas.

3.1.2. This sub-subgroup contains the thiolester hydrolases, like palmitoyl-CoA, succinyl-CoA, and hydroxyaclglutathione hydrolase.

3.1.3. Enzymes of this sub-subgroup catalyze the hydrolysis of a variety of phosphate esters of alcohols, sugar alcohols, phenols, and amines. Phosphodiesters are not hydrolyzed by this group of enzymes, but the linkage between organic and inorganic phosphate groups is. Alkaline-phosphatase catalyzes transphosphorylation reactions and can be found in milk, liver, and *Escherichia coli*. The acid-phosphatases remove inorganic phosphate groups from a variety of phosphomonoesters and phosphoproteins. They also attack the pyrophosphate link, and some of them hydrolyze ATP and ADP. The 5'-and 3'-nucleotidases attack different ribonucleotides and deoxyribonucleotides, like AMP, IMP, and UMP, splitting them to nucleosides and orthophosphates. The different sugarphosphatases, like glycerol-2-phosphatase, are very specific in their catalytic activities.

3.1.4. Some of the phosphate-splitting enzymes attack phosphoricdiesters, yielding phosphoricmonoesters and an alcohol. The phosphodiesterase I and II attack polyribonucleotides and oligodeoxyribonucleotides, liberating nucleotide-5'-phosphates or nucleotide-3'-phosphates, respectively. Phospholipase C and phospholipase D split phosphatidylcholine at different points. The different deoxyribonucleases and nucleases attack DNA and RNA, respectively. Much research has been done recently with the 3', 5'-cyclic-nucleotide-phosphodiesterase, which opens up the cyclic phosphate ring in 3', 5'-cyclic-AMP.

3.1.5. This sub-subgroup contains the dGTPase that produces dG and triphosphate from dGTP.

3.1.6. Sulfuric ester hydrolases attack sulfuric acid esters. The arylsulfatase, sterol sulfatase, and glycosulfatase belong to this sub-subgroup.

3.2. Enzymes in this subgroup are capable of splitting glycosyl compounds.

3.2.1. Glycosidases of this sub-subgroup catalyze the hydrolysis of the glycosidic bond of simple glycosides and oligo- and polysaccharides. They are water-soluble, slightly acidic proteins, (except for lysozyme), with an optimum pH between 4.5 and 7.0. The hydrolysis occurs according to different mechanisms, since some reactions proceed with inversion of configuration (e.g., α-amylase), whereas retention of configuration occurs with others (β-amylase). Dextranases are enzymes capable of hydrolyzing the α-1, 6-glycosidic linkages of the bacterial polysaccharide dextran. Endodextranases can be found in mammalian tissues (e.g., spleen, liver, kidney, lung, and brain).

Pectinases hydrolyze pectin (methylated α-1, 4-polygalacturonic acid) and can be found in vegetables and microörgainisms. Many marine bacteria contain chitinase.

The α-1, 4-glycosidic linkage of polysaccharides, such as starch and glycogen, is broken down by the different α-amylases, either randomly or from the nonreducing end in a regular manner. The β-amylases attack alternate glycosidic bonds, starting from the nonreducing end of the starch chain. Amalyses are distributed widely in plants and animals, but the β-amylases seem to be restricted to higher plants.

Invertases catalyze the hydrolysis of sucrose and related glycosides. All mammalian tissues and body fluids contain β-glucuronidase, with highest activity found in the liver, kidney, and spleen. This enzyme can hydrolyze *in vitro* β-glucuronides formed in the mammalian body from alcohols, phenols, and carboxylic acids and excreted in the urine and bile.

β-Galactosidase is the first oligosaccharide-splitting enzyme to be obtained in crystalline form. Other widely distributed enzymes of this sub-subgroup are the lysozymes, which are capable of dissolving bacterial cell walls. They are found in tears, saliva, blood serum, in different animal tissues, and in plants and bacteria. They are often used in the pharmaceutical industry to disrupt bacteria so that intracellular bacterial products may be obtained. The product of the cell-wall hydrolysis is β-(1, 6)-N-acetylglucosaminyl-N-acetylmuranic acid.

The hyaluronidases are capable of breaking the glycosidic bonds of hyaluronic acid. They have been utilized in studies of the chemical structure of acid mucopolysaccharides and, in medicine, to facilitate the absorption of parenterally administered agents. Hyaluronidases are not strictly specific. Some of them can split, at the same rate, N-acetylhyalobiuronidylgalacto-saminyl and N-acetylhyalobiuronidylglucosaminyl groups linked to hyaluronic acid. These enzymes are widely distributed in nature.

Neuraminidase was first found in the influenza virus. This enzyme splits β-D-N-acetylneuramic acid (sialic acid) from mucoproteins. It is widely distributed in bacteria and viruses.

3.2.2. Enzymes belonging to this sub-subgroup hydrolyze N-glycosyl compounds. For example, nucleosidase splits N-ribosylpurine to purine and D-ribose. Adenosyl monophosphate nucleosidase splits AMP to adenine and D-ribose 5-phosphate.

3.2.3. Enzymes of this sub-subgroup hydrolyze 5-glycosyl compounds.

3.3. Enzymes belonging to this subgroup hydrolyze ether bonds.

3.3.1. A thioether-hydrolyzing enzyme is adenosylhomocysteinase, which forms adenosine and L-homocysteine from 5-adenosyl-L-homocysteine.

3.4. Enzymes of this subgroup act on peptide bonds.

3.4.1. This sub-subgroup of enzymes hydrolyzes α-aminoacylpeptide groups. L-Leucylpeptide is split specifically by leucine aminopeptidase, whereas a variety of aminoacyloligopeptides is split by different aminopeptidases.

3.4.2. Carboxyterminal amino acids are split by carboxypeptidase A. Carboxypeptidase B splits only carboxyterminal-L-lysine.

3.4.3. Enzymes of this sub-subgroup, like glycylglycine dipeptidase and iminopeptidase, split dipeptides.

3.4.4. Many well-known enzymes belong to this sub-subgroup. Pepsin, the principal enzyme of gastric juice, was perhaps the first enzyme to be recognized and the second to be crystallized. It hydrolyzes different internal peptide bonds, provided both fragments are in the L-configuration. Trypsin, which is activated from trypsinogen, can be found in the pancreatic juice of most mammals. It hydrolyzes peptides, amides, or esters at bonds involving L-arginine or L-lysine carboxyl groups.

Chymotrypsin A and B hydrolyze peptides or amides, especially at aromatic L-amino acid carboxyl sites. Papain, another well-known enzyme, hydrolyzes peptides, amides, and esters, mostly at basic amino acid or leucine or glycine sites. Studies of this enzyme contributed much to our present knowledge of enzyme action, especially to the sulfhydryl theory of enzymes. Thrombin converts fibrinogen to fibrin, and plasmin hydrolyzes fibrin to soluble products.

Some other enzymes of this sub-subgroup are subtilopeptidase A, aspergillopeptidase A (converts trypsinogen to trypsin), keratinase, urokinase (converts plasminogen to plasmin), and clostridiopeptidase A and B.

3.5. Enzymes of this subgroup act on C—N bonds other than peptide bonds.

3.5.1. Enzymes of this sub-subgroup hydrolyze linear amides. The most interesting member of this sub-subgroup is L-asparaginase. This enzyme has been used with some success against acute lymphoblastic leukemia. It hydrolyzes L-asparagine to L-aspartate and ammonia, and thereby deprives the malignant cell of its essential nutrient, L-asparagine. A similar enzyme, glutaminase, hydrolyzes L-glutamine to L-glutamate and ammonia.

Monocarboxylic acid amides are split by amidase, and urea is split by urease. Aminoacylase splits N-acyl amino acids in general. Acetylornithinedeacetylase splits α-N-acetyl-L-ornithine. Acyllysine deacylase is specific in splitting ε-N-acyllysine.

3.5.2. This sub-subgroup of enzymes splits cyclic amides. Barbiturase hydrolyzes barbiturate to malonate and urea. The most important members of this subgroup are the penicillinases. Extensive research is being done to

synthesize penicillins that resist the penicillinases. These enzymes are produced by bacteria in self-defense against penicillin. Penicillinases differ in
their hydrolyzing ability, hence the specific resistance by certain bacteria,
especially staphylococci, to different penicillins.

3.5.3. Enzymes of this sub-subgroup split linear amidines. For example,
L-arginine is split to L-ornithine and urea by arginase.

3.5.4. This sub-subgroup of enzymes splits cyclic amidines. Cytosine is split
to uracyl and ammonia by cytosine deaminase. Adenosine deaminase requires ATP for its function. Other members of this group are AMP
deaminase, methenyl-tetrahydrofolate cyclohydrolase, pterin deaminase, and
dCMP deaminase.

3.5.5. Members of this sub-subgroup attack cyanides. For example, nitri-
lase catalyzes the reaction:

$$\text{Nitrile} + H_2O \rightarrow \text{carboxylate} + NH_3$$

3.6. Enzymes of this subgroup hydrolyze acid anhydride bonds.

3.6.1. This sub-subgroup of enzymes hydrolyzes pyrophosphate to orthophosphate. A notable member of this subgroup is inorganic pyrophosphatase.
ATPase, isolated from mammalian sources, and apyrase, isolated from plants,
hydrolyze ATP, ADP, and orthophosphate. Other members of the sub-subgroup are nucleoside diphosphatase, acylphosphatase, and CTPase.

3.7. Enzymes of this subgroup act on C—C bonds.

3.7.1. These enzymes, such as fumarylacetoacetase, act on ketonic substances.

3.8. This subgroup of enzymes acts on halide bonds.

3.8.1. Flouroacetate flourohydrolase, a member of this sub-subgroup, hydrolyzes C—F bonds.

3.8.2. P-halide bonds are hydrolyzed by enzymes like DFPase. These
enzymes are capable of hydrolyzing a variety of phosphorofluoridates.

Lyases

Members of this group catalyze the addition of a radical to a double bond, or
conversely, remove a group from the substrate with the formation of a double
bond. The group removed or added may be carbon dioxide, formaldehyde,
acetaldehyde or other aldehydes, hydrogen cyanide, a ketoacid, pyruvate,
water, ammonia, an amine hydrogen sulfate, mercaptan, hydrochloric acid,
etc. These reactions are not characteristic of any single class of biochemically
important substances. The main function of lyases is participation in the
intermediary metabolism of carbohydrates, fats, and certain amino acids.

Only a few lyases will be mentioned specifically in the following subclassi-
fication, since numerous enzymes catalyze identical or very similar reactions.

4.1. This subgroup of enzymes catalyzes the removal of carbon dioxide. The substrates include 2-oxoacids, oxalate, acetoacetate, L-aspartate, L-glutamate, L-histidine, L-arginine, and UDPglucuronate.

4.1.2. This sub-subgroup of enzymes catalyzes the removal of an aldehyde group. For example, deoxyriboaldolase produces D-glyceraldehyde-3-phosphate and acetaldehyde from 2-deoxy-D-ribose 5-phosphate. Threonine aldolase splits L-threonine to glycine and acetaldehyde.

4.1.3. Enzymes of this sub-subgroup act on ketoacids. Malate synthetase catalyzes the reaction:

$$\text{L-Malate} + \text{CoA} = \text{acetyl-CoA} + H_2O + \text{glyoxylate}$$

This enzyme is highly specific and can be found in many living species. Also, acetyl-CoA is formed from 3-hydroxy-3-methyl-glutaryl-CoA plus Coa and from citrate plus CoA by hydroxymethylglutaryl-CoA-synthase and citrate synthase, respectively. A complex reaction is catalyzed by ATP-citrate lyase:

$$\text{ATP} + \text{citrate} + \text{CoA} = \text{ADP} + \text{orthophosphate} + \text{acetyl-CoA} + \text{oxaloacetate}$$

4.2.1. This sub-subgroup of enzymes hydrolyzes the general reaction:

$$X \rightleftharpoons Y + H_2O$$

One of the most interesting members of this sub-subgroup is carbonic-anhydrase, which catalyzes the reaction:

$$H_2CO_3 \rightleftharpoons CO_2 + H_2O$$

The enzyme can be found in erythrocytes, gastric mucosa, and renal cortex. Its various physiological roles center around the diffusion of carbon dioxide and transport of hydrogen and sodium ions. Other members of the group are fumarase, aconitate hydrotase (enzymes of the citric acid cycle), dihydroxy-acid dehydratase, phosphopyruvate hydratase, and serine, homoserine, and threonine dehydratases. Synthetases, like tryptophan, cysteine, and porphobilinogen synthetase, are also members of this sub-subgroup.

4.3. Members of this enzyme subgroup are carbon-nitrogen lyases.

4.3.1. The ammonia lyases, like aspartate ammonia lyase, histidine ammonia lyase, and phenylalanine ammonia lyase, belong to this sub-subgroup.

4.3.2. Amidine lyases, members of this sub-subgroup, include adenylosuccinate lyase and argininosuccinate lyase.

4.4. This subgroup of enzymes comprises the carbon-sulfur lyases.

4.4.1. A member of this sub-subgroup is S-alkylcysteine lyase, which catalyzes the cleavage of many 5-alkylated L-cysteine molecules.

4.5. Enzymes of this subgroup are the carbon-halide lyases.

Isomerases

Isomerases catalyze isomerization of molecules. The reaction involved could be racemization, epimerization, *cis–trans* isomerization, intramolecular oxidoreduction, and intramolecular group transfer. The last reaction could be a simple transfer of an acyl or other group or reactions involving addition to double bonds (lyases).

5.1. Racemases and epimerases. Enzymes of this subgroup catalyze racemization, i.e., conversion of an L form to the optically opposite D form, and epimerization, i.e., conversion of the optical configuration of an optically active carbon atom.

5.1.1. Enzymes of this sub-subgroup catalyze racemization reactions of amino acids and amino acid derivatives. For example, alanine racemase catalyzes the reaction:

L-Alanine⇌D-alanine

and hydroxyproline epimerase catalyzes the reaction:

L-Hydroxyproline⇌D-allohydroxyproline

5.1.2. Enzymes of this sub-subgroup act on hydroxy acids and their derivatives. Lactate racemase catalyzes the reaction:

L-Lactate⇌D-lactate

5.1.3. Enzymes of this sub-subgroup act on carbohydrates and on carbohydrate derivatives. UDPglucose epimerase, a member of this sub-subgroup, catalyzes the reaction:

UDPglucose⇌UDPgalactose

Other members of this group are ribulophosphate 4-epimerase, aldose mutarotase, and *N*-acylglucosamine 2-epimerase.

5.2. 5.2.1. *Cis–trans* isomerases catalyze *cis–trans* conversion reactions. Maleylacetoacetate isomerase catalyzes the reaction:

4-Maleylacetoacetate⇌4-fumarylacetoacetate.

5.3. Enzymes of this subgroup catalyze intramolecular oxidation–reduction reactions.

5.3.1. Intramolecular oxidoreductases of this sub-subgroup interconvert aldoses and ketoses. Arabinose isomerase, for example, catalyzes the reaction:

D-Arabinose → D-ribulose

Triosephosphate isomerase acts on glyceraldehyde:

D-Glyceraldehyde-3-phosphate → dihydroxyacetonephosphate

Other members of this sub-subgroup are L-arabinose-, xylose-, glucose-phosphate-, mannose-, glucuronate-, and glucosamine-phosphate isomerases.

Ligases or Synthetases

These enzymes catalyze the synthesis of two molecules and simultaneously cleave the pyrophosphate bond of ATP or of a similar triphosphate. Further classification is based on the bond formed, which can be C—O, C—S, C—N, or C—C.

6.1. Enzymes of this subgroup catalyze the synthesis of two molecules with the formation of a new C—O bond. Many enzymes, namely the amino-acid-tRNA synthetases, belong to this subgroup.

Tryosyl-tRNA synthetase, for example, catalyzes the reaction:

$$ATP + \text{L-tyrosine} + tRNA = AMP + \text{pyrophosphate} + \text{L-tyrosyl-tRNA}$$

Each amino acid has its corresponding enzyme that catalyzes the condensation of it to a tRNA, as in the equation above.

6.2. Enzymes of this subgroup catalyze the fusion of two molecules, with the formation of a C—S bond.

Acetyl-CoA synthetase (6.2.1.) catalyzes the reaction:

$$ATP + \text{acetate} + CoA = AMP + \text{pyrophosphate} + \text{acetyl-CoA.}$$

Acyl-CoA sulfatase (which utilizes a wide variety of acids), the specific succinyl-CoA synthetase, and choloyl-CoA synthetase also belong to this sub-subgroup.

6.3. Ligases of this subgroup form C—N bonds.

6.3.1. Enzymes of this sub-subgroup link ammonia or amines with carboxylic acids.

For example, asparagine synthetase catalyzes the reaction:

$$ATP + \text{L-aspartate} + NH_3 = AMP + \text{pyrophosphate} + \text{L-asparagine}$$

6.3.2. Acid–aminoacid ligases or amide synthetases link amino acids with acids or with amino acids, according to the general formula:

$$ATP + \text{acid} + \text{amino acid} = AMP + \text{pyrophosphate} + \text{acid–amino acid}$$

Pantothenate synthetase, γ-glutamyl-cysteine, glutathione, and D-alanyl-alanine synthetase belong to this sub-subgroup.

6.3.4. Enzymes of this sub-subgroup form C—N bonds other than those specified in the previous 6.3. sub-subgroups. CTP synthetase, for example, catalyzes the reaction:

$$ATP + UTP + NH_3 = ADP + \text{phosphate} + CTP$$

Tetramethyl-tetrahydrofolate synthetase, GMP synthetase, adenylsuccinate synthetase, and argininosuccinate synthetase belong to this sub-subgroup.

6.3.5. Enzymes of this sub-subgroup form C—N bonds and utilize glutamine as N-donor. NAD synthetase, for example, catalyzes the reaction:

ATP + deamino-NAD + L-glutamine + H_2O = AMP + pyrophosphate

$$+ NAD + L\text{-glutamate}$$

6.4.1. Ligases of this sub-subgroup form C—C bonds. For example, acetyl-CoA-carboxylase catalyzes the reaction:

ATP + acetyl-CoA + CO_2 + H_2O = ADP + phosphate + malonyl-CoA

Propionyl-CoA carboxylase and pyruvate carboxylase also belong to this sub-subgroup.

Reference

1. Florkin, M., and Stotz, F., eds., "Comprehensive Biochemistry, 13." Elsevier, Amsterdam, 1965.

PART II

SPECIFIC ENZYMES

Chapter 8

Enzymes of Acetylcholine Metabolism

Francis F. Foldes

Introduction

Acetylcholine (ACh) and two of its regulatory enzymes, namely, choline acetyltransferase (EC 2.3.1.6; choline acetylase, ChAc) and acetylcholinesterase [EC 3.1.1.7; acetylcholine acetyl hydrolase; red cell cholinesterase; (ChE); specific ChE; AChE] have important physiologic functions. Thus, for example, they are essential for neuromuscular and ganglionic transmission, interneuronal transmission in certain parts of the central nervous system (CNS) (83, 126, 215, 341a, 401) regulation of membrane permeability and perhaps also for axonal conduction (353, 358).

The physiologic role, if any, of the third regulatory enzyme of ACh, butyrylcholinesterase (EC 3.1.1.8; acetylcholine acyl hydrolase; plasma ChE; nonspecific ChE; BuChE) has not been clarified as yet (182). BuChE, however, has important pharmacologic functions. It is responsible for the enzymatic breakdown of ester type neuromuscular blocking agents and local anesthetic agents and therefore it is of great significance in anesthesiology (141).

The activities of one or more of the enzymes of ACh metabolism, and especially that of BuChE, may be influenced by physiologic states [e.g., age and sex (396), pregnancy (154)], inherited (genetically determined) conditions such as atypical BuChE (253), pathologic conditions (e.g., liver disease) (161), and drugs (141).

Alterations of the activity of one or more of these enzymes may have significant influence on the course of anesthesia. Thus, for example, decreased BuChE activity encountered in liver disease may increase considerably (161) and in the presence of atypical or other abnormal variants of

BuChE [e.g., fluoride resistant (203) or "silent" (315)] excessively the intensity and duration of action of succinylcholine chloride (suxamethonium, Anectine, SCh). Conversely, many drugs used in anesthesiology may influence the activities of the enzymes of ACh metabolism (30, 97, 121, 133, 148, 152, 156).

In this chapter I will attempt to summarize the information presently available on the distribution, physical and chemical characteristics, kinetics, and physiologic role of ChAc, AChE, and BuChE. Whenever applicable I discuss the influence of normal and abnormal activities of the various enzymes of ACh metabolism on pharmacologic actions and interactions of drugs used in anesthesiology, and the effects of drugs administered to patients before, during, and after anesthesia on the activities of these enzymes. Limitations of space only permit the presentation of a small fraction of the information available on many aspects of the topics to be discussed. The two guiding principles in the selection of the material to be included in this chapter are: (a) the presentation of the current understanding of the interaction of anesthetic and adjuvant drugs with the enzymes of ACh metabolism; and (b) supplying background information for those who wish to embark on the investigation of one or the other of the many unexplored areas of this important field.

Choline Acetyltransferase

ChAc catalyzes the synthesis of ACh from acetyl-coenzyme A (AcCoA) and choline (354):

$$ChAc + AcCoA + CH \rightleftharpoons ACh + CoA + ChAc$$

AcCoA in turn is synthetized by other enzymes, the most important of which is AcCoA synthetase (EC 6.2.1.1.) (423).

Distribution

ChAc is widely distributed in the animal kingdom and is also present in at least one microorganism, *Lactobacillus plantarum* (411, 434) and in a plant, the nettle (34). The presence of ACh in several plants (242) indicates that other plants may also contain this enzyme. With few exceptions (211) ChAc is only found in multicellular animals with specialized conducting tissue. In various mammals highest ChAc activities were found in certain parts of the brain (e.g., caudate nucleus, putamen, cerebral cortex (211, 376) and peripheral nervous system (e.g., ventral spinal roots, mixed spinal roots, sciatic nerve, superior cervical ganglion, retina) (211). High concentrations of ChAc are also present in the placenta (57), but its functional significance in this organ has not been clarified as yet.

Subcellular Distribution

The subcellular distribution of ChAc was studied most thoroughly in brain tissue (90, 92, 167, 196, 213, 215). When brain tissue is gently homogenized in iso-osmotic sucrose and subjected to a combination of differential and density gradient centrifugation (196), three particulate (P_1, P_2, P_3) fractions and a solute (S_3) fraction can be separated. With this technique most of the ChAc was found in the P_2 fraction that contained the detached nerve terminals called boutons or synaptosomes. Depending on the region of the brain analyzed 70 to 90% of the ChAc was found in occluded form in the particulate fractions and the remainder in the soluble fraction (167). In homogenates of regions rich in cholinergic terminals, the proportion of the particulate and, in those containing many cholinergic cell bodies (e.g., spinal cord), the proportion of the soluble fraction of ChAc is greater (167).

When the synaptosomal layer (P_2) is treated with hypoosmotic solutions the synaptosomes burst and their contents can be separated by density gradient centrifugation (438). Part of the ChAc can then be recovered in particulate and part in soluble form. The particulate ChAc is attached to fractions containing larger membrane particles (166) and is not occluded. The ratio of the nonoccluded particulate and the soluble ChAc depends on the pH and ionic strength of the medium. At physiologic pH (7.0 to 7.4) and ionic strength $(1.5 \times 10^{-1} M$ KCl or NaCl) expected to exist in the synaptosomes *in vivo* most, if not all, of the ChAc may be expected to be present in nonoccluded, soluble, and therefore functional form in the synaptosomes (167).

Synthesis

The site of synthesis of ChAc is controversial. Most investigators believe that ChAc is synthesized in the cell body of the neurons and is transported from here in the exoplasm to the site of its action in the nerve terminal (127, 214, 271). Local axonal synthesis of various proteins, including AChE, however, was reported [for references, see Rosenberg et al. (381)] and therefore the possibility of the axonal synthesis of ChAc cannot be definitely excluded. ChAc is also synthesized in neuroblastoma cell cultures (329). The synthesis of ChAc in these cultures was depressed by about 45% by chronic exposure to $3 \times 10^{-6} M$ morphine.

Purification

In recent years considerable advances have been made in the purification of ChAc. Because of the different sources of the enzyme and variations in the methods of purification the reported results are not strictly comparable. Potter and Glover (376) using bovine caudate nucleus and putamen obtained

420- to 1300-fold purification with about 20% yield. The specific activities of the partly purified extracts were between 0.1 to 0.3 μg of ACh synthesized/mg protein/min. The extract contained no deacylases or cholinesterases (ChE), but it did contain significant amounts of carnitine acetyltransferase. Chao and Wolfgram (71) using bovine caudate nuclei obtained 400-fold purification with about the same specific activity as Potter and Glover (376) with a 50% yield. White and Wu (437) reached 100- to 380-fold purification from autopsied human caudate nuclei and putamen. The specific activity reported by these investigators was significantly higher, about 7.7 μg ACh synthesized/mg protein/min. They, however, calculated specific activity from the maximum velocity obtained by extrapolating to saturating concentrations of both AcCoA and choline (Ch).

Physical and Chemical Properties

ChAc is a relatively basic protein. Its molecular weight was reported by various investigators to be 65,000 (376), 100,000 (436) and 120,000 (71), respectively. According to Chao and Wolfgram (71) in dilute buffer ChAc dissociates into two subunits with a very significant loss of activity. The molecular weight of the minor unit was about 69,000 and that of the major unit about 51,000.

Investigation of the electrical properties (e.g., isoelectric point, net surface charge) of ChAc obtained from different species (167, 327, 328, 437) revealed physicochemical differences. ChAc obtained from rat and cat (167) had relatively strong positive net surface charge and a high isoelectric point (pH 7.8) and adsorbed readily to negatively charged membranes. ChAc of guinea pig and pigeon (167) had a weak surface charge and low isoelectric point (pH 6.8) and did not adsorb readily to membranes. Multiple forms of ChAc were obtained subsequently from rat (327, 328, 437) and human (437) brain. These differences in the electrical properties of ChAc obtained from different species, or even from different tissues of the same species, may help to explain the occasionally variable results obtained in kinetic studies by different investigators.

The partially purified enzyme is stable at −20°C for months and at 4°C for weeks.

Kinetics

Only some of the essential points of the complex kinetics of the two-substrate ChAc will be discussed.

ChAc is a rather selective enzyme. Thus, for example, it will not catalyze the transfer of acetate from AcCoA to carnitine. Although at a slower rate it will catalyze the formation of propionylcholine (PrCh) from propionyl-CoA and Ch, but not that of butyrylcholine (BuCh) (45). If Ch (trimethylammoniumethylalcohol) is replaced by dimethylamino or monomethyl-aminoethylalcohol the rates of synthesis of the resulting compounds are decreased to 8 to 2% of the ACh rate, respectively (44). More recent studies (376) also showed that ChAc is highly specific for Ch, but not for AcCoA which could be substituted by propionyl-CoA without significantly affecting the speed of synthesis. Acetylthiocholine (ATCh), however, was synthesized at a rate comparable to that of ACh.

According to Potter and Glover (376) partially purified ChAc is inactive below pH 5, but has a broad pH optimum between pH 7 and 10. Others (71) found that the reaction proceeds at optimal rate at pH 7.

The effect of salts (KCl or NaCl) on ChAc activity depends on the purity of the preparation (376). The greater the purity of the enzyme, the lower the salt concentrations one needs to produce maximal rates. Independent of purity maximal rates can be obtained with 1.5×10^{-1} M KCl.

The Michaelis constants (K_m) of ChAc obtained from bovine caudate nuclei and putamen (376), from human putamen and caudate nucleus, human sciatic nerve, rat brain, rabbit brain, and rabbit sciatic nerve (437) were found to be very similar for both Ch and AcCoA. Depending on the source of ChAC they ranged from 4×10^{-4} M to 1.2×10^{-3} M for Ch and from 6×10^{-6} to 1.8×10^{-5} for AcCoA. Over a very wide range neither substrate (Ch or AcCoA) inhibited or facilitated the synthesis of ACh by ChAc (376).

Kinetic studies with varying concentrations of each substrate in the presence of fixed concentrations of the other indicate a sequential enzyme mechanism (437). This means that both substrates, Ch and AcCoA, combine with ChAc before either of the products, ACh and CoA, are released. Similar results were reported by others (376) for central nervous system (CNS) ChAc. In contrast, Schuberth (392) reported a nonsequential, "ping pong" mechanism for the synthesis of ACh by human placental ChAc. According to this mechanism one of the products, probably CoA, leaves the enzyme before the second substrate, Ch, combines with it. Subsequent studies by others (343, 344, 385) indicate that the synthesis of ACh by placental ChAc is also a sequential reaction.

The synthesis of ACh by ChAc is inhibited by high concentrations of ACh. According to Kaita and Goldberg (251) greater than 10^{-2} M concentrations of ACh cause progressive inhibition of ChAc. The inhibition with 10^{-1} M ACh was 45% and it increased with increasing ACh concentrations. The inhibition was competitive for Ch, but not AcCoA. Potter and Glover (376), and Glover and Potter (181) confirmed the inhibitory effect of ACh on ChAc, but they were of the opinion that ACh is a noncompetitive inhibitor of ChAc.

ACh, PrCh, and benzoylcholine (BeCh) in 5×10^{-2} M concentrations all produced about the same 30% inhibition of ChAc (251). The inhibitory effect of the same concentration of acetyl-β-methylcholine (MeCh) was only 15%.

The second product of the enzymatic synthesis of ACh, coenzyme-A (CoA), also inhibits ChAc (181). The inhibition is competitive for the first

substrate, AcCoA and noncompetitive for the second substrate, Ch (437).
For the backward reaction:

$$E + ACh + CoA \rightarrow CH + AcCoA + E$$

the K_m for each substrate decreased as the concentration of the other increased (376). Or, in other words, in contrast to the situation with the forward reaction, the substrates of the backward reaction reciprocally influence each other's affinity to the enzyme. The K_m values observed were 7.5×10^{-4} to 5×10^{-3} for ACh and 2.5×10^{-5} to 1.5×10^{-4} for CoA.

Inhibitors

The influence of substrates (AcCoA and Ch) and products (CoA and ACh) on the rate of synthesis of ACh by ChAc had been discussed in the preceding paragraphs. The inhibitory effects of other compounds on ChAc are discussed briefly below.

Soon after the discovery of ChAc by Nachmansohn and Machado (361), weak inhibitory activities were reported for certain naphtoquinones (359) and α-keto acids (363). The most potent of these compounds 2-methyl-1,4-naphtoquinone-8-sulfuric acid had an I_{50} of the order of 10^{-4} M. In 1963, Nachmansohn (354) closed his review on ChAc with the following statement: "A really potent and specific inhibitor of ChAc has not been found as yet. Such a compound obviously would be of great interest."

Based on the assumption that some degree of similarity might exist between the active sites of the AcCh synthesizing (ChAc) and hydrolyzing (AChE) enzymes, Cavallito and his associates (64–67, 406, 407, 435, 436) synthesized and tested *in vitro* a large number of compounds that were expected to bind to a negatively charged group at the active site of ChAc that interacts with the quaternary ammonium group of Ch.

The first potent inhibitor of ChAc synthesized was hexamethylene-1-4-(1-naphtylvinyl)-pyridinium-6-trimethylammonium dibromide (406) which had an I_{50} value of 9×10^{-7} M for ChAc. This compound, however, was not selective since it also had significant inhibitory effect on AChE ($I_{50} = 6 \times 10^{-7}$ M) and BuChE ($I_{50} = 3 \times 10^{-6}$ M). It also had significant neuromuscular (n.m.) blocking activity (63a). Further investigation of different styrylpyridine derivatives and analogues of the general structure shown in Figure 8-1 revealed that ChAc inhibitory activity may be expected to be present in molecules in which an aromatic ring system (a) is conjugated to a pyrido ring (c) through an exocyclic unsaturated (ethenylic or ethynylic) bond (b). Such molecules are flat, rigid, and coplanar. There should be no substituents on a, b, or c (except on N) that would add significant third dimension to the molecule. Ionization of the cationic center of c enhances ChAc inhibitory effect by increasing the intensity of ionic bonding and by π electron deficient

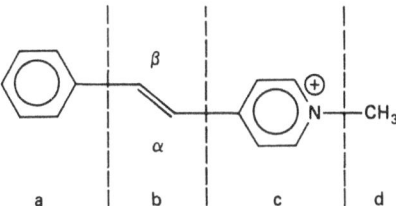

Figure 8-1. The general structural formula of styrylpyridine-type choline acetyltransferase inhibitors.

1 (Monomer)

2 (Dimer)

Table 8-1. The influence of photodimerization on the N.M., anti-ChE, and anti-ChA activity of styrylpyridine-type choline acetyltransferase inhibitors

Compound	Type	Aryl group	Experiment condition	N.M.Block[a] ED50[c]	Anti-ChE[b] I50	Anti-ChA[b] I50
GSA-I-9	1	Phenyl	Dark	1.6×10^{-4}	1.4×10^{-4}	5.0×10^{-6}
GSA-III-5	2	Phenyl	Light	7.6×10^{-6}	4.6×10^{-6}	0 at 10^{-4}
GSA-I-17A	1	Naphthyl	Dark	3.4×10^{-5}	6.4×10^{-5}	8.5×10^{-7}
GSA-II-61	2	Naphthyl	Light	2.8×10^{-6}	3.5×10^{-7}	5.0×10^{-3}
GSA-I-7	1	Phenanthryl	Dark	7.3×10^{-6}	4.4×10^{-5}	5.5×10^{-5}
	1+2	Phenanthryl	Light	2.9×10^{-6}	2.0×10^{-5}	23% 10^{-4}
GSA-I-11	1	p-Biphenyl	Dark	2.1×10^{-5}	2.6×10^{-5}	23% at 10^{-4}
	1+2	p-Biphenyl	Light	1.2×10^{-5}	2.1×10^{-5}	18% at 10^{-4}

[a]Note that photodimerization increases n.m. and anti-ChE and decreases anti-ChA activity.
[b]Rat diaphragm.
[c]Rat brain.
[d]Moles/liter.

character and electron accepting properties. The π electron donor and hydrophobic properties of a also increase ChAc inhibitory activity. Substitutions on d have little influence on anti-ChAc activity, but have significant influence on anti-ChE potency. Hydroxyethyl substitution on d, however, markedly increases water solubility.

The *trans* isomers of the compounds, obtained in synthesis, when exposed to light are transformed to their *cis* isomers if b is an ethylenic (—CH=CH—) bond (436). This transformation results in a marked decrease of anti-ChAc and an increase of anti-ChE and n.m. blocking activity (143) (Table 8-1). Thus, for example, in the dark the I_{50} values of the anti-ChA and anti-AChE activity of bromide-N-trimethylammoniumhexyl-[4-(naphtylethenyl)-pyridinium] were 8.5×10^{-7} and 6.4×10^{-5} M, respectively, and the I_{50} anti-ChA/I_{50} anti-AChE ratio was 0.013. When exposed to light the I_{50} values of the anti-ChA and anti-AChE of the resulting dimer were 5×10^{-5} and 3.5×10^{-7}, respectively and the I_{50} anti-ChA/I_{50} anti-AChE ratio was 142.8. Similarly to anti-ChE activity, n.m. blocking potency was also increased by light-induced dimerization.

The potency and specificity anti-ChAc activity of the best compounds of this series is presented in Table 8-2.

In vitro the styrylvinylpyrydinium type inhibitors cause reversible, noncompetitive inhibition of ChAc (406). It has been suggested that they interfere with a catalytic histidine moiety in ChAc and block the transfer of acetyl group from the enzyme complex to Ch (64).

Table 8-2.   The potency and specificity of representative choline acetyltransferase inhibitors investigated

Compound	I_{50} (M)[a] for		Specificity
	ChAc	AChE	(ChAc/AChE)
N-hydroxyethyl-[4-(1-naphthylethenyl)-pyridinium] bromide	6.0×10^{-7}	1.5×10^{-3}	2500
N-trimethylammoniumhexyl-[4-(1-naphthylethenyl) -pyridinium] bromide	8.5×10^{-7}	6.4×10^{-5}	75
N-methyl-[4-(1-naphthylethenyl)-pyridinium] iodide (IN 1633)	4.7×10^{-7}	4.2×10^{-4}	890
N-acetamide-[4-(1-naphthylethenyl)-pyridinium] iodide	3.8×10^{-7}	2.5×10^{-3}	6580
N-methyl-[4-(naphthylethynyl)-pyridinium] iodide	1.0×10^{-6}	9.0×10^{-4}	900

[a]Inhibitory activities determined in dark.

The kinetic studies with the various styrylpyridine type inhibitors of ChAc were carried out with crude or partially purified tissue homogenates. Further investigations with the now available more highly purified preparations (71, 376, 437) will be required to elucidate the mode of action and kinetics of these compounds.

Other recently investigated inhibitors of ChAc include bromoacetyl-CoA (72), bromoacetonyl-trimethylammonium bromide (371), and haloacetylcholine derivatives (343, 344). Bromoacetyl-CoA has been reported to be a potent inhibitor of carnitine acetyltransferase and ChAc but its I_{50} value for the latter enzyme is not stated (72). Bromoacetonyltrimethylammonium bromide has an I_{50} of 5×10^{-4} M for placental ChAc (371) and the I_{50} values of the most potent of the halogenated ACh derivatives, chloroacetylcholine was of the order of 10^{-5} M in AcCoA regenerating systems and somewhat lower (3×10^{-6} M estimated) in systems containing synthetic AcCoA (344). The *in vitro* inhibitory effect of chloroacetylcholine is short-lasting, probably because of its rapid hydrolysis in the incubating medium. Chlorocholine itself is a very weak inhibitor of ChAc. In 4×10^{-2} M concentration it only causes 29% inhibition.

Preliminary screening of the *in vivo* pharmacologic effects of the styrylpyridine type ChAc inhibitors was carried out by Hemsworth and Foldes (221). The compounds caused a nondepolarization or a biphasic depolarization block both *in vivo* and *in vitro*. *In vitro* the block caused by the bisquaternary and monoquaternary derivatives on the rat's phrenic nerve-diaphragm preparation could be readily reversed by washing. In contrast, the block caused by the tertiary inhibitors was difficult or impossible to reverse. All compounds also depressed the twitch tension of the directly stimulated muscles. Neuromuscular blocking doses of some of the compounds caused transient severe fall of blood pressure and slowing of the respiratory rate in cats. These studies indicated that the pharmacologic effects of the styrylpyridine analogues investigated cannot be attributed to their ChAc inhibitory effects alone.

Studies by other investigators also indicated that the *in vivo* ChAc inhibitory effect of the styrylpyridine type inhibitors is less than what would be expected from their *in vitro* efficacy. Nonspecific inhibitors of ChAc (*n*-methyl- and *n*-ethylmalaimide, *p*-hydroxymercuribensoate, iodoacetate, copper sulfate, 5,5'-dithiobis-(2-nitrobenzoic acid) has been reviewed by Hebb (212). Their action is mainly due to the production of conformational changes in the enzyme (435).

Physiologic Role

The physiologic role of ChAc is the regulation of the synthesis of ACh. Synthesis usually occurs close to the sites of action of ACh. Since ChAc, in comparison to the catabolizing enzymes of ACh, AChE, and BuChE is a slowly acting enzyme there are special storage mechanisms for ACh in the organisms. The compartmentalization of ChAc, ACh, and AChE is so designed that relatively large quantities of ACh can accumulate in close vicinity of AChE without being broken down by it. In the CNS this is achieved by storing most of the ACh ($>70\%$) inside the presynaptic nerve terminals (synaptosomes) (196, 215, 443). About 50% of the synaptosomal ACh is contained in the synaptic vesicles (444). The synaptic vesicles have a diameter of about 500 Å and contain about 2000 molecules of ACh (445). Electrophysiologic investigations (125, 274) indicated that ACh is released in discrete packages called "quanta." The morphologic equivalent of quanta are the synaptic vesicles (89–92). Most of the remaining synaptosomal ACh is in the cytoplasmic sap. A small portion, however, is localized in the external synaptosomal membrane or in vesicles close to it (32).

Fortuitously, there is little or no AChE within the synaptosomes. This enzyme is assumed to be synthesized in the endoplasmic reticulum (32, 314). From here it is probably excreted to the outer surface of the cell and migrates along the axon and accumulates at the site of its physiologic action, on the outer surface of the synaptosome. By this arrangement stores of ACh can accumulate in the nerve terminal.

Interestingly, it is not the more labile cytoplasmic ACh, but the ACh segregated in the synaptic vesicles that is released by the nerve impulse. The cytoplasmic ACh seemingly serves the purpose of filling up newly formed vesicles. The ACh released in excess is adsorbed partly to AChE present on the outer surface of the synaptosome and in the subsynaptic membrane, and partly to the subsynaptic cholinergic receptors. Within a very short period (2 to 3 msec) all the released ACh is hydrolyzed into acetic acid and Ch. These are reabsorbed into the synaptosome and resynthesized with the help of AcCoA-synthetase and ChAc to ACh.

Recent studies indicate that brain tissue cannot synthesize Ch from precursors (13, 56, 78). The Ch utilized for the synthesis of ACh by brain tissue is obtained either from the plasma (393) or it is reabsorbed after the hydrolysis of ACh (13, 394). Experimental evidence also indicates that brain can obtain Ch necessary for the synthesis of ACh from phospholipids (78). Hemicholinium-3 is a competitive inhibitor of the uptake of choline (394).

Most of the observations made on the compartmentalization of ChAc, ACh, and AChE and on the synthesis, storage, release, and metabolism of ACh in the brain are also applicable to other synaptic transmission sites, such as the autonomic ganglia and the n.m. junction (135).

Interaction with Anesthetics and Adjuvant Drugs

There is little information available on the *in vitro* or *in vivo* effect of drugs used in anesthesiology on ChAc activity. ChAc does not directly influence the pharmacologic effects of anesthetic drugs. Conceivably, a significant decrease of ChAc activity, by reducing the amount of available ACh, would increase the effect of those anesthetic agents which depress ACh synthesis and/or release. Such a situation may arise in neurologic or neuromuscular disorders which cause degeneration of the presynaptic elements at a synaptic junction. Selective degeneration of the presynaptic elements has been described in the "weaver mouse" (223) in whom an inherited anomaly causes disintegration of the external granular cells of the cerebellum. ChAc activity and ACh content was found to be decreased and AChE activity increased in these animals (330).

Conversely, anesthetic agents and adjuvant drugs are capable of influencing ChAc activity both *in vivo* and *in vitro*. It has been reported that the concentration of ACh in rat brain varies inversely with the degree of cerebral activity (397). Thus, anesthetic agents increase (178, 378, 420a) and CNS stimulants (e.g., pentylinetetrazol) decrease (178, 296) brain ACh content. It was reported recently (397) that the brain of rats injected intraperitoneally with 1 g/kg urethane or 0.35 g/kg chloralhydrate contained significantly higher concentrations of ACh than those of control animals. Morphine was also reported (82) to increase the ACh content of the brain in addicted rats by inhibiting its release. Withdrawal was followed by a sudden increase of free (released) ACh. The withdrawal symptoms could be attenuated by atropine.

The increased brain ACh concentration under the influence of anesthetic agents has been attributed to inhibition of release (37, 67, 81, 341, 427). The increase in the intracellular (synaptosomal) ACh concentration inhibits ChAc activity (251, 376) by a feedback mechanism. It appears, however, that pentobarbital also has a direct depressant effect on ChAc (394).

Atropine in low concentrations stimulates ACh release from cortical brain slices and the above-mentioned feedback mechanism increases ACh synthesis (374, 397). Other mechanisms, such as interference by atropine with the adsorption of ACh to muscarinic sites had also been suggested for the explanation of the seemingly increased free ACh present under the influence of atropine (68, 397).

Cholinesterases

In this section the distribution, synthesis, purification, physical and chemical properties, and kinetics of AChE and BuChE will be considered separately. For practical purposes the inhibitors, physiologic roles, and genetically determined anomalies of AChE and BuChE; the influence of physiologic conditions and pathologic states on these enzymes; and their interaction with anesthetic and adjuvant drugs are discussed together.

Acetylcholinesterase

AChE catalyzes the hydrolysis of ACh and that of certain other Ch and thiocholine (TCh) esters. AChE can also hydrolyze many non-Ch esters (25).

Distribution. AChE is widely distributed in various species and in the different tissues of mammalian species. Specifically, it is present in all conducting fibers of nerve and muscle, in the pre- and postsynaptic membranes of junctional sites in both the central and peripheral nervous system (356, 357), in red cells (335), in placenta (369), and in variable concentration in most other tissues (282).

In general, AChE, in contrast to BuChE, is a membrane-bound enzyme, localized in close proximity to the sites of action of ACh but separated from the ACh stores within the membranes (32, 215, 443). This arrangement allows AChE to fulfill its regulatory role on the functions of released ACh without interfering with the accumulation of adequate reserves of ACh.

AChE is present in all parts of the central nervous system in variable concentrations. Its concentration is relatively high at sites rich in synapses (e.g., the basal ganglia) and in the area of the primary neurons of motor and autonomic fibers (282). The AChE concentration in the region of the primary afferent neurons (e.g., dorsal roots) and their termination in the spinal cord is low (282). In the human brain (164), the decreasing order of AChE activity in representative areas was: caudate nucleus > globus pallidus > thalamus > cerebellar cortex > cerebral cortex. AChE activity in the various cerebral cortical areas is 20 to 30 times lower than in the thalamus or globus pallidus. AChE activity of the white matter of the internal capsule is of the same magnitude as that of the gray matter of different cerebral cortical areas. In the subcortical white matter AChE activity is about five to 10 times lower than in cortical gray matter.

In the somatic portion of the peripheral nervous system relatively high concentrations of AChE are present in the motor neurons and nerve trunks (282). In contrast, the concentration of AChE in the sensory neurons, nerve trunks, and their terminals is relatively low (282).

The areas of origin of sympathetic preganglionic fibers in the cord have moderate to high AChE activity (280). The cervical sympathetic trunk and sympathetic and parasympathetic ganglia also have high AChE activity (288, 388). In those ganglia where the postganglionic fiber is adrenergic (e.g., stellate) most AChE is localized on the outer surface of the terminal presynaptic fibers (282, 287, 388). When the postganglionic fiber is cholinergic (e.g., ciliary ganglion) there is high AChE activity in both the pre- and postsynaptic fibers. AChE activity is also high within the postganglionic fibers of parasympathetic ganglia (282, 287).

At cholinergic neuroeffector sites most of the AChE activity is localized in the ganglia, which usually are in close proximity to the effector cells and in the axonal terminations of the short postganglionic fibers (417). In the heart, AChE activity is usually higher in the atria than in the ventricles (19) and AChE staining is the most intensive in the sino-auricular and atrioventricular nodes which may be considered peripheral vagal ganglia (260a).

In other organs, such as in the blood vessels, the respiratory, digestive, and urogenital systems, AChE is localized predominantly in parasympathetic ganglia and postganglionic fibers (282). An exception to this is the placenta. In this organ, which has no innervation, there is a high concentration of AChE confined almost exclusively to the fetal portion (194).

Histochemical studies indicate that most of the AChE at the n.m. junction is postsynaptic. The presynaptic membrane contains significantly less AChE (80, 282). Similar observations were made with methods based on the combination of histochemistry and electron microscopy (35). With this technique the presence of AChE was also demonstrated in the synaptic vesicles of the axonal terminals. The AChE activity per unit volume of endplates is about 50 times greater than that of the surrounding muscle (177). The number of AChE molecules in a single endplate has been estimated to be about 10 to 20 million (430).

It was suggested by Zupancic (474, 475) that the anionic site of AChE is identical with the cholinergic receptor. *In vitro* studies with purified AChE (267) and *in vivo* experiments on the monocellular electroplax preparation (268) appear to contradict this assumption.

Subcellular Distribution. The AChE distributed throughout the entire length of all types of nerve fibers of various species (356, 357) may be divided into an external, functional and an internal, reserve portion (59, 284, 323). AChE is also present at various synaptic sites. At the n.m. junction most of the AChE is localized in the postsynaptic membrane (79, 80, 87). It was believed earlier that in autonomic ganglia the localization of AChE is mostly presynaptic (288, 388). However, recent studies combining electron microscopic and histochemical techniques, allowing better visualization of the synaptic areas, revealed that similarly to the n.m. junction, most of the AChE is located postsynaptically in autonomic ganglia (284, 286).

AChE is a membrane-bound enzyme (266, 339). It can be bound not only to extracellular, but also to intracellular membranes (282).

In biopsied, homogenized human muscle, subjected to differential ultracentrifugation, the highest concentration of AChE was found in the microsomal fraction (408). The AChE activity of this fraction was 19 times greater than that of the muscle homogenate.

Synthesis. Relatively little information is available on the sites of synthesis of AChE. In nerve tissue it is probably synthesized near to its final localization in the endoplasmic reticulum (286, 288). From here it migrates to the surface of the cell membrane where it exerts its physiologic function. In the course of ontogenesis AChE appears in both skeletal (295, 352) and heart muscle (292) before innervation. Or, in other words, AChE of muscle originates, at least in part, in nonneural elements, possibly in connective tissue fibers, which later become parts of the sarcolemma (262, 398). The AChE present in human erythrocytes probably originates from their precursors, the erythroblasts (462). Human platelets and their precursors the megakaryocytes contain no AChE (462). The situation is reversed in the cat: Megakaryocytes and platelets do, erythoblasts and red cells do not contain AChE. In other mammals there is an inverse relationship between the AChE content of red cells and platelets (463). The fetal portion of the placenta is rich in AChE. It was suggested (194) that ACh is synthesized in the hemopoietic elements of the placenta.

Purification. The first active solution of AChE was obtained from the electric tissue of *Torpedo marmorata* in 1939 by Nachmansohn and Lederer (360, 360a). Several hundredfold purification of AChE was achieved later from enzyme solutions extracted from the electric tissue of *Electrophorus electricus* (383). One milligram of this extract hydrolyzed 440 mmol of ACh per hr. Further purification of extracts of the electric tissue of the *Electrophorus electricus* yielded a homogenous protein (309) with a specific activity of 750 mmol of ACh hydrolyzed by 1 mg protein in 1 hr. About 60 mg pure AChE was obtained from 10 kg toluene pretreated material. Disc electrophoresis and high-speed centrifugation studies indicated that the purified enzyme preparation is a homogenous protein (308). Subsequently, the purified enzyme was obtained in crystalline form (309).

Partially purified red cell and brain AChE extracts were also prepared [for references, see Usdin (424)]. Partially purified electric eel (Sigma Chemical Co., St. Louis, Mo.; Worthington Biochemical Corp., Freehold, N.J.), and bovine erythrocyte (Winthrop Laboratories, New York, N.Y.) AChE are commercially available.

Physical and Chemical Properties. The molecular weight of AChE is about 260,000 (310); its isoelectric point is at pH 5 (306, 309). The ultraviolet absorbtion curve of purified AChE has a minimum at 250 and a maximum at 280 nm (308).

Physicochemical (310) and electron microscope (70) studies revealed that AChE is a dimer consisting of two identical protomers of about 130,000 molecular weight (307). Each protomer contains two, nonidentical polypeptide chains of about 65,000 molecular weight. Each protomer has one active site located on one of the polypeptide chains. The role of the second polypeptide chain is not known; it is possible that its presence is essential for AChE activity (310, 310a).

Based on indirect evidence obtained from analogy with other hydrolytic enzymes and from the interaction of substrates and inhibitors with the enzyme at different pH values (424), it is generally accepted that the active center of AChE, contains an anionic and an esteratic site (455, 468). Based on the inhibitory effects of trimethylammonium (457) and urethane type (163) inhibitors with varying quaternary ammonium-carbonyl carbon distances on electric eel and human red cell AChE, respectively, the distance between the anionic and esteratic site of AChE was estimated to be about 6.0 Å. It was also suggested (41) that in AChE, for each esteratic site there are two anionic sites in close proximity to one another. Since both the substrates and inhibitors of AChE are stereoselective it is probable that the active site of AChE is three-dimensional (294).

The structure of the active sites of AChE is a much debated topic. There is considerable evidence for the presence of serine (75c, 390a), histidine (346, 455), and aminocarboxylic acids (40a, 454). A glutamine–serine–alanine sequence was described in the active surface of ACh by Oosterbaan (245, 366). For a more extensive discussion of the composition of the active center of AChE the reader is referred to Usdin (424).

AChE of various species (e.g., human, bone, insect) may consist of isoenzymes with catalytically similar, but allosterically different characteristics (105, 248, 332, 395).

Kinetics. It is generally accepted (290) that the positively charged quaternary nitrogen of ACh is attracted by coulombic forces, probably to strongly negative phosphate groups (62, 63) at the anionic site or to a negative ester group (171a, 171b), and the carbonyl carbon to a positively charged nucleophilic group consisting of histidine (455) and serine (75c) at the esteratic site of the active center of AChE (33, 453) (Figure 8-2). In addition, covalent bond formation between the ester group and esteratic site (452) and dipolar forces in the ester oxygen atom may also be involved in the binding (290).

The hydrolysis of the substrate (e.g., ACh) occurs in at least three steps. At first, a high-energy unstable complex is formed (Step I) which can dissociate into free enzyme and unchanged substrate, or goes on to form a more stable acetylated enzyme and yields Ch (Step II). Finally, (Step III) the acetylated enzyme reacts with water and dissociates into free enzyme and acetic acid. The reaction between AChE and ACh may be illustrated as follows:

$$\text{Step I} \qquad\qquad \text{Step II} \qquad\qquad \text{Step III}$$

$$E + ACh \underset{k_{-1}}{\overset{k_1}{\rightleftarrows}} EACh \quad \overset{k_2}{\longrightarrow} \quad \underset{\overset{|}{CH}}{EA} \quad \overset{k_3}{\longrightarrow} \quad E + A$$

In this equation E = free enzyme; ACh = substrate; EACh = enzyme substrate complex; Ch = choline; A = acetate; k_1 and k_{-1} are the forward and backward reaction rate constants of Step I, k_2 and k_3 are the reaction rate constants of steps II and III.

The hydrolysis of ACh by AChE occurs very rapidly. The turnover time (i.e., the time necessary for one active center to break down one molecule of substrate) of AChE for ACh had been estimated to be of the order of 30 to 60 μsec (285, 299) or even less (306). This means that the turnover number (i.e., the number of molecules of substrate broken down by one active center of

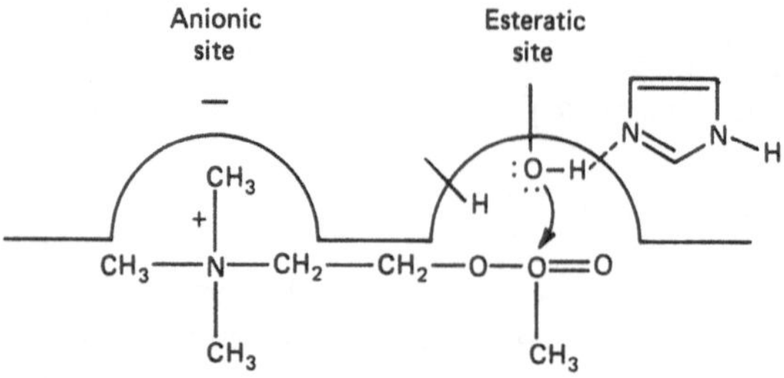

Figure 8-2. The hypothetical interaction between acetylcholine and the active sites of acetylcholinesterase [From Wilson (453)].

enzyme in 1 min) of AChE for ACh, assuming that there are two active centers in each molecule of AChE (307), is of the order of 2,000,000 to 3,000,000. The specific activity (i.e., millimoles of substrate hydrolyzed by 1 mg of protein in 1 hr) of the pure enzyme is 750 (306).

In comparison to BuChE, AChE is a relatively selective enzyme and hydrolyzes relatively few substrates. Of the various choline esters propionylcholine (PCh) (362) and acetyl-β-methylcholine (MeCh) (5) are hydrolyzed relatively rapidly by AChE, but at a slower rate than ACh. In contrast acetylthiocholine (ATCh) and acetyl-β-methylthiocholine (MeTCh) are hydrolyzed more rapidly by human red cell AChE than their corresponding oxyesters, ACh and MeCh (133) (Table 8-3). Certain neutral compounds, such as triacetin (2, 20, 442) or β, β-dimethylbutyl acetate (2) are hydrolyzed relatively rapidly by AChE. Optical isomerism may markedly influence the rates of hydrolysis of the same substrates by AChE. Thus, for example, only the D-isomer of MeCh is hydrolyzed by various AChE (27, 233, 234). The L-isomer of MeCh acts as an inhibitor (233). With other compounds the L-isomer may be the preferred substrate of AChE [for references, see Usdin (424)].

Table 8-3. The hydrolysis rates of various substrates by human red cell acetylcholinesterase

Substrate	Substrate concentration (m)	Hydrolysis rate ($\mu M/ml/hr$)
Acetylcholine	3×10^{-3}	308
Acetylthiocholine	3×10^{-3}	385
Acetyl-β-methylcholine	2×10^{-2}	148
Acetyl-β-methylthiocholine	2×10^{-2}	193
Propionylcholine	3×10^{-3}	197

AChE activity is inhibited by excess substrate (5, 19, 468). The optimal substrate concentration, however, varies with different substrates. Thus, for example, the optimal substrate concentration for partially purified AChE is about 3×10^{-3} M with ACh or PrCh (20), but about an order higher for MeCh. In agreement with this, by using nonpurified human red cell (133, 164), brain or muscle (408) AChE the optimal substrate concentrations of ACh and MeCh were found to be of the order of 3×10^{-3} M and 2×10^{-2} M respectively. AChE activity is not inhibited by high concentrations of MeCh (164).

Despite wide variations in the specific activity of AChE with various substrates there is relatively little difference in the affinity of these substrates to the enzyme. This affinity as expressed by the K_m, the substrate concentration at which the enzymatic reaction proceeds with half the maximal rate (V_{max}), is of the order of 10^{-4} M for electric eel AChE (103).

Changes in pH may affect the substrate, enzyme, or both (76, 77). Theoretically, pH changes influence the degree of dissociation of ionizing groups. In the case of ChE this would be of particular importance for the quaternary nitrogen of Ch. This group, however, is completely ionized over a wide pH range and therefore influence of pH changes on ChE activity cannot be attributed to their effects on the enzyme. The pH optimum for AChE activity is between eight and nine, probably at about 8.25 (43a, 455). The effect of pH and AChE activity had been reviewed by Cohen and Oosterbaan (77).

The optimal temperature range for AChE activity is between 37° and 40°C. Reducing the reaction temperature by 10°C caused an about 15% decrease of nonpurified red cell AChE activity (133). Lowering the temperature from 37° to 17°C caused an about 50% reduction of the hydrolysis rate of ACh and MeCh by homogenized rat muscle (428).

The K_m of ACh and other substrates for electric eel AChE (456) and that of ACh for red cell AChE (400) seem to be temperature independent. Consequently, the effect of temperature changes must be attributed to their influence on the turnover numbers.

The influence of various monovalent (Na^+, K^+, Li^+) and divalent (Ba^{2+}, Ca^{2+}, Mg^{2+}) cations on AChE activity is controversial (19, 77). NaCl (0.15 M to 0.5 M) (403) and $MgCl_2$ (0.01–0.04 M) has been recommended as the optimal ionic medium (77).

Recently three genetically determined variants of AChE have been reported (75a, 402). The variations observed represented the phenotypic expression two codominant alleles at a single locus.

Butyrylcholinesterase

BuChE catalyzes the hydrolysis of BuCh at a faster rate than that of ACh. It also hydrolyzes many other Ch and nonCh esters not hydrolyzed by AChE, but in contrast to AChE it does not hydrolyze MeCh.

Distribution. BuChE is less widely distributed among various species than AChE. It has not been detected in any invertebrate species studied (25). In man and other mammals the highest BuChE activity is present in the plasma (5) and in the liver (104). Moderate BuChE activity was also demonstrated in the brain (164, 171), retina (414), cerebrospinal fluid (136), muscle (88, 408), heart (408), intestines (101, 280), pancreas (19), kidney (317), skin (317), and milk (414).

BuChE, in contrast to AChE which is membrane bound, is a solubilized enzyme present in solution in plasma, cerebrospinal fluid and milk and most BuChE is recovered after differential centrifugation of muscle homogenates from the supernatant (408).

Synthesis. It is generally accepted that plasma BuChE is synthesized in the liver (387). Plasma BuChE activity is markedly decreased in liver disease. The decrease is proportional to the severity of the liver involvement (161). It is possible that BuChE may also be synthesized in other tissues. Variations in the properties of BuChE obtained from different tissues of the same animal (42, 279, 387) supports this assumption.

The influence of hormones on the synthesis of plasma BuChE is controversial. The controversial findings may be due to differences in the species of experimental animals and methodology. In man, plasma BuChE was found to be increased in hyperthyroidism (7, 420) and decreased in hypothyroidism (420), or treatment with antithyroid drugs (7). In female rats, thyroidectomy had no effect on plasma or liver BuChE (424) but castration decreased it (386). In male rats thyroidectomy (424) or castration (386) increased it. In male guinea pigs, however, castration decreased and testosterone increased plasma BuChE (222). In female mice injected with sublethal doses of diisopropylfluorophosphate (DFP), castration, hypophysectomy, or estrogen treatment (9), thyroidectomy, or thyroid treatment (8) had no effect on the rate of regeneration of BuChE in plasma or liver which returned to control levels in about 3 weeks. The daily subcutaneous injection of large doses of corticoids, 10 mg/kg methylprednisolone sodium succinate (Solu-Medrol) or 2 mg/kg dexamethasone sodium phosphate (Hexadrol phosphate) caused about a 40 percent decrease of plasma BuChE activity in 12 days in male and female mongrel dogs. After discontinuation of the corticosteroids, BuChE activity returned to control levels in about 4 to 6 weeks (145). Similar reversible decrease of plasma BuChE was observed after the oral administration of 60 to 100 g prednisone or prednisolone in patients (464, 465).

Purification. A variety of methods have been employed for the purification of BuChE [for references, see Augustinsson (25); Usdin (424)]. A 50- to 100-fold purification of horse serum BuChE was achieved with ammonium sulfate precipitation by Steadman and Steadman as early as 1935 (409). Subsequently, 3400- (326, 413) to 10,000- (207) fold purification of human plasma BuChE and 5000 (412) to 14,000 (414) of horse plasma BuChE was

Table 8-4. The degree of purification and specific activity of plasma butyrylcholinesterase preparations

Source	Approximate purification[a]	Specific activity[b]	Reference
Horse	500	12,000	Strelitz (412)
Horse	14,000	—	Svensmark (414)
Man	3400	9000	Surgenor and Ellis (413)
Man	—	20,000	Malmström *et al.* (326)
Man	10,000	—	Haupt *et al.* (207)

[a]Indicates the ratio of the amount of protein in the starting material to the amount of purified protein.

[b]nmol of ACh hydrolyzed by 1 mg of protein per hr.

obtained. Purified BuChE is very stable and stored at room temperature retains most of its activity for several years (77a). Partially purified human plasma BuChE preparation is commercially available from Behringwerke, A.G. (Marburg-Larr, DBR) and a similar horse plasma BuChE preparation from Worthington Biochemical Corporation (Freehold, N.J.).

The degree of purification and specific activity of plasma BuChE preparations are summarized in Table 8-4.

Physical and Chemical Properties. The molecular weight of human BuChE was estimated to be 300,000 (413) to 348,000 (207). Much higher molecular weights (750,000) for horse serum BuChE were determined by inhibitor kinetic (216) and gel filtration (415) studies. It is conceivable that the molecule of horse (367) and perhaps also that of human BuChE, similarly to AChE (70, 310) consists of two or more protomers with several active sites. The concentration of BuChE in human plasma is about 2×10^{-8} M (182). The isoelectric point of human BuChE is about 3.0 (414, 416).

The active center of BuChE, similarly to AChE, has an anionic and an esteratic site (451, 454). The suggested amino acid sequence of the active site of BuChE is:

Phe-Gly-Glu-Ser-Ala-Gly–(Ala, Ala, Ser) (366)*

It seems probable that serine (75b) and histidine (346, 455) are the basic groups at the esteratic site of BuChE. The negative charge at the anionic site is probably carried by an ω-carboxyl group of a dicarboxilic amino acid (e.g., glutamic acid) (113) or phosphate groups (62, 63).

*Abbreviations: Phe=phenylalamine; Gly=glycine; Glu=glutamic acid; Ser=serine; Ala= alanine.

BuChE exists in the form of several different isoenzymes. Isoenzymes are all derived from the same organ and have the same catalytic action but have different physical and/or chemical properties (331). The differences in the chemical structure of the isoenzymes usually do not involve the active sites (24). The isoenzyme with the greatest mobility toward the anode on electrophoresis is numbered 1 and the rest are numbered consecutively according to their decreasing order of mobility (431).

Human plasma BuChE has been separated electrophoretically (204) into 4 isoenzymes (C_1, C_2, C_3, and C_4). Of these isoenzymes, C_4 contains about 90% of the total activity (189). Depending on the population sampled, in 5 to 17% of the subjects, a fifth isoenzyme (C_5) of BuChE is also present (205) in the vicinity of C_4. Sera containing C_5 usually have relatively high BuChE activity (205a). All the BuChE isoenzymes are determined by the same structural genes (317). Liddel (317) separated five isoenzymes from serum, liver, kidney, brain, ileum, and skin of individuals who were normal or atypical homozygotes (see above) or heterozygotes for BuChE and found that the inhibition characteristics of all isoenzymes were very similar within each group. Recently, as many as seven (298) and 12 (249) isoenzymes of BuChE were identified in human plasma.

Kinetics. The active center of BuChE, as already mentioned, contains an anionic site with a strong negative charge (451, 454), probably carried by the ω-carboxyl groups of dicarboxylic amino acids (e.g., glutamic acid) (113) which have the necessary negative charge at physiologic pH (250). The high reactivity of the esteratic site is the result of the interaction of the hydroxyl group of serine with the imidazole group of histidine (250). The functional groups of the active sites of AChE and BuChE are probably very similar. It is possible, however, that BuChE has only one and AChE two anionic sites (40, 41). In addition to the anionic site, another nonesteratic site was postulated to be present in BuChE but not in AChE (26). The dominant forces at this site were assumed to be Van der Waal's forces. It was also suggested recently that there are differences in the hydrophobic areas of AChE and BuChE (250) and that in the vicinity of the esteratic site of AChE there is one, and on that of BuChE there are two such sites.

The mechanism of the hydrolysis of ACh by BuChE is similar to that of AChE (see p. 105). The turnover number of human plasma BuChE is considerably lower than that of AChE. It is estimated to be about 50,000 (424) and that of horse plasma BuChE from about 50,000 (412) to about 84,000 (244) with ACh substrate.

Although BuChE hydrolyzes some of the substrates of AChE (e.g., MeCh) (22, 210) at only a negligible rate is it capable of splitting a much greater variety of substrates that AChE. Thus, in addition to ACh, BuChE hydrolyzes a number of fatty acid and ω-amino fatty acid esters of choline (e.g., PrCh, BuCh, valerylcholine, heptanoylcholine, 6-aminocapoylcholine, 7-aminoheptanoylcholine) (138), dicarboxylic acid esters of choline (e.g., SCh) (180, 182, 422, 446), succinylmonocholine (SMCh) (129, 141, 162, 187, 446), suxethonium (161, 162), aromatic esters of choline such as benzoylcholine (BeCh) (335, 335a), ester-type local anesthetic agents [e.g., procaine (151, 209, 252), tetracaine (18), meprylcaine, isobucaine (149), their halogenated analogs (e.g., 2-chloroprocaine) (139, 151)], and many other esters. The thio-analogs of ACh and BuCh are hydrolyzed at a faster rate than their oxy-analogs, both by human (133) and horse plasma BuChE (210). It is of interest that, in contrast to MeCh, which is hydrolyzed extremely slowly by human plasma BuCh, its thio-analog is split at the rate of 113 μmol/ml plasma/hr (about 40 percent of the ACh rate) by this enzyme (133). There is a wide variation in the absolute and relative hydrolysis rates of various substrates by BuChE of different species (25, 49). Thus, for example, although horse plasma BuChE hydrolyzes ACh at about the same (22, 23) or faster (467) rates than human plasma BuChE, it hydrolyzes procaine (141) or succinylcholine (22, 138) at significantly lower rates than human BuChE. Procaine is hydrolyzed by various mammalian plasmas at 3 to 20% of the human rate (141) and of all mammalian plasmas investigated, with the exception of the Mangabey monkey (22), human plasma BuChE hydrolyzed SCh the most rapidly. The marked species variation of plasma BuChE activity is exemplified by the finding that the enzymatic hydrolysis rates of ACh and SCh by plasma BuChE are about eight times lower in the Macaque than in the Mangabey monkey (22). It is of interest that the plasma BuChE of the cat, an animal widely used for the pharmacologic testing of n.m. blocking agents, does not hydrolyze SCh (138).

Steric configuration of the substrate is also important (424) for BuChE. When the asymmetric atom is in the alcohol moiety usually the D-isomers are hydrolyzed at a faster rate than the L-isomers (179, 234). When the acyl moiety of the molecule is asymmetrical, sometimes the L- (10, 17), at other times the D- (17) isomers are hydrolyzed more rapidly.

In contrast to ACh, excess substrate does not inhibit BuChE. The substrate concentrations at which BuChE activity is maximal (V_{max}) is about 2×10^{-2} M for aliphatic (164) and 5×10^{-5} M for aromatic substrates (e.g., procaine, BeCh) (151, 154, 252, 405).

There is considerable variation between the affinity of various substrates of BuChE to the enzyme. Thus, for example, for partially purified human plasma BuChE (cholase) the following K_m values were obtained: ACh, 1.8×10^{-3}; SCh, 1.8×10^{-3}; SECh, 1.3×10^{-4}; and SMCh, 2.9×10^{-2} (162). The K_m of procaine was about three orders lower, 2.6×10^{-6} than that of ACh (151). Goedde et al. (188) using a sensitive radio-isotope technique for the determination of the hydrolysis rates of SCh and SMCh and a different plot for the calculation of K_m found significantly lower K_m values, 4×10^{-5} M for SCh and 8.4×10^{-3} for SMCh.

The optimal temperature for BuChE activity is between 37° and 40°C.

BuChE is much more temperature dependent than AChE. Decreasing the temperature by 10°C causes about a 50% decrease of the reaction rate of BuChE and only about 15% decrease of that of red cell AChE (133). The enzymatic hydrolysis rate of ACh by muscle ChE (a mixture of AChE and BuChE) decreases parallel with the lowering of temperature (110).

The pH optimum of BuChE for most substrates is between 8 and 9 (188). Interestingly, the pH optimum of SMCh is between 5 and 6 (188).

The 5-fold dilution of human plasma causes an about 25% increase in the hydrolysis rate of ACh by BuChE (142). Since dilution also causes a similar increase of the activity of purified human plasma BuChE (cholase) this phenomenon cannot be explained by the adsorption of ACh to other plasma proteins. It is conceivable that the slower hydrolysis rate observed in the presence of relatively high enzyme concentrations is caused by an interaction of enzyme molecules with one another that interferes with the adsorption of ACh to the active sites (133).

Genetically Determined Variants. Since the frequently used muscle re-laxants, SCh and ester-type local anesthetic agents (e.g., procaine or tetracaine), are metabolized by BuChE (141), inherited abnormalities of this enzyme may adversely affect the course of anethesia. For this reason this topic is discussed in some detail.

The discovery of genetically determined variants of BuChE started with the observations (51, 122) that after the administration of the usual doses of SCh, prolonged apnea was occasionally encountered in apparently healthy individuals. Further investigations revealed that not only the patients in-volved, but also some of their family members had abnormally low plasma BuChE activities (6, 168, 301). Kalow and associates, who made similar observations in a large number of patients who received SCh in conjunction with electroshock therapy (257) attributed the excessively increased sensitivity of these patients to SCh to the presence of qualitatively different, "atypical" plasma BuChE (258, 260). Important differences were found between the properties of normal and atypical plasma BuChE. Thus, for example, plasmas containing the atypical enzyme hydrolyzed ACh and various other substrates 2 to 10 times more slowly than the normal enzyme (86, 154). The decreased rate of hydrolysis by atypical BuChE was primarily attributed to the much lower affinity (higher K_m value) of the various substrates to the atypical than to the normal variant (86, 254). The ratios of the K_m values of the various substrates to the normal and atypical enzyme are 1.9, 3.2, 5.5, and 6.4 for BuCh, PrCh, BeCh, and ACh, respectively, (86), 20 for SMCh (188), and 40 for SCh (241). Tetracaine hydrolchloride (Pontocain) is the only known substrate that is hydrolyzed by atypical plasma BuChE as well as or better than by the normal enzyme (404, 405).

Besides the lower affinity of the various substrates, there are also other differences between the atypical and the normal BuChE. Thus, for example, increasing the ionic strength of the assay system from 7×10^{-4} M to $1.25 \times$ 10^{-1} M NaCl increased the activity of normal BuChE two- to threefold, but had no influence on the atypical enzyme (74, 416a). Various cationic activators (e.g., Mg^{2+}, Ca^{2+}, tetramethylammonium, Ch) also had less effect on the atypical than on the normal enzyme (74). It was suggested (74, 256) that the different kinetics of the normal and atypical BuChE are due to differences in the structure of the anionic site, which in turn affect the hydrolytic process at the esteratic site.

Others (36) are of the opinion that there is no difference in the amino acid composition of the active site of the normal and atypical enzyme, but the configuration of this site is distorted by structural changes at secondary sites in a way that interferes with the absorption of the substrates.

The interaction of atypical BuChE with various inhibitors is also different from that of the normal enzyme. The decreased inhibitory effect of 10^5 dibucaine hydrochloride (Nupercain) was used originally for the differentiation of atypical from normal plasma BuChE (256, 260). While 10^{-5} M dibucaine causes a $>71\%$ inhibition of normal BuChE, it inhibits the atypical variant by $<30\%$ (254). The percent of inhibition caused by 10^{-5} M dibucaine was termed by Kalow the "dibucaine number" (DN) (256, 260). Atypical BuChE is also resistant to the inhibitory effect of other inhibitors (154, 255, 317) except for the organophosphorous type anti-ChE. The normal and atypical enzymes also behave differently in the presence of various buffers (257a, 259).

Based on the trimodal distribution of the DN in large population samples, Kalow and Staron (259) postulated that the type of BuChE is determined by nondominant, allelic, autosomal genes. Other investigators suggested that the gene expressivity of the atypical varient is inbetween dominant and recessive (6, 303). Kalow and Staron (259) designated the normal gene by "N" and the atypical, dibucaine resistant gene by "D." Those individuals who have two N genes (NN) are normal homozygotes, with qualitatively and quantitatively normal BuChE; those with two atypical genes (DD) are atypical homozygotes; and those with one N and one D gene (ND) are heterozygotes. The normal and atypical enzymes have different electrophoretic mobility and can be separated from the plasma of heterozygotes by paper electrophoresis or column chromatography (316).

Besides the atypical dibucaine resistant BuChE, other genetically determined abnormal variants had also been described (185). Most of these are characterized by decreased (or absent) activity and/or altered sensitivity to various inhibitors. One of these is the fluoride resistant variant (203). Both fluoride-resistant homozygotes (FF) and heterozygotes (NF) have been described (302, 439). While 5×10^{-5} M sodium fluoride inhibits normal BuChE about 64%, the fluoride-resistant homozygote is only inhibited about 34% (438). Another abnormal variant is the "silent" BuChE (99, 100, 315). Subjects who are homozygotes for the silent gene have no (315) or insignificant

(189, 190) BuChE activity. Yet another variant described by Whittaker (440, 441) is excessively inhibited by 1% n-butyl alcohol.

A BuChE variant characterized by markedly increased activity was described by Neitlich (364). This variant, subsequently termed E Cynthiana (461) is very rare and its frequency was estimated to vary from 1 : 770 (276) to 1 : 2600 (364). So far, no homozygotes of this variant have been encountered. In the four heterozygotes of the only family studied (364), plasma BuChE activity was about 300% of the normal mean in three and about 150% in the fourth. The DN was normal in all. The specific enzyme activity is normal and the increased activity is due to overproduction of the variant enzyme (461).

Based on the demonstrated existence of normal, atypical, fluoride-resistant, and silent BuChE variants, Lehmann et al. (302) proposed that the type of BuChE present in any individual is controlled by four allelic autosomal genes acting at the same site. The four genes were originally designated as: N (normal), D (dibucaine resistant), F (fluoride resistant), and S (silent) genes (300). When both genes are identical the individual (in genetic terms often called the "propositus") is a homozygote; when the two genes are different, the propositus is a heterozygote. There are at present four known types of homozygotes (NN, DD, FF, and SS) and six possible combinations of heterozygotes (ND, NF, NS, DF, DS, and FS). Subsequently two other nomenclatures were proposed, one by Motulsky (345) and the other by Goedde and Baitsch (183, 184). These systems of classification also distinguish between genotype, which is based on the presence of one or two of the genetically determined variants (genes) and the phenotype which is the manifestation of the genes in the individual. Although, the elaborate classifications of Motulsky (345) and Goedde and Baitsch (183, 184) are more meaningful for geneticists, they complicate the understanding of the involved principles for clinicians. For this reason, in this chapter, the more simple nomenclature of Lehmann et al. (302) is employed. Since none of the nomenclatures is universally accepted and are used interchangeably in various publications for the sake of comparison, the three nomenclatures and the most important characteristics of the different variants are summarized in Table 8-5.

Clear-cut differentiation of the various homozygotes from one another is relatively simple. The discrimination between the various heterozygotes and between heterozygotes and homozygotes is much more difficult and requires detailed genetic investigation of the family members of the propositus. Thus, for example, because of the wide variation of BuChE activity among normal subjects this parameter may be very similar in normal homozygotes and in normal–atypical (ND) or normal–silent (NS) heterozygotes. The DN of these heterozygotes may be so close to that of normal homozygotes that differentiation between them is not possible. Determining the ratio of the hydrolysis rates of procaine and tetracaine by BuChE, and multiplying the ratio of the

Table 8-5. Human plasma butyrylcholinesterase variants and their characteristics[a]

	Genotype			Phenotype			Approximate				
	Lehmann and Liddel[b]	Motulsky[c]	Goedde and Baitsch[d]	Old designation	Motulsky	Goedde and Baitsch	Relative BuChE activity	DN	FN	P/T Ratio[e]	Frequency
NN	$E_1^u E_1^u$	$Ch_1^u CH_1^u$	Usual (normal)	U	$Ch_1(UU)$[a]	100	72^e	64^a	364	$96/100^a$
ND	$E_1^u E_1^a$	$Ch_1^U CH_1^D$	Intermediate	I	$Ch_1(UD)$	57^e	56^e	48^a	203	$1/33^g$
DD	$E_1^a E_1^a$	$Ch_1^D CH_1^D$	Atypical	A	$Ch_1(DD)$	20^e	12^e	23^a	43	$1/3650^e$
SN	$E_1^u E_1^s$	$Ch_1^S Ch_1^U$	Usual	U	$Ch_1(US)$	55^e	69^e	64^a	365	$1/200^e$
SS	$E_1^s E_1^s$	$Ch_1^S Ch_1^S$	Silent	S	$Ch_1(SS)$	—	—	—	—	$1/170,000^e$
SD	$E_1^s E_1^a$	$Ch_1^S Ch_1^D$	Atypical	A	$Ch_1(DS)$	10^e	11^e	23^a	52	$1/11,000^e$
FN	$E_L^L E_1^u$	$Ch_1^F Ch_1^U$	U_1	UF	$Ch_1(UF)$	80^a	76	52^a	?	?
FF	$E_1^L E_1^L$	$Ch_1^F Ch_1^F$	—	F	$Ch_1(FF)$	50^a	67	34^a	?	?
FD	$E_1^L E_1^a$	$Ch_1^F Ch_1^D$	I_1	IF	$Ch_1(DF)$	60^a	50^a	30^a	?	?
FS	$E_1^L E_1^s$	$Ch_1^F Ch_1^S$	—	F	$Ch_1(FS)$	61^f	67^f	43^f	?	?

[a]Adapted from Usdin (424).
[b]Lehmann and Liddell (300).
[c]Motulsky (345).
[d]Goedde and Baitsch (183, 184).
[e]Smith and Foldes (405).
[f]Whittaker 1967 (439).
[g]Davies et al. (86).

two rates by 100 gives a number termed the P/T ratio (404, 405). In one study, the P/T ratio was found to be 364 ± 7.7 for normal homozygotes, 203 ± 5.9 for normal–atypical (ND) heterozygotes and 43.2 ± 2.8 for atypical homozygotes (405). This wide divergence of the P/T ratios allows unequivocal differentiation between the most frequent NN or ND and also between the ND and DD variants, but not between the NN and NS or the DD and DS variants. For similar reasons, differentiation between the FN, FF, FD, and FS variants, may be very difficult without detailed genetic studies of family members. Both the DN number and the P/T ratio may be determined rapidly with simple ultraviolet spectrophotometric techniques (260, 405).

For the determination of the DN number, stock solutions of 10^{-3} M benzoylcholine chloride (24.4 mg in 100 ml), and 2×10^{-4} M dibucaine hydrochloride (7.6 mg in 100 ml), and a pH 7.4 phosphate buffer, containing 6.07 g Na_2HPO_4 and 2 g $NaH_2PO_4.H_2O$ in 1 liter of distilled water, are prepared. In well-stoppered bottles the BeCh and dibucaine solutions may be stored at 4°C for several months and the phosphate buffer for several weeks. At first the enzymatic hydrolysis rate of BeCh is determined by observing the change of optical density (extinction coefficient) at 37°C in a system consisting of 2.8 ml phosphate buffer, 0.2 ml of heparinized plasma diluted 1:9 (10-fold dilution in phosphate buffer), 0.2 ml of the BeCh stock solution and 0.8 ml distilled water. The buffer, diluted plasma, and distilled water are mixed first and the BeCh substrate is added at 0 time. Before the addition of the substrate all components of the system should be heated to 37°C. A blank in which the substrate is replaced by 0.2 ml distilled water is used. The absorption chamber of the spectrophotometer is kept at 37 ± 0.2°C throughout the assay. The micromoles of BeCh hydrolyzed per milliliter of plasma per hour is calculated from the equation:

$$\mu\text{mol/ml Plasma/60 min} = \frac{\Delta E}{t} \times \frac{50 \times 60}{720} = \frac{\Delta E}{t} \times 4.17$$

In this equation ΔE is the observed change of the extinction coefficient at 246 nm wavelength; t is the number of minutes during which the measured change of the extinction coefficient occurred; 50 is the factor that converts the results from the 0.02 ml plasma present in the reaction medium to 1 ml; 60 is the factor that converts the results to 60 min; and 720 is the change in the extinction coefficient caused by the hydrolysis of 1 μmol of BeCh.

The extinction coefficient observed at 30 sec or 1 min intervals after the start of the hydrolysis and the observed change is plotted against time. A straight portion of the line (indicating constant hydrolysis rate) is used for the calculation of $\Delta E/t$.

After determining the hydrolysis rate of BeCh, the rate is again determined in the presence of 10^{-5} M dibucaine. This time 0.2 ml of 2×10^{-4} M dibucaine is added to both the assay medium and the blank and the volume of distilled water is reduced by 0.2 ml to 0.6 ml in both. It is advisable to allow the plasma to interact with dibucaine for 5 to 10 min before the start of the assay. In every other respect the determination of the inhibited hydrolysis rate is carried exactly the same way as that of the uninhibited rate.

The DN can be calculated from the equation:

$$DN = 100\left(1 - \frac{R_i}{R_u}\right).$$

In this equation R_i is the inhibited and R_u the uninhibited hydrolysis rate of BeCh.

For the calculation of the procaine/tetracaine ratio (P/T ratio), the hydrolysis rates of procaine and tetracaine are determined. The concentrations of the stock solutions of procaine and tetracaine are the same as that of BeCh, 10^{-3} M. Instead of diluted, undiluted plasma is used giving a final plasma dilution of 1:20. The change of the extinction coefficient is determined at 292 and 313 nm for procaine and tetracaine, respectively. The hydrolysis rates of procaine and tetracaine, respectively, are calculated from the equations:

$$\text{nmol Procaine/ml plasma/60 min} = \frac{\Delta E}{t} \times \frac{5 \times 60}{3300} = \frac{\Delta E}{t} \times 0.091.$$

$$\text{nmol Tetracaine/ml plasma/60 min} = \frac{\Delta E}{t} \times \frac{5 \times 60}{4700} = \frac{\Delta E}{t} \times 0.064.$$

The P/T ratio $= \frac{R_p}{R_t} \times 100$, where R_p and R_t are the hydrolysis rates of procaine and tetracaine, respectively.

Acceleration, Inhibition, and Reactivation of Activity of Cholinesterases

Various chemical compounds may accelerate or inhibit the activity of ChE or reactivate the inhibited enzyme. The effect of a given compound will depend not only on the type of the enzyme (ACh or BuChE) but also on the chemical structure of the substrate.

Acceleration of the Activity of Cholinesterases. Divalent cations increase the activity of AChE. Nachmansohn (352) found that the increasing order of the activating effect was $Ba^{2+}, Mg^{2+}, Ca^{2+}, Mn^{2+}$. 2×10^{-5} M Mn^{2+} caused a 14-fold increase in the hydrolysis rate of ACh by electric eel AChE. Ca^{2+}, 5×10^{-3} M, increased the rate of hydrolysis of ACh by red cell hemolysates by 50% (351), and that of purified red cell AChE by 100% (426). As a rule, relatively high concentrations of monovalent ions (e.g., Na^+, K^+, Li^+) are required for accleration of enzymatic activity (352). Their effect is usually nonspecific and may be attributed to the increase of the ionic strength of the medium. Thus, for example, the hydrolysis rate of 3×10^{-3} M ACh or 2×10^{-2} M MeCh was found to be the highest in the presence of 0.2 to 0.5 M NaCl (351).

Kato (272) found that atropine and eserine are adsorbed to different sites on purified squid AChE and that atropine and related compounds accelerate the hydrolysis rate of ACh by this enzyme (273) without interfering with its adsorption to the enzyme.

The hydrolysis of acetyl fluoride by electric eel AChE can be accelerated by 0.12 M tetramethyl- or 0.003 M tetraethylammonium (338). The increased rate of hydrolysis was attributed to conformational changes of the enzyme that facilitated the access of the substrate to the esteratic site.

Various amines, such as trimethyl- or triethylamine, tryptamine, benzylamine, histamine (117, 118, 120), various narcotic analgesics (61, 121, 170, 208), amphetamine (153) and quaternary ammonium compounds such as tetramethyl-, tetraethyl-, tetrapropyl-, tetrabutylammonium (120), or choline (202) are capable of significantly accelerating the hydrolysis of aromatic substrates (e.g., BeCh or procaine) by plasma BuChE. 10^{-3} M 14-hydroxydihydromorphinone and 14-hydroxydihydrocodeinone caused the greatest acceleration (350 to 400%) of the hydrolysis of procaine (121) and 3×10^{-4} M tetrapropylammonium accelerated the hydrolysis of BeCh by 240% (120). None of the amines or ammonium compounds accelerate the hydrolysis of aliphatic substrates such as ACh by plasma BuChE or red cell AChE (121). Certain alcohols, however, accelerate the hydrolysis of ACh by cholinesterases (421).

The accelerating effect of compounds containing positively charged nitrogen atom (N^+) on the hydrolysis of aromatic substrates by BuChE may be explained by the following hypothesis: It is generally accepted that the N^+ of aromatic substrates is attracted by coulombic forces to the anionic and their ester group by coulombic and other forces to the esteratic site of cholinesterases. BuChE also contains a second nonesteratic site where Van der Waal's forces supply the dominant reactive force (26). This site may play an important role in the binding of aromatic substrates such as BeCh or procaine containing a relatively large flat benzene ring. It is conceivable that if the N^+ of an aromatic substrate interacts with the anionic site and its benzene ring with the above nonspecific site by Van der Waal's forces, the ester group of the substrate will not be in a position favorable for its interaction with the esteratic site and the rate of hydrolysis will be relatively slow. When the anionic site is occupied by another group containing N^+, it is conceivable that the aromatic substrate will be adsorbed by Van der Waal's forces to a nonspecific site of BuChE in a way that will favor the interaction of its ester group with the esteratic site resulting in increased rate of hydrolysis observed in the presence of such compounds.

Inhibitors of Cholinesterases. The inhibitors of cholinesterases (ChE) may be reversible or irreversible. The reversible inhibitors dissociate readily from the enzyme and can be easily removed (e.g., by dialysis) with full restoration of enzyme activity. Or, in other words, there is a reversible equilibrium between enzyme and inhibitor: $E + I \rightleftharpoons [EI]$. In contrast, with irreversible inhibitors there is no such equilibrium and the reaction between enzyme and inhibitor is unidirectional: $E + I \rightarrow [EI]$. The binding of enzymes and irreversible inhibitors can only be broken up by the use of "reactivators" (449, 450), causing the return of enzyme activity (243, 243a).

The reversible inhibitors may be competitive, noncompetitive, and uncompetitive (424). Competitive inhibitors compete with the substrate for the same site of the enzyme and are usually structurally similar to the substrate. Their presence increases the K_m of the substrate but has no effect of V_{max}. If the reciprocal of the substrate concentration (1/S) is plotted against, the reciprocal of the rate of enzymatic reaction (1/v) in the absence and presence of the inhibitor, the two lines intersect with the ordinate (y axis) at the same point. Noncompetitive inhibitors combine with the enzyme at another site than the substrate. They do not influence the K_m, but decrease V_{max}. If 1/S plotted against 1/v in the absence and presence of the inhibitor, the two lines intersect on the abscissa (x axis) (95). Uncompetitive inhibitors combine with the enzyme–substrate (ES) complex.

In this chapter the effects of reversible inhibitors of ChE are discussed primarily. Irreversible inhibitors are only discussed briefly. Those wishing to obtain more information on the complex kinetics and structure action relationships of ChE inhibitors are referred to the handbook edited by Koelle (281), Usdin's (424) monograph, and the exhaustive list of references cited by these authors.

Kinetics of Cholinesterase Inhibition. The complex kinetics of cholinesterase inhibition are discussed only to the extent necessary for the understanding of the pharmacologic and clinical use of inhibitors. Their inhibitory effect on various ChE will be expressed in terms of their I_{50} values (i.e., the concentration required to cause 50% inhibition of the uninhibited hydrolysis rate of a substrate).

The effect of competitive inhibitors on ChE depends on numerous factors. These include: (a) the type of ChE; (b) the choice of the substrate; (c) the time of equilibration between the enzyme and inhibitor before the addition of substrate; (d) the relative concentrations of the inhibitor and substrate; (e) whether the inhibition is observed *in situ* or in a homogenate or solution (109); (f) the purity of the enzyme or, in other words, the presence of substances (e.g., proteins) in the assay system to which the inhibitor may be bound reversibly or irreversibly (142); (g) the methods used for the purification of the enzyme; and (h) the pH, ionic strength, and temperature of the assay system. Most of the above considerations also apply to noncompetitive and uncompetitive, reversible, and irreversible inhibitors.

Because of the multiplicity of factors that influence enzyme inhibition, the comparison of the relative potency of different inhibitors determined under variable experimental conditions is often difficult.

Most, but not all, anticholinesterases (anti-ChE) have relatively greater inhibitory effect on BuChE than on AChE.

The inhibitory effect of various inhibitors on the hydrolysis of substrates with a low K_m (high affinity to the enzyme) is less than on those which have a high K_m.

The optimal incubation time varies with the type of enzyme and the inhibitor. With BuChE it is less than 10 min for edrophonium and ambenonium, 1 to 2 hrs for neostigmine and about 3 hrs for pyridostigmine (142). The optimal incubation time of these inhibitors with red cell AChE is similar (142).

In the presence of relatively high concentration of proteins much of the inhibitor may be adsorbed to them and only a relatively small fraction is present in free form and available for interaction with the enzyme. Thus, for example, in human plasma diluted 1 : 1.1 (practically undiluted) the I_{50} values of neostigmine, pyridostigmine, and ambenonium are 2.1×10^{-7}, 2.5×10^{-6}, and 2.5×10^{-5} M, respectively. In plasma diluted 1 : 20 or in purified human plasma BuChE (Cholase) the corresponding I_{50} values are 10^{-8}, 1.9×10^{-7}, and 2.6×10^{-6} M. The addition of 4.5% human serum albumin to the system containing purified human plasma BuChE decreases the inhibitory effect (increases the I_{50} value) of the anti-ChE to that observed in undiluted plasma. This indicates that in undiluted plasma 90 to 95% of these inhibitors is reversibly adsorbed to serum albumin (142).

As already discussed, the inherited abnormal variants of plasma BuChE are, as a rule, significantly less sensitive to the inhibitory effects of various compounds. Thus, for example, dibucaine, neostigmine, RO2-0683 [(2-hydroxy-5-phenylbenzyl) trimethylammonium (bromide) dimethylcarbamate], and SCh inhibit the atypical, much less than the normal variant (see Table 8-6).

The interaction between ChE and inhibitors depends on the chemical structure of the latter. Inorganic ions, such as F, are uncompetitive inhibitors of ChE (73, 293). They become attached to sites different from the active

Table 8-6.   The inhibition of the hydrolysis of benzoylcholine by normal and atypical plasma butyrylcholinesterase

Inhibitor	Concentration (M)	Normal Plasma	Atypical Plasma
Dibucaine	10^{-5}	73.2 ± 1.2[a]	25.3 ± 3.6
Neostigmine	10^{-7}	72.2 ± 1.9	8.7 ± 2.3
RO2-0683	3.3×10^{-9}	78.4 ± 1.9	9.5 ± 3.2
Succinylcholine	3.3×10^{-4}	76.6 ± 1.3	12.7 ± 3.0

[a]Mean ± S.E.M. of inhibition expressed as percent of uninhibited rats.

center of ChE and can get adsorbed, not only to the free enzyme, but also to the enzyme–substrate complex and interfere with both the acylation and deacylation of ChE. F inhibits AChE much more than BuChE (73).

Neostigmine and other carbamates containing both an ester and a quaternary ammonium group are adsorbed to both the esteratic and anionic sites of ChE (153, 163, 350, 351). Inhibitors like simple tetraalkylammonium compounds or edrophonium, which contain a quaternary ammonium group but no ester group, are adsorbed only to the anionic site of ChE. Organophosphorus and selenophosphorus compounds, and organosulfonates interact with the esteratic site. Some organophosphorus compounds such as echothiophate iodide (phospholine) also have a quaternary ammonium group and interact with both the esteratic and the anionic site (165). There are several theories on the interaction of bisquaternary inhibitors with ChE. They may interact with the anionic site of the active center and thereby inhibit the adsorbtion of the substrate, interact with two anionic sites in the immediate vicinity of the active center, cover its surface and thereby make it inaccessible for the substrate, or they may interact with more than one ChE molecule (320).

Inhibitors which interact primarily with the anionic site of ChE, for example, tetraethyl compounds (43, 153) or edrophonium (148) have a greater inhibitory effect on AChE than on BuChE. These findings are in agreement with the observation that the negative charge of the anionic center of AChE is greater than that of BuChE and the interaction between the anionic site and the quaternary ammonium groups is relatively more important for the binding of inhibitors (and also substrates) to AChE than to BuChE (451).

Interference, by steric hindrance, with the adsorption of the ester group of an inhibitor to ChE causes a greater loss of BuChE than of AChE activity (163), indicating that binding at the esteratic site is relatively more important for the inhibition of BuChE than for that of AChE.

With the relatively simple carbamate-type (urethane-type) inhibitors both AChE and BuChE activity is optimal when the distance between N^+ and $C=O$ groups is about 5Å (135). This and similar observations made with hydroxyphenyltrimethylammonium compounds (457) indicate that the distance between the anionic and esteratic sites of ChE is about 5Å.

Representative Cholinesterase Inhibitors. Compounds capable of inhibiting ChE include simple tetraalkyl ammonium compounds, phenolic hydroxy derivatives, carbamates, bisquaternary compounds, and miscellaneous compounds including narcotic analgesics, local anesthetics, endolalkylamines, tranquilizers, atropine, and its derivatives.

The various compounds to be discussed may be arbitrarily divided into three groups: The first of these includes those agents which are clinically employed because of their anti-ChE activity (see Table 8-7). The I_{50} values in these and subsequent tables unless otherwise indicated refer to human erythrocyte AChE and human plasma BuChE, respectively. As already mentioned, similar to the wide variations in the activity of ChE obtained from the same tissues of different species, their relative and absolute sensitivity to various inhibitors may also differ widely. This and other circumstances (e.g., substrate concentration, ionic composition of the assay system, and duration of incubation of the inhibitor with the enzyme before the addition of the substrate) result in the wide variation of the I_{50} values reported by different investigators and occasionally on different occasions by the same investigator. For example, in our laboratory, using a 20 min preincubation period, the I_{50} values of neostigmine were 2.3×10^{-7} and 5.6×10^{-8} M for AChE and BuChE, respectively. When neostigmine was incubated with AChE and BuChE for the optimal periods (1 to 3 hr) (142), the I_{50} values were 1.2×10^{-9} and 1.0×10^{-8} M, respectively.

It is of interest that hexafluorenium is the first compound used in anesthesiology to intensify and prolong the action of an anesthetic drug by deliberately inhibiting its enzymatic breakdown (156, 159).

Table 8-7 also indicates the kinetic classification and the primary clinical uses of the ChE listed.

The second group listed in Table 8-8 includes a wide variety of drugs used by anesthesiologists and also other compounds of theoretical interest. If uniform distribution is assumed in total body water, extracellular fluid, or plasma, only relatively few of these compounds (e.g., pancuronium, chlorpromazine) would be capable of producing clinically significant inhibition of AChE or BuChE. It is conceivable, however, that if, on the basis of their physicochemical properties (e.g., lipid/water distribution coefficient) they are distributed selectively, they may cause significant inhibition of ChE at certain sites (e.g., central nervous system, peripheral nerves). These considerations apply especially to nonquaternary compounds such as narcotic analgesics and local anesthetics that are capable of penetrating the blood–brain barrier or the nerve fiber membrane.

The compounds included in the third group (see Table 8-9) were chosen primarily on the basis of their selectivity toward AChE and BuChE, respectively. The data included in Table 8-9 should be useful for those who, for experimental purposes, wish to inhibit AChE or BuChE alone, or both together. As in the other tables presented in this chapter, the I_{50} values shown apply to human erythrocyte AChE and human plasma BuChE under the indicated experimental circumstances. With ChE from other human tissues and from ChE of other species, these values can only be considered guidelines and the corresponding I_{50} values must be determined individually for each enzyme. This can be done by determining the inhibitory effect of a 10^{-4} or 10^{-6} M concentration of moderately or highly potent inhibitors, respectively. Depending on the degree of inhibition obtained, the inhibitory effect of 5 or 6 other concentrations of the inhibitor is determined. These concentrations are so selected that about half of them should be below and half above the

Table 8-7. Anticholinesterases in clinical use

Generic name	Proprietary name	Chemical classification	I$_{50}$ (M) Values AChE $(3.1\times10^{-3}\ M)^a$	BuChE $(2.2\times10^{-2}\ M)^a$	Kinetic classification	Clinical uses	References
Edrophonium[b,c,d] chloride	Tensilon	Phenolic quaternary ammonium	1.1×10^{-5}	1.4×10^{-3}	Reversible, competitive	Diagnosis of myasthenia gravis (MG). Residual n.m. block	Foldes et al. (140a)
Neostigmine[e,f] Methyl sulfate	Prostigmine	Aromatic carbamate	1.2×10^{-9}	1.0×10^{-8}	Irreversible, noncompetitive	Treatment of MG. Reversal of residual n.m. block	Foldes and Smith (142)
Pyridostigmine bromide[e,f]	Mestinon	Heterocyclic carbamate	6.8×10^{-9}	1.9×10^{-7}	Reversible, competitive	Same as those of neostigmine	Foldes and Smith (142)
Ambenonium[c,e] chloride	Mytalase	Aromatic bisquaternary ammonium	7.5×10^{-10}	2.6×10^{-6}	Reversible, competitive	Treatment of MG	Foldes and Smith (142)
Hexafluorenium bromide[c,e]	Mylaxen	Aromatic bisquaternary ammonium	2.2×10^{-6}	1.5×10^{-7}	Reversible, competitive	Prolongation of the n.m. effect of succinylcholine	Foldes et al. (159)
Echothiophate iodide	Phospholine	Organophosphate	1.5×10^{-8}	1.0×10^{-9}	Irreversible	Treatment of glaucoma and MG	Foldes et al. (155); Leopold and Krishna (305)

[a] ACh concentration in assay system.
[b] Source of BuChE was purified human plasma (Cholase).
[c] Inhibitor incubated with enzymes for 20 min before addition of substrate.
[d] Manometric method.
[e] Titrimetric method.
[f] Inhibitor incubated with enzymes for optimal period (several hours).

Table 8-8. Clinically used compounds with anticholinesterase activity

Type of compound	Generic name	Proprietary name	Chemical classification	I_{50} (M) Values		References
				AChE[a] $(3.1 \times 10^{-3}\ M)$[c]	BuChE[b] $(2.2 \times 10^{-2}\ M)$[c]	
Nondepolarizing n.m. blocking agents	d-Tubocurarine chloride	—	Bisquaternary ammonium compound	7.2×10^{-4}	3.0×10^{-4}	Foldes et al. (147)
	Dimethyl-d-tubocurarine chloride	Metubine Mecostrin	Bisquaternary ammonium compound	3.2×10^{-3}	4.8×10^{-4}	Foldes et al. (147)
	Gallamine triethiodide	Flaxedil	Bisquaternary ammonium compound	4.5×10^{-4}	2.4×10^{-4}	Foldes et al. (147)
	Alcuronium chloride	Alloferine	Bisquaternary ammonium compound	6.7×10^{-4}	5.4×10^{-5}	Foldes and Deery (137)
	Pancuronium chloride	Pavulon	Bisquaternary ammonium compound	3.0×10^{-4}	5.6×10^{-8}	Foldes and Deery (137)
	Benzoquinonium	Mytolon	Bisquaternary ammonium compound	2.2×10^{-7}	2.1×10^{-5}	Foldes et al. (147)
Depolarizing n.m. blocking agents	Succinylcholine chloride	Anectine	Bisquaternary ammonium compound	1.3×10^{-3}	6.4×10^{-4}	Foldes and Deery (137)
	Suxethonium bromide	Brevedil-E	Bisquaternary ammonium compound	2.6×10^{-4}	4.3×10^{-4}	Foldes and Deery (137)

124

Category	Drug	Trade name	Structure type			Reference
Narcotic analgesics	Decamethonium chloride	Syncurine	Bisquaternary ammonium compound	3.9×10^{-5}	2.4×10^{-5}	Foldes and Deery (137)
	Morphine phosphate	—	Phenanthrene derivative	1.0×10^{-3}	4.0×10^{-3}	Foldes et al. (152)
	Oxymorphone hydrochloride	Numorphan	Semisynthetic morphine derivative	4.6×10^{-3}	4.6×10^{-3}	Foldes et al. (152)
	Levorphan tartrate	1-Dromoran	Morphinan derivative	5.6×10^{-5}	7.1×10^{-4}	Foldes et al. (152)
	Meperidine hydrochlorides	Demerol	Piperidine derivative	$> 10^{-2}$	2.2×10^{-3}	Foldes et al. (152)
	Alphaprodine hydrochloride	Nisentil	Piperidine derivative	3.4×10^{-3}	2.8×10^{-3}	Foldes et al. (152)
	Nalorphine hydrochloride	Nalline	Morphine derivative	1.1×10^{-3}	2.0×10^{-4}	Foldes et al. (152)
Narcotic antagonists	Naloxone hydrochloride	Narcan	Semisynthetic morphine derivative	6% at 10^{-3}	10% at 10^{-3}	Foldes and Deery (137)
	Levallorphan tartrate	Lorfan	Morphinan derivative	4.0×10^{-5}	8.5×10^{-5}	Foldes and Deery (137)
Local anesthetic	Procaine hydrochloride	Novocaine	p-Aminobenzoic acid ester of an amino alcohol	2.8×10^{-4}	1.9×10^{-4}	Baart et al. (30)
	2-Chloroprocaine hydrochloride	Nesacaine	p-Aminobenzoic acid ester of an amino alcohol	3.5×10^{-5}	2.7×10^{-5}	Baart et al. (30)

(Continued)

Table 8-8. (*Continued*)

Type of compound	Generic name	Proprietary name	Chemical classification	I$_{50}$ (M) Values		References
				AChE[a] $(3.1\times10^{-3}\,M)^c$	BuChE[b] $(2.2\times10^{-2}\,M)^c$	
Local anesthetic	Tetracaine hydrochloride	Pontacaine	p-Aminobenzoic acid ester of an amino alcohol	2.0×10^{-5}	1.3×10^{-6}	Baart et al. (30)
	Dibucaine hydrochloride	Nupercaine	Aromatic amide	1.6×10^{-3}	1.0×10^{-5}	Foldes and Deery (137a)
	Lidocaine hydrochloride	Xylocaine	Aromatic amide	1.2×10^{-2}	4.0×10^{-4}	Foldes and Deery (137a)
	Mepivacaine hydrochloride	Carbocaine	Aromatic amide	1.5×10^{-2}	1.2×10^{-3}	Foldes and Deery (137a)
	Bupivacaine hydrochloride	Marcaine	Aromatic amide	1.0×10^{-2}	3.0×10^{-5}	Foldes and Deery (137a)
	Etidocaine hydrochloride	Duranest	Aromatic amide	1.9×10^{-2}	4.4×10^{-5}	Foldes and Deery (137a)
Miscellaneous compounds	Chloropromazine	Thorazine derivative	Phenothiazine	3.0×10^{-4}	2.0×10^{-5}	Erdos et al. (117)
	Tryptamine	—	Indolalkylamine	4.2×10^{-3}	2.0×10^{-3}	Erdos et al. (117)
	5-Hydroxytryptamine	—	Indolalkylamine	3.2×10^{-3}	7.1×10^{-4}	Erdos et al. (117)
	Bufotenine	—	Indolalkylamine	1.6×10^{-4}	7.8×10^{-5}	Erdos et al. (117)
	1-Epinephrine hydrochloride	Adrenaline	Indolalkylamine	2.2×10^{-2}	2.0×10^{-5}	Zsigmond (472)
	d-Lysergic acid	—	Indolalkylamine	4.9×10^{-5}	1.4×10^{-6}	Zsigmond et al. (471)

[a]Human plasma BuChE (Cholase), specific activity about 1000 μmol/mg/hr. [b]Washed hemolyzed human red cells. [c]Substrate concentration.

Table 8-9. Selectivity of anticholinesterases

Type of compound	Generic name or code	I_{50} (M) values for		Inhibition ratio		References
		AChE	BuChE	AChE/BuChE	BuChE/AChE	
Selective inhibitors of AChE	3116 CT	—	—	250,000	—	Funke et al. (174)
	B.W.284C51	1.0×10^{-7}	1.0×10^{-3}	10,000	—	Austin and Berry (28)
	Ambenonium	7.5×10^{-10}	2.6×10^{-6}	3,500	—	Foldes and Smith (142)
	R02-1250	1.3×10^{-8}	1.1×10^{-5}	850	—	Foldes et al. (163)
Selective inhibitors of BuChE	Pancuronium	3.0×10^{-4}	5.6×10^{-8}	—	5360	Foldes and Deery (137)
	RO2-0683	9.0×10^{-7}	3.2×10^{-9}	—	280	Foldes and Deery (137)
	DFP	1.0×10^{-7}	7.2×10^{-10}	—	140	Foldes et al. (155)
	Mipafox	2.2×10^{-5}	3.9×10^{-7}	—	56	Aldridge (4a)
Nonselective inhibitors of cholinesterases	Physostigmine	2.3×10^{-7}	3.6×10^{-8}	0.16	6.39	Wills (447)
	Neostigmine	2.3×10^{-7}	5.6×10^{-8}	0.25	4.00	Foldes et al. (163)
	RO2-9147	1.1×10^{-7}	4.5×10^{-8}	0.40	2.40	Foldes et al. (163)
	Phospholine	1.5×10^{-8}	1.0×10^{-9}	0.07	15.00	Foldes et al. (155)

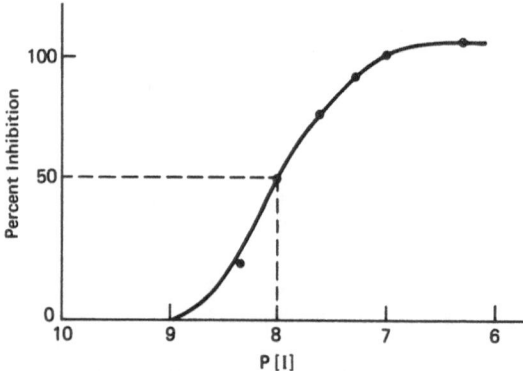

Figure 8-3. The inhibition curve of neostigmine obtained by plotting the negative logarithm of the inhibitor concentration ($p[I]$) against percent inhibition.

expected I_{50} value. From the plot of the negative logarithm of the inhibitor concentration ($p[I]$) against percent inhibition the $p[I]_{50}$ and from this the I_{50} value can be determined. For example, from Figure 8-3 the $p[I]_{50}$ of neostigmine is 8. This means that the $I_{50} = 10^{-8}$ M. When the $p[I]$ is not a round number, as in the case of edrophonium for AChE (4.96), the I_{50} value can be calculated as follows:

$$p[I]_{50} = 4.96 = 5.00 - 0.04.$$

the antilogarithm of $5.00 - 0.04 = 10^{-5} \times 10^{0.04} = 10^{-5} \times 1.1 = 1.1 \times 10^{-5}$, or in other words, the I_{50} of edrophonium for AChE is 1.1×10^{-5} M.

For an inhibitor to be considered truly "selective," its concentration causing complete inhibition of one enzyme (I_{100}) should have no measurable inhibitory effect on any other enzyme. This is indeed the case for BW.248C51 with human erythrocyte AChE (28) and for pancuronium and succinylcholine with human plasma BuChE with BuChE (137).

Reactivation of Inhibited Cholinesterases. As discussed earlier, ChE activity is inhibited when the esteratic site of the enzyme is occupied reversibly or irreversibly by a group that prevents its interaction with the ester group of ACh or that of other substrates. In the case of reversible inhibitors (e.g., edrophonium, ambenonium) the binding between the enzyme and the inhibitor depends on the concentration of the inhibitor in contact with the enzyme, and it occurs rapidly. When this concentration is decreased, *in vivo* by excretion and redistribution, or *in vitro* by dialysis, below a critical level enzyme activity will be reestablished. In the case of irreversible inhibitors, the inhibition occurs in two steps (325). At first a reversible complex is formed between them and the enzyme. The rate of formation of this complex is relatively rapid, and similar to the complex formation between the enzyme and reversible inhibitors, depends on the concentration of the inhibitor and its affinity to the enzyme. The second step occurs more slowly and consists of the formation of a relatively irreversible complex between a part of the inhibitor and the esteratic site (51a). During this second step, similar to the reaction between ACh and AChE, the inhibitor is split. One part containing a carbamyl (e.g., in neostigmine), phosphoryl (e.g., in DFP), sulfonyl (e.g., in methanesulfonyl fluoride), or selenyl group becomes "irreversibly" attached to the esteratic site. The second part, which is usually the most acidic group of the inhibitor, is called the leaving group (e.g., in DFP it is fluoride). In agreement with the slow reaction rate of the second step, maximal inhibition with irreversible inhibitors usually occurs slowly. For example, the maximal inhibitory effect of neostigmine with human erythrocyte ACh only develops in about 3 hr (142).

Reversible inhibitors can be recovered intact from the enzyme–inhibitor complex whatever their duration of interaction. In contrast, after the completion of the second step of their interaction only the breakdown products of the irreversible inhibitors can be recovered.

During reactivation, the reacting group of the inhibitor is removed and enzymatic activity is restored. To what extent the enzyme regains its preinhibition activity depends on the extent of physical and chemical changes that occurred in the enzyme molecule during inhibition and reactivation (38, 39). These changes are referred to as "aging" and are discussed subsequently.

Reactivation may occur spontaneously or through interaction with a nucleophylic reagent, with high affinity to the group attached "irreversibly" to the esteratic site of ChE. During spontaneous reactivation the nucleophylic group is the hydroxyl ion of water. Both spontaneous and induced reactivation depends on numerous factors including: the type (AChE or BuChE) and source of ChE, the type of inhibitor, the pH and temperature, and in the case of induced reactivation, the chemical and physiochemical properties and concentration of the reactivator. The reactive group of the reactivator is usually the nucleophylic anion ($=N—O^-$) (367).

Spontaneous reactivation of human plasma BuChE is significant after inhibition with neostigmine (142, 146) and the closely related carbamate, RO-2063 (146), but not with the heterocyclic carbamate, pyridostigmine (142). In the presence of high inhibitor concentrations (e.g., 10^{-7} to 3.3×10^{-6} M) of neostigmine or RO2-0683 after 24 hr incubation, there was no reactivation (142, 146). With low concentrations (10^{-8} to 1.6×10^{-7} M), the reactivation was complete. There was also significant reactivation after 24 hr incubation in heat-inactivated plasma or phosphate buffer of pH 7.4, but reactivation was less complete than in normal plasma (146). This indicates that in addition to the reaction between these inhibitors and water, enzymatic breakdown of the inhibitors also plays a significant role in the reactivation process (146). Phenolic quaternary ammonium (e.g., edrophonium) or bis-quaternary aromatic ammonium (e.g., ambenonium) type inhibitors did not undergo spontaneous reactivation after 20 hr incubation in plasma (142).

After 21 hr incubation with human erythrocytes there was only moderate, 20 percent, reactivation of neostigmine (142), indicating that it was probably caused by interaction with the solvent and not by enzymatic hydrolysis of the substrates.

The spontaneous reactivation of ChE inhibited by organophosphates is also due to the nucleophilic effect of the hydroxyl group of water. In the case of human plasma BuChE, however, enzymatic hydrolysis may also be a contributing factor (336).

Most of the presently known reactivators contain a quaternary ammonium group separated from the nucleophylic group by about the distance (5 Å) between the anionic and esteratic site of ChE (424). There are numerous compounds capable of reactivating irreversibly inhibited ChE. Historically, Ch (449a) was one of the first compounds found to be capable of reactivating phosphorylated AChE. Other compounds introduced or synthetized as re-activators include hydroxylamine (449a), hydroxamic acids (456a), ketoximes, and aldoxymes (72a) such as pyridine-2-aldoxine methiodide (2-PAM); pyri-dine-4-aldoxime methiodide (4-PAM), and double aldoxime molecules in which 2 molecules of 4-PAM are joined by a methylene chain (TMB-4) (228, 377), or by an oxygen, sulfur, or ether link (e.g., toxogonin (LuH6) (112, 116). Of these compounds, 2-PAM is used most widely in the treatment of intoxication caused by irreversible ChE inhibitors. Of the newer compounds both TBM-4 and toxogonin *in vitro* are about 20 times more potent reactiva-tors of human red cell AChE than 2-PAM (219, 227). Toxogonin, however, is relatively less toxic *in vivo* than 2-PAM or TBM-4 (115).

After relatively short (20 min) incubation with neostigmine, human plasma BuChE can also be reactivated by simple quaternary ammonium compounds (e.g., tetraethyl, tetramethyl- , tetrapropyl- , or tetrabutylammonium) and aromatic amines, such as morphine, oxymorphone, or amphetamine (156). The concentrations of these compounds required for the reactivation of neostigmine, when used alone cause moderate inhibition of both human plasma BuChE and red cell AChE. The reactivating effect of these com-pounds is probably due to the displacement of the quaternary nitrogen of neostigmine from the anionic site of BuChE (153), before the esteratic site becomes carbamated. In contrast, instead of reactivation, these compounds increase the inhibitory effect of neostigmine on AChE. The difference in the influence of these compounds on neostigmine inhibited BuChE and AChE may be due to the anionic site of ACh having a greater negative charge than BuChE, or AChE having two and BuChE one anionic site (41, 451), and that adsorption of the quaternary nitrogen of ACh to the anionic site is more important for its hydrolysis by AChE than by BuChE (163).

There is no significant difference between the *in vitro* and *in vivo* efficacy of reactivators. The choice of the reactivator to be used *in vivo* is made on the basis of the therapeutic index. Thus, for example, TMP-4 is twice as active as

2-PAM *in vivo* but its acute toxicity is at least four times greater (321) and its chronic toxicity is also much greater (4). Toxogonin may have a more favorable therapeutic ratio than TMP-4 (111), and may be even superior in this respect to 2-PAM. More clinical data will be required to confirm the validity of this assumption.

The biologic half-life of 2-PAM in man is about 1.7 hr (289). Since effective reactivation of phosphoryllated ChE requires the sustained presence of reactivators, the relatively short half-life of 2-PAM should be taken into consideration in the management of organophosphorus intoxication.

Aging. Human plasma BuChE (226) and AChE (340) inhibited with organophosphates, depending on the duration of the interaction between the enzymes and the inhibitors, become gradually refractory to reactivation by nucleophylic compounds. This phenomenon is referred to as "aging." Aging is caused by the transformation of the enzyme by a chemical reaction, probably dealkylation (368, 425), between the phosphoryl (or other similar) group and the enzyme, to an inactive form (218). In addition to the chemical structure of the inhibitory group, and the time elapsed before the application of the reactivator, the type of enzyme, the temperature and the pH also have considerable influence on aging.

Aging occurs extremely rapidly with saman (pinacolylmethyl-phosphonofluoridate) (341a) with a half-life* of less than 90 sec (46) and much more slowly with tabun or sarin (217). Organophosphates containing a quaternary ammonium group (e.g., phospholine) cannot be reactivated by oximes (210) and in this sense they age instantaneously.

Plasma BuChE tend to age more rapidly than AChE. Thus, for example, the half-life of aging for DFP inhibited human plasma BuChE is 20 min (224) and that of human erythrocyte AChE is 266 min (225). The corresponding data for tabun are 336 to 355 min for BuChE and 798 min for AChE (217). Human plasma BuChE inhibited by organophosphates ages much more rapidly than horse plasma BuChE (217). The situation is reversed with erythrocyte AChE of these two species (225).

The rate of aging decreases with decreasing temperatures (225), or with increasing pH (218). Aging seems to occur more slowly in purified than in unpurified enzyme preparations (217).

Aging does not occur with all types of enzymes. In addition to ChE, aging was also observed with atropinesterase (3), but not with trypsin or chymotrypsin (197).

Aging can occur not only *in vitro*, but also *in vivo*. Adsorption to other enzymes and to nonenzymatic sites by altering the inhibitor concentration

---

*The half-life of aging may be defined as the duration of interaction between the inhibitor and the enzyme, before the addition of the reactivator, that results in loss of 50 percent of activity.

available for interaction with ChE, may influence the aging process *in vitro* (206).

Species Variation of Cholinesterase. There is a marked variation in the specific activity with ACh and in the substrate specificity of ChE obtained from the same tissues of different species. There is a similar variation in the sensitivity of ChE of different species to various inhibitors.

With ACh substrate plasma BuChE activity is relatively high in man, horse, and certain species of monkeys, such as chimpanzee (186), and Mangabey, but not in others, such as the Macaque. BuChE activity is much lower in other mammals (e.g., dog, guinea pig, cat, rat) and is the lowest in the cow (186). There can be significant variation in plasma BuChE activity not only from species to species, but also within different inbred strains of the same species. For example, a more than fourfold variation of BuChE activity was found in 23 strains of mice (12). In mammalian plasmas the order of the relative hydrolysis rates of various substrates is usually BuCh > PrCh > ACh > BeCh > SCh > procaine (22, 23, 150). Turtle plasma, however, hydrolyzes PrCh about four times faster than BuCh (22, 23). None of the mammalian plasmas hydrolyze MeCh at an appreciable rate, but chicken (22, 23) and electric eel plasmas (467) do.

There is considerable variation in the erythrocyte AChE activity of various mammals (60a). Of all the species investigated, it was found to be the highest in man. The relative activities (man = 100) were: chimpanzee 71, rhesus monkey 64, pig 16, guinea pig 14, goat 12, horse 12, dog 8, rabbit 5, and cat 2. Significant differences are also present in the AChE activity of various fractions of the cerebellum of different species (193).

There is a wide variation in the sensitivity of ChE of various species against different inhibitors. The dibucaine number of rhesus monkey (98.5) and horse (91.4) is higher, and that of cat (20.8) and dog (68.4) is lower than that of man (80.2) (186), and sodium fluoride inhibits significantly the BuChE of man and chimpanzee only. The inhibitory effect of mipafox on human erythrocyte AChE ($I_{50} = 2.2 \times 10^{-5}$) is about seven times greater than on horse erythrocyte AChE ($I_{50} = 1.5 \times 10^{-4}$). Its relative inhibitory effect on the plasma BuChE of these species is the reverse: The I_{50} values are 3.9×10^{-7} for human and 3.8×10^{-8} for horse BuChE. Because of this, mipafox is a highly selective inhibitor of horse (I_{50}AChE/I_{50}BuChE = 3950) but not of human (I_{50}AChE/I_{50}BuChE = 56) (25, 210a) BuChE.

Influence of Physiologic States on Human Cholinesterases. Age, sex, and pregnancy have a significant effect on human plasma BuChE activity. In contrast, human erythrocyte AChE is influenced less significantly by physiologic variables.

The plasma BuChE activity of neonates and infants of less than 6 months of age is 40 to 50 percent lower than that of young adults (334, 469). There was no significant sex difference in BuChE activity in this age group (469). BuChE of neonates and young infants showed similar sensitivity to inhibition by dibucaine and SCh as adult BuChE, but was extremely resistant to inhibition by neostigmine (469). The erythrocyte AChE activity of the fetus and the newborn was also reported to be much lower than that of adults (261). In healthy young adult males, plasma BuChE activity with five different substrates (ACh, BuCh, BeCh, SCh, and procaine) is significantly higher than in females of the corresponding age group (396). In aged males, plasma BuChE activity is significantly lower than that of young males (396) and approaches that of young adult females. There is no difference in BuChE activity of young and aged females.

Human plasma BuChE of the same individual may vary ± 6 percent within a 2-year period (21). Significant diurnal variation of BuChE activity with a maximum around noon and a minimum around midnight was also reported (94).

Except for the lower activity in the newborn (261), there is no age or sex variation of human erythrocyte AChE (21, 396).

It is of interest that in the rat the influence of sex on plasma BuChE activity is the opposite of that observed in man. Thus, in mature female rats, BuChE activity is more than three times higher than in mature males or in immature females (347). Castration or thyroidectomy in male rats causes a significant increase of plasma BuChE activity (123).

Plasma BuChE activity decreases in pregnancy (154, 312, 399, 433), but the enzyme is qualitatively normal. This diminished activity probably represents a decreased rate of synthesis in the liver. BuChE activity returns to normal levels within a few weeks after delivery.

Influence of Pathologic Conditions on Cholinesterases. Plasma BuChE activity may be significantly decreased in a number of pathologic conditions. Erythrocyte AChE activity is only affected significantly in terminal conditions.

Changes of plasma BuChE activity have the greatest diagnostic and prognostic significance in various pathologic conditions of the liver (14, 97). Since the half-life of plasma BuChE in man is about 8 to 10 days (246, 364) its diagnostic usefulness is greater in subacute and chronic, than in acute liver disease. Thus, for example, in acute hepatitis or obstructive jaundice the elevation of the SGOT activity and icterus precedes the fall of BuChE activity (57).

In incipient cirrhosis of the liver low plasma BuChE activity may be the first sign of decreased liver function (195). BuChE activity closely parallels the clinical course, and its progressive fall is an ominous prognostic sign. The decreased synthesis of BuChE in pathologic changes of the liver can also be demonstrated histochemically. Under normal conditions BuChE is evenly distributed in the lobules and in the cytoplasm of the liver (201). In contrast, in obstructive jaundice, cirrhosis, or congestive liver disease, BuChE staining is markedly diminished or absent in the involved areas (47).

Decreased liver function caused by right-sided congestive heart failure is also accompanied by fall of BuChE activity (324). After digitalization, BuChE activity returns toward normal levels. Reversible fall of BuChE activity has also been observed after acute myocardial infarction (220). The maximal changes usually occur at the end of the first week after infarction.

In chronic kidney insufficiency, plasma BuChE activity is markedly reduced; erythrocyte AChE activity, however, is normal (465). The characteristic sex difference of BuChE activity (396) is not present in these patients. Single hemodialysis has no effect on BuChE activity but causes a small clinically unimportant, but statistically significant elevation of erythrocyte AChE activity. Regular hemodialysis of 6 months duration caused a significant increase of BuChE, but had no effect on AChE activity. After renal transplantation there was a sharp fall of BuChE activity that coincided with the administration of large doses of glucocorticoids (464, 465). After tapering the dose of glucocorticoids, BuChE activity returned, over several weeks, to pretransplant levels. Investigation of the various possibilities of decreased BuChE activity in chronic kidney insufficiency indicates that, in all probability, it is due to diminished synthesis of the enzyme in the liver.

In nephrosis, plasma BuChE activity is normal or greater than normal (410, 466). This is because the synthesis of BuChE and serum albumin in the liver are parallel processes (98, 124, 312, 313, 410, 432). To compensate for its loss through the urine, the synthesis of serum albumin in nephrosis is markedly increased. This is accompanied by an increased synthesis of BuChE. Because of its large molecular weight, however, only insignificant amounts of BuChE are lost in the urine and there is a consequent increase of plasma BuChE activity. In terminal nephrosis, when increased protein synthesis cannot keep pace with the progression of the disease, BuChE activity decreases. This decrease of BuChE activity is an ominous prognostic sign.

Plasma BuChE activity was found to be decreased in burnt children (60). The decrease is usually proportional to the extent of the burn and reaches its lowest level in about 2 weeks when the patients are ready for skin grafting. The diminished BuChE activity decreases the rate of enzymatic breakdown of SCh. This intensifies and prolongs the pharmacologic effects of SCh at the n.m. junction and on the myocardium. There is significant increase [for references, see Schaner et al. (391)], in the amount of K^+ released by SCh from the muscle (278, 333). The hyperkalemia combined with the vagotonic effect of the relatively high concentrations of SCh reaching the myocardium may result in a high (1:200 to 1:250) incidence of cardiac arrest (60). The incidence of this complication can be decreased by limiting the dose of SCh to 0.4 mg/kg (200).

Plasma BuChE activity may decrease significantly postoperatively (97). It usually reaches a minimum between the third and sixth postoperative day and returns to its preoperative value by the end of the second week. Continued fall of BuChE activity is usually a sign of complications (59a, 98). The postoperative fall of BuChE activity can be attributed to the transient impairment of liver function caused by anesthesia, surgery, and impaired nutrition in the immediate postoperative period.

Malnutrition (51) due to protein poor diet, esophageal stenosis, absorption anomalies, severe anemias (51), and cachexia (379) is also associated with decreased BuChE activity.

BuChE activity was found to be significantly lower in 177 cancer patients than in their sex matched controls (380). The decrease was the lowest in untreated patients with localized tumors, greater after radiation or anticancer drug therapy and lowest in terminal cases. Some of the alkylating agents used in the treatment of malignant conditions cause 50 percent inhibition of the *in vitro* hydrolysis of BuCh by BuChE in concentrations ranging from $3..3 \times 10^{-4}$ for triethylmelamide to 9.1×10^{-3} for triethylmethiophosphoramide (473). In the dose range used these compounds are unlikely to produce clinically significant inhibition of ChE. The hydrolysis product of another anticancer agent AB-132, however, inhibits BuChE significantly both *in vitro* ($I_{50} = 5.6 \times 10^{-5} M$) (144). and *in vivo* (429) and may cause prolonged apnea in patients who received SCh.

BuChE activity is increased in hyperthyroidism (232, 420) and decreased in hypothyroidsm (420). Extremely high BuChE values may be encountered in gliomas (460, 460a).

Physiologic Functions of Cholinesterases

Despite extensive investigations, dating back to the classical studies of Hunt (235, 236, 237) Hunt and Taveau (238, 239), Dixon (96), Winterberg (458a), Dale (85), Loewi (318), Loewi and Mansfield (318a), Loewi and Navratil (319), Fuhner (172, 173), and others (1, 373, 263), the physiologic role of ChE is still highly controversial.

The physiologic effect of ChE does not necessarily depend on their ability to hydrolyze ACh (182). Thus, for example, in *in vitro* experiments removing by repeated washing with Ringer's solution, the ChE adsorbed loosely to the myocardium interferes with the normal activity of the frog heart (47). Normal myocardial activity can be reestablished by adding of purified AChE or BuChE, but not γ-globulin or human serum albumin, to the system (291). The positive inotropic effect of ChE was not abolished by inhibition with $10^{-4} M$ eserine that completely inhibits their enzymatic activity.

After the washout of ChE the intracellular sodium concentration was found to be significantly elevated, despite increased activity of the "sodium pump." Because of this it was suggested that the physiologic role of ChE on the frog heart is the lowering of the permeability of the muscle membrane to sodium (47).

The most widely investigated and least controversial role of ChE is that of AChE in synaptic transmission in the peripheral nervous system and at cholinergic neuroeffector sites (282). AChE located primarily, but not exclusively, on the postsynaptic membrane (269, 270) hydrolyzes the relatively large amounts of ACh released by the nerve impulse, makes possible the repolarization of the postjunctional membrane depolarized by ACh and thereby completes the depolarization-repolarization sequence essential for junctional transmission. Even this role of AChE has been questioned (106), on the basis that the concentration of ACh at the synapses is far below (10^{-8} to 10^{-6} M), the concentration found to be optimal (10^{-3} M) for the *in vitro* hydrolysis of ACh. Others have estimated the ACh concentration at synapses from 5×10^{-6} to 5×10^{-5} (384, 458) to 10^{-3} M (365). The resolution of this problem is further complicated by the fact that the kinetics of the intact enzyme in its natural environment may be quite different from those of extracted or purified enzyme (107, 108, 110).

The possibility that BuChE also contributes to the removal of ACh released by the nerve impulse at synaptic junctions cannot be excluded. About 5 to 10 percent of the ChE activity of biopsied human muscle is attributable to BuChE (408) and BuChE was also localized histochemically at the n.m. junction (88, 114, 232a).

The transmitter functions of ACh in the central nervous system and in axonal conduction, and consequently the importance of ChE on these processes, are highly controversial. There is much circumstantial evidence (e.g., the high ACh, AChE, and ChAc content of many brain areas) in favor of the transmitter functions of ACh and AChE in the central nervous system (126, 264). Until now, however, the role of ACh and AChE had only been proven for the transmission of impulses from the motor neuron to its spinal interneuron, the Renshaw cell (84, 102).

If possible, even more controversy, manifested by numerous publications and animated discussions at meetings (265, 283, 356, 357) surrounds the role of ACh and AChE in axonal conduction. It was proposed by von Muralt (348, 349) and expanded by Nachmansohn and his collaborators (58, 93, 355) that ACh plays an important role in the alteration of the permeability of the nerve membrane to K^+ and Na^+, and thereby to the changes of the resistance to the nerve fiber to conduction of the nerve impulse. The experimental evidence accumulated in favor of this hypothesis by Nachmansohn and his colleagues (93) has been criticized by others (106, 283). At present the evidence for the role of ACh in axonal conduction must be considered inconclusive and the solution of the problem one way or another will require more research.

ACh and ChE have also been implicated in the control of the passive permeability of and/or active transport through various cell membranes (31,

230, 231, 277, 297). The widespread distribution of ACh and ChE in tissues not related to transmission of impulses, for example, in the placenta (69, 311, 369) favors this assumption.

Until recently no definite physiologic role had been attributed to BuChE. It was suggested that its primary function is the protection of the physiologically important AChE from naturally occuring inhibitors (304). BuChE achieves this by its ready availability in the liver, through which ingested, and in the plasma, through which both ingested and parenterally administered inhibitors pass in the course of their distribution to various parts of the body. The fact that the affinity of eserine and most other inhibitors to BuChE is greater than to AChE increases its ability to protect the functionally important AChE. Thus, for example, in patients treated for glaucoma (370) or myasthenia gravis (155) with echothiophate, plasma BuChE activity may be completely inhibited without any sign or symptom, of anti-ChE intoxication. The same applies to subjects (e.g., farmers, chemical workers, exterminators) with subclinical anti-ChE intoxication.

BuChE, however, may also be involved, together with erythrocyte AChE, in the maintenance of the optimal Ch:ACh balance in plasma (175), fat metabolism in the liver (31, 75), and the metabolism of certain tissues (e.g., blood vessels, myelin sheath, neuroglia) in which it is present in relatively high concentrations (176). Another physiologic function of BuChE is the hydrolysis of BuCh formed by butyryl-CoA in the course of the metabolism of fatty acids in the liver (75). In addition to its already mentioned possible role in synaptic transmission BuChE (and also AChE) that persists in denervated motor end plates may be involved in the re-innervation of these structures (114).

The Role of Cholinesterases in Anesthesiology

As discussed earlier in this chapter, many drugs used before, during, and after anesthesiology may interact with ChE and these interactions may have clinical significance. From the practical point of view these interactions may be considered under two headings: (a) the influence of anesthetic and adjuvant drugs on ChE; and (b) the influence of ChE on the metabolism of anesthetic drugs.

The Influence of Anesthetic and Adjuvant Drugs on Cholinesterases. Inhibition and occasionally acceleration (121) of ChE may be an unwanted side effect of drugs used in conjunction with anesthesia. Anti-ChE may be used deliberately to prolong and intensify the effect of depolarizing (156, 159), or to antagonize nondepolarizing (2a, 130, 382) n.m. blocking agents.

The influence of anesthetic and adjuvant drugs on ChE had been observed both *in vivo* and *in vitro* studies. In *in vivo* studies on surgical patients (50) thiopental–nitrous oxide–oxygen, ether–oxygen, and cyclopropane–oxygen anesthesia or the administration of as much as 1000 mg SCh over a 10 min period had no significant effect on plasma BuChE. In other studies on anesthetized volunteers relatively small increases or decreases of plasma BuChE activity were encountered with thiobarbiturates or propanidid (Epontol) alone or combined with 1% halothane and nitrous–oxide–oxygen, and also after the intravenous injection of 25 mg droperidol and 0.5 mg fentanyl (97). The changes of plasma BuChE encountered in these studies, however, were not large enough to have any clinical significance.

In evaluation of the possible clinical significance of the interaction of anesthetic and adjuvant drugs and ChE, in addition to the anti-ChE potency of the compound, many other variables, such as route of administration, absorption, plasma binding, distribution, metabolism, and excretion, must be taken into consideration. The detailed discussion of this complex problem is beyond the scope of this monograph and only a few examples of practical importance will be considered briefly.

Of the various drugs administered occasionally to patients before surgery, phospholine may have the greatest effect on ChE and on the course of anesthesia. Phospholine is being used extensively for the treatment of glaucoma (305) and occasionally for the treatment of myasthenia gravis (155, 418). In patients on chronic phospholine therapy, plasma BuChE may be almost completely inhibited and recovery of normal activity may take 3 months or more (145). In such patients prolonged apnea may be encountered after the intravenous administration of conventional doses of SCh (370). Prolonged apnea (429) caused by inhibition of BuChE (144) may also occur after the adminstration of SCh to patients with an anticancer drug, AB-132.

Because of the relationship of their inhibitory effect on BuChE and their dose range used in anesthesiology, procaine (158) ($I_{50} = 1.9 \times 10^{-4}$ M), tetracaine ($I_{50} = 1.3 \times 10^{-6}$ M), 2-chloroprocaine ($I_{50} = 2.7 \times 10^{-5}$ M), pancuronium ($I_{50} = 5.6 \times 10^{-8}$ M), and chloropromazine (119) ($I_{50} = 2.2 \times 10^{-5}$ M) may interfere with the metabolic transformation of ester type n.m. blocking agents and local anesthetics. When any of these drugs are used together with SCh the duration and intensity of action of the latter may be increased. From the clinical point of view the effect of these compounds on the enzymatic breakdown and systematic toxicity of ester type local anesthetic agents, may be much more important. These factors must be considered when large doses of ester type local anesthetic agents are administered parenterally (e.g., for peridural block or block of large nerve plexuses) or topically into the tracheobronchial tree (e.g., bronchoscopy). Tetracaine, which is both a substrate and an inhibitor of BuChE, inhibits its own hydrolysis. In patients, who for some reason (e.g., malnutrition) already have low plasma BuChE activity and/or decreased plasma volume, the tracheobronchial administration of relatively small doses of tetracaine may cause severe systemic reactions (11).

Deliberate inhibition of AChE by edrophonium (16, 240, 322), neostigmine (2a, 382), and pyridostigmine (470) is widely employed for the reversal of the residual n.m. block at the end of surgery. The pharmacologic basis of the safe use of these compounds had been discussed elsewhere (382). It bears emphasis, however, that the muscarinic effects of neostigmine and pyridostigmine should be antagonized by the concomitant use of atropine and that the optimal dose of anti-ChE should be determined by careful administration of fractional doses. Excessive doses of anti-ChE may not only cause severe arrhythmias, but may also cause prolonged depolarization block (134).

Deliberate inhibition of plasma BuChE by hexafluorenium is employed for the prolongation and intensification of the neuromuscular effects of SCh. This technique, the details of which had been described elsewhere (156, 159) may be the method of choice for the production of surgical relaxation under special circumstances. This applies when the use of nondepolarizing relaxants are relatively contraindicated (e.g., in renal shutdown, need to avoid anti-ChE for the antagonism of residual n.m. block), or when the side effects of conventional doses of SCh (e.g., rapid endplate depolarization that may cause increased intragastric or intraocular pressure, potassium release from the muscles, bradyarrhythmias) must be prevented. The n.m. block produced by the combined administration of the recommended doses of hexafluorenium and SCh (132) will wear off within 30 min after the last fractional dose of SCh, without the need for any antagonist, in the vast majority of cases (132). Since the termination of the block is due to a large extent to the alkaline hydrolysis of SCh, its duration is not affected significantly by quantitative or qualitative changes of BuChE activity or diminished urinary excretion.

The Influence of Cholinesterases on the Metabolism of Anesthetic Drugs. While the physiologic role of BuChE is unclarified and highly controversial, its pharmacologic significance is universally accepted. In addition to various esters of Ch it also hydrolyzes many other esters. From the point of view of the anesthesiologist the hydrolysis of the ester type n.m. and local anesthetic agents has the greatest importance.

In general the speed of onset, intensity, and duration of action of both the desired effects and the unwanted side effects of drugs depends on their "free" (unabsorbed to plasma ingredients, such as proteins or cellular elements of the blood) plasma level. The only exceptions to this rule are compounds which become irreversibly bound to receptors (e.g., organophosphates to ChE) or which on contact destroy receptor sites (52). The free plasma level of drugs depends on absorption from sites of administration, binding to blood ingredients, distribution to the various body compartments, tissue binding, metabolism in the blood and at other sites (e.g., in the liver) and excretion (e.g., in urine, feces, sweat). Under physiologic circumstances, with the exception of metabolic transformation, the fate of drugs is very similar in most mammalian species (52). As a rule the activity of the microsomal enzymes of the liver, responsible for the metabolic transformation of most drugs, is greater in most mammalian species than in man (52). In contrast, the ability of human BuChE to hydrolyze ester type n.m. and local anesthetic agents is much greater in man than in other mammals (29, 138, 141). The high BuChE activity of human plasma has important implications: It will decrease the relative potency and even more so, the duration of action of hydrolyzable relaxants and decrease the systemic toxicity of ester type local anesthetic agents. In agreement with this there is an inverse relationship between the enzymatic hydrolysis rates of ester type local anesthetic agents, such as SCh or its diethyl analogue (160, 161) (SECh) or the various ω-amino fatty acid ester of choline (138). In species (e.g., cat) which hydrolyze these esters very slowly or not at all (138), the intensity and duration of the n.m. effect of the various compounds is significantly greater than in man (54, 55).

There is no consistent relationship between potency and duration of action of ester type local anesthetic agents and their enzymatic hydrolysis rates. Tetracaine, which is about eight to 10 times as potent, and has a significantly longer duration of action than procaine, is hydrolyzed five times slower than the latter (151). 2-Chloroprocaine which is about one and one-half times more potent and has about the same duration of action as procaine (139) is hydrolyzed four to five times faster than procaine (151). In contrast, except for rapid intravenous injection, when it parallels local anesthetic potency (141), systemic toxicity is inversely proportional to the hydrolysis rate by plasma BuChE (149, 159a). In proportion to their clinical potency the systemic toxicity of the amide-type local anesthetic agents is greater than that of the ester type agents (149, 159a).

The qualitative and quantitative species variation in the metabolism of compounds hydrolyzed by BuChE, on one hand, and those broken down by microsomal enzymes of the liver, on the other hand, may result in significant variation in the pharmacologic effects of ester type muscle relaxants and local anesthetic agents in different mammals. These factors must be taken into consideration when attempting to transfer the findings of pharmacologic and toxicologic data obtained in laboratory animals to man. It is essential that the clinical investigation of not only these compounds, but of drugs in general, should be preceded by clinico–pharmacologic and metabolic studies in man. Such studies should include the determination of the optimal dose and/or concentration and the toxicity by the administration of graded doses and the "half life" of the optimal dose. This approach will decrease the dangers and reduce the time and effort expended between the synthesis and animal testing and the introduction of therapeutic agents into clinical practice.

It is evident from the foregoing that any qualitative or quantitative alteration of plasma BuChE activity may significantly influence the intensity and duration of action and toxicity of ester type n.m. blocking and local anesthetic agents.

After the administration of conventional doses, the intensity and duration of action of hydrolyzable relaxants is (e.g., SCh) markedly, but not excessively, increased when plasma BuChE activity is low, but the enzyme is qualitatively normal (122, 161), and decreased when BuChE activity is abnormally high (364). The systemic toxicity of SCh (e.g., potassium release from muscles (131, 274, 333) bradyarrhythmias), however, may become alarming even in the presence of a quantitative decrease of BuChE activity (60, 391).

In the presence of atypical BuChE variants, the duration of action may be excessively prolonged after the administration of conventional doses of SCh (154, 253, 257).

The relationship between plasma BuChE activity and duration of action of SCh, however, is complicated by the findings that, in some patients, prolonged apnea was encountered in the presence of normal BuChE activity (15, 154, 459) and that one subject (161) with lower than normal BuChE activity was found to be resistant to SCh. These findings indicate that, in addition to qualitative and quantitative variation of BuChE activity, other presently unknown factors may also cause increased or decreased sensitivity to SCh.

Qualitative and/or quantitative changes in BuChE activity have little or no influence on the intensity and duration of action of hydrolyzable local anesthetic agents. Such changes, however, may markedly increase systemic toxicity of these compounds. Thus, for example, systemic reaction was encountered in a young female operated on for the relief of intestinal obstruction in the sixth month of pregnancy (154). About 20 min after the peridural administration of 22 ml 3 percent 2-chloroprocaine containing 1:200,000 epinephrine, she developed muscular twitching which was treated by the administration of oxygen and small doses of intravenous thiopental. Subsequently her plasma BuChE activity was found to be less than half of the mean of the corresponding age and sex group. One month after delivery her BuChE activity was normal and a few months later she had a gynecologic operation under peridural anesthesia with the same dose of 2-chloroprocaine, without any untoward effect.

Another case of severe systemic toxicity caused by a hydrolyzable local anesthetic agent was encountered in a 57-year-old male patient who had stenosing carcinoma of the esophagus. This patient, who lost 17 lb in a relatively short time, was scheduled for esophagoscopy under general anesthesia. When the patient was positioned for endoscopy the surgeon decided to bronchoscope the patient, awake, before esophagoscopy. To facilitate this, the mouth and pharynx were sprayed with 4 ml of 0.1 percent tetracaine and 2 ml of 0.1% tetracaine was instilled into the tracheobronchial tree. Within 1 to 2 min the patient developed severe tonic–clonic convulsions and lost consciousness. The patient received 100 mg SCh intravenously, was intubated, and ventilated with 100 percent oxygen. Subsequently, 75 mg thiopental was also administered intravenously, and when the vital signs became stabilized, esophagoscopy was performed. The SCh induced apnea lasted 20 min. After completion of the esophagoscopy and the return of adequate spontaneous respiration, the endotracheal tube was removed and the patient was bronchoscoped. Recovery was uneventful. The plasma BuChE activity determined 3 days postoperatively was less than one-third (with AChE substrate 72 μmol/ml plasma/hr) of the normal mean for the corresponding age and sex (396). One week later, uneventful palliative surgery was performed under neuroleptanesthesia (157). On this occasion the intravenous administration of 0.2 mg/kg SCh provided satisfactory muscular relaxation for endotracheal intubation and apnea of 4 min duration.

To minimize the occurrence of systemic toxic reactions caused by local anesthetic agents the lowest effective volume and concentration of the least toxic compound should be used and unless contraindicated 1:400,000 to 1:200,000 epinephrine should be added to its solution to minimize vascular absorption. Of the presently available local anesthetic agents 2-chloroprocaine has the highest therapeutic index and should be the agent of choice (148a) for techniques (e.g., continuous peridural block) requiring the repeated administration of relatively large doses of local anesthetic agents. It should also be used in preference to other local anesthetic agents for the blocking of nerve plexuses (e.g., axillary block) and nerve trunks (e.g., sciatic block) requiring large doses of local anesthetic agents, provided that the required duration of the block does not exceed 90 min.

The preferential use of 2-chloroprocaine is especially important in obstetric practice. Although plasma BuChE activity is usually reduced in pregnancy (312, 399, 433), 2-chloroprocaine is still metabolized faster than the more slowly hydrolyzable ester type and the nonhydrolyzable amide type local anesthetic agents. This is not only important from the point of view of the mother, but also from that of the fetus. In the fetus and the newborn the drug metabolizing microsomal enzyme systems of the liver are incompletely developed (53, 169, 247) and plasma BuChE activity is about 50 percent lower than in young adults (334, 471). Urinary excretion is also less efficient in the fetus and newborn (365a). For this reason the disposition of the amide-type local anesthetic agents (all of which readily penetrate the placental barrier) by the fetus and newborn is slow (342, 375, 419). Consequently any toxic effect of local anesthetic agents (e.g., fetal hypotension, bradycardia, arrhythmias) will persist and will not only affect the immediate well-being of the fetus and newborn (389, 390) but may also have delayed adverse effects on physical and mental development.

Because of its relatively rapid hydrolysis by BuChE of the maternal plasma and the placenta, little or no unchanged 2-chloroprocaine reaches the fetal circulation (128). For this reason, 2-chloroprocaine should be the agent of choice not only for lumbar or sacral peridural, but also for pudendal and paracervical block in obstetrics.

What could be done to minimize the incidence and severity of untoward reactions to drugs metabolized by plasma BuChE? The most important single factor is awareness of the possibility of such reactions. Ideally, it would be desirable to determine routine plasma BuChE activity of all patients in whom the use of such drugs are contemplated. From the practical point of view this would require the adaptation and inclusion of the determination of BuChE activity among the automated techniques used routinely on hospitalized patients. Barring this, some method of reliable determination of plasma BuChE activity should be available at major hospitals. The modification of the ultraviolet spectrophotometric method (229, 256, 258, 260) for the determination of BuChE activity with BeCh substrate described earlier in this chapter (see p. 116) is relatively simple and accurate, and can easily be carried out in any laboratory. BuChE activity should be determined with this or some other suitable method whenever there is a high degree of suspicion that qualitative and/or quantitative changes in the activity of this enzyme may be present. Such changes are most likely to be encountered in relatives of individuals who are known homozygotes or heterozygotes of one of the abnormal (atypical, fluoride-resistant silent) variants of BuChE. Quantitative changes of BuChE may also be present in certain pathologic conditions, such as liver disease (161), hypothyroidism (420), malnutrition and cachexia (379), malignancies (380) and burns (60), or after the iatrogenic [e.g., echothiophate for the treatment of glaucoma (305)] or accidental intoxication with organophosphates (447).

While the knowledge of plasma BuChE activity is very useful it should be remembered that the normal level of this enzyme does not necessarily exclude prolonged SCh induced apnea (154) or systemic reaction to hydrolyzable local anesthetic agents.

Beyond what has been discussed earlier, nothing more can be done for the prevention of systemic reactions that may accompany the use of hydrolyzable local anesthetic agents. Excessively prolonged apneas, reported to occur with the use of SCh in patients with abnormal variants of BuChE, however, can be usually, but not always prevented by adhering to two simple rules: The first of these is the limitation of the initial dose of SCh to the minimum (about 0.5 to 0.6 mg/kg) that will provide optimal conditions for endotracheal intubation; the second is the delaying of the administration of any more SCh until the start of the return of spontaneous respiratory activity, or if this parameter is monitored, the return of the indirectly stimulated twitch tension. If there is any reason to believe that the patient may be an abnormal homozygote or heterozygote for BuChE and no facilities are readily available for the determination of the activity of this enzyme, a test dose of 0.2 mg/kg SCh should be used. The treatment of prolonged apnea caused by SCh is a controversial subject. Most clinicians prefer to control and later assist ventilation until the return of adequate spontaneous respiratory activity. Others (187, 191) recommend the intravenous administration of highly purified (1200- to 1300-fold) human plasma BuChE (Behringerwerke, Marburg-Lahn, DBR) in amounts corresponding to the expected BuChE content of the patient's plasma volume (about 50 mg of the purified preparation/liter). The administration of these amounts of purified BuChE to five subjects who were atypical homozygotes (*DD*) and to one who was an atypical–silent hetero-zygote (*DS*) increased the enzymatic hydrolysis rate of BeCh to normal levels and also caused a significant elevation of the dibucaine number (*DN*) from 16 and 27 to 49 and 64. Furthermore, following the administration of 1 mg/kg SCh, neuromuscular transmission in the diaphragm and in the skeletal muscles returned to normal in about the same time (8 min) after the injection of purified BuChE as in normal subjects. In some cases the injection of the purified BuChE preparation was followed by mild foreign protein reaction (188).

Whether or not the advantages of the use of purified human plasma BuChE in the therapy of SCh induced prolonged apnea justify the risk of possible adverse reactions will have to be determined in larger groups of patients.

The therapy of systemic reactions caused by local anesthetic agents had been discussed elsewhere (140). The first and most important measure is the prompt administration of 100% oxygen at the first sign of toxicity. Subsequent therapy depends on the clinical course. If the patient develops tonic–clonic convulsions and loses consciousness, 1.0 to 1.5 mg/kg SCh should be administered rapidly to control muscular hyperactivity and facilitate endotracheal intubation and adequate ventilation. Since n.m. blocking agents do not influence the increased irritability of the motor cortex, 1 to 1.5 mg/kg thiopental should be administered intravenously next. If the reaction was caused by a hydrolyzable local anesthetic agent usually no more SCh and/or thiopental will be required. If signs of muscular hyperactivity recur more SCh, and if the still paralyzed patient regains consciousness, more thiopental should be given.

If the clinical course is milder and the patient only manifests muscular twitchings and becomes disoriented without losing consciousness, thiopental should be administered first and muscle relaxants should only be used if the patient subsequently develops convulsions.

The Therapy of Anticholinesterase Intoxication

An anesthesiologist may be called upon to assist or to manage the treatment of anti-ChE intoxication. Such intoxications may occur as complications of the therapy of myasthenia gravis and are referred to as "cholinergic crisis" (140a, 448). They may also be the results of accidental or intentional intoxication with organophosphates. The management of anti-Ch intoxication has been discussed extensively by Wills (447).

The signs and symptoms of anti-ChE intoxication are caused primarily by the muscarinic and nicotinic actions of the accumulated ACh. The muscarinic signs and symptoms are due to the effects of ACh at cholinergic effector sites and include stimulation of glandular activity resulting in tearing, salivation, increased tracheobronchial and gastric secretions, stimulation of the smooth muscle of the respiratory, gastrointestinal, and genitourinary tracts, causing bronchospasm, intestinal cramps, diarrhea, involuntary urination, uterine cramps, inhibition of the smooth muscle of the vascular system (85a) resulting in decreased peripheral resistance, and bradycardia.

The nicotinic signs and symptoms are caused by stimulation in mild and depression in severe intoxication of n.m. and ganglionic transmission. They include variable degrees of hypertension followed by hypotension in severe intoxication, muscular fasciculation and twitching that may be followed, depending on the severity of the intoxication, by muscle weakness or paralysis.

A third group of signs and symptoms referable to the central nervous system include anxiety, confusion, dizziness, ataxia progressing in severe cases to coma, respiratory depression, and generalized convulsions with an EEG pattern similar to that seen in epilepsy. The central nervous system manifestations of anti-ChE intoxication are probably due to the muscarinic actions of ACh since they are amenable to treatment with large doses of antimuscarinic agents (e.g., atropine).

In addition to the above signs and symptoms organophosphate intoxication may cause various coagulation disturbances that may result in increased or decreased coagulation time (275) and delayed central (372) or peripheral (48) muscular paralysis.

The therapy of anti-ChE poisoning depends on the severity of the intoxication and the presenting signs and symptoms. In severe intoxication, when adequate spontaneous ventilation is impaired by central respiratory depression, paralysis of the respiratory muscles, convulsions or profuse tracheobronchial secretions, endotracheal intubation should be performed as quickly as possible. The muscle relaxant (usually SCh) used to facilitate intubation will also control the convulsions temporarily. After a brief period of oxygenation the accumulated secretions should be removed by suctioning and artificial ventilation should be continued. After the establishment of adequate ventilation an intravenous infusion should be started partly to support the circulation and partly to have ready access for the intravenous administration of drugs. Next, 1 to 2 mg increments of atropine should be administered intravenously 1 to 2 min apart until atropinization (dry skin, decrease of myosis, improvement of bradycardia) is achieved and maintained. This may require as much as 50 mg, or more atropine in the first 24 hr (447). If the convulsions should recur during the process of atropinization, nondepolarizing n.m. blocking agents (e.g., *d*-tubocurarine, pancuronium, gallamine) should be administered as required for their control. If available, pancuronium and gallamine are preferable to *d*-tubocurarine. They both have a tendency to increase pulse rate and are less likely to cause endogenous histamine release, that may increase the bronchoconstrictor effect of anti-ChE, than *d*-tubocurarine.

When reactivation of the inhibited ChE with oximes is indicated, it should be carried out with as little delay as possible to avoid "aging" of the enzymes.

As discussed earlier after variable length of phosphorylation (or carbamation) with anti-ChE, enzymes cannot be reactivated any more by oximes. Effective reactivation by oximes require the maintenance of relatively high concentrations for prolonged periods. The most widely used reactivator is 2-PAM. Its initial dose in severe anti-ChE intoxication is about 25 to 30 mg/kg injected over 5 to 10 min (198). Since the half-life of 2-PAM in man is about 100 min (289) following the injection of the initial dose, the administration of 2-PAM should be continued at a rate that will deliver the equivalent of the initial dose in 3 to 4 hr. Thus, for example, if the initial dose was 2100 mg in a 70 kg patient, 700 mg (70 ml of a 1 percent solution) should be administered per hr for the first 24 hr. Of the other presently available reactivators toxogonin (Lü H6) (116) may have some advantages over 2-PAM. It is considerably more potent than 2-PAM and its recommended dose is 3 to 5 mg/kg (447).

Oximes had also been recommended for the treatment of anti-ChE poisoning that may complicate the treatment of myasthenia gravis by these compounds (199). The use of oximes in such patients, however, is seldom indicated. Usually assisted or controlled ventilation, removal of secretions, and the administration of atropine is sufficient to tide the patient over until the spontaneous reactivation of ChE.

In several anti-ChE intoxication the support of circulation and correction of any acid–base or electrolyte disturbances may also be necessary. Both respiratory and metabolic acidosis may occur in anti-ChE poisoning (192), and there may be significant loss of sodium and potassium (447). Congestive failure may also accompany severe anti-ChE intoxication (337).

Whenever there is reason to believe that still unabsorbed anti-ChE is present on the skin or in the stomach, parallel with the described therapeutic measures, the patient should be decontaminated. Anti-ChE on the skin can be chemically destroyed by weak alkalis (including soap), hypochlorides (e.g., Chlorox) or hydrogen peroxide. After oral ingestion the removal of anti-ChE from the stomach by gastric lavage with weak alkalis may be attempted. To avoid aspiration, gastric lavage should be carried out only after endotracheal intubation with a cuffed tube.

The management of mild anti-ChE intoxication should be expectant. It should consist of decontamination, the oral or intramuscular administration of atropine, removal of accumulated secretions and above all, careful observation and readiness to institute other measures if the worsening condition of the patient warrants them.

References

1. Abderhalden, E., Paffrath, H., and Sickel, H., Beitrag zur Frage der Inkre-(Hormon) Wirkung des Cholins auf die motorischen Funktionen des Verdauungst kanales. II. Mitteilung. *Pfluegers Arch. Gesamte Physiol. Menschen Tiere* **207**, 241 (1925).

2. Adams, D. H., The specificity of the human erythrocyte cholinesterase. *Biochim. Biophys. Acta* **3**, 1 (1949).

2a. Adams, R. C., Curare as aid to relaxation in anesthesia. *Surg. Clin. North Am.* **25**, 735 (1945).

3. Adie, P. A., The reactivation of inhibited atropinesterase (EC 3-1-1-10). *In* "Proceedings of the Conference of Structure and Reactions of DFP Sensitive Enzymes" (E. Heilbronn, ed.), p. 167. Research Institute of National Defence, Stockholm, 1967.

4. Albanus, L., Järplid, B., and Sundwall, A., On the toxicity of TMB-4, a reactivator of inhibited cholinesterase. *Biochem. Pharmacol.* **12**, 111 (1963).

4a. Aldridge, W. N., The differentiation of true and pseudo cholinesterase by organophosphorus compounds. *Biochem. J.* **53**, 62 (1953).

5. Alles, G. A., and Hawes, R. C., Cholinesterases in the blood of man, *J. Biol. Chem.* **133**, 375 (1940).

6. Allot, E. N., and Thompson, J. D., The familial incidence of low pseudocholinesterase level. *Lancet* **2**, 517 (1956).

7. Ambrus, C. M., and Ambrus, J. L., Effect of hyperthyroidism and treatment with thiouracil on cholinesterase levels. *Z. Vitam.-, Horm.- Fermentforsch.* **2**, 464 (1948).

8. Ambrus, P. S., and Ambrus, J. L., Effect of thyroid hormone and thyroidectomy on the synthesis of cholinesterases. *Res. Commun. Chem. Pathol. Pharmacol.* **2**, 118 (1971).

9. Ambrus, M. S., and Black, J., Role of the pituitary and sex hormones in the synthesis of cholinesterases. *Life Sci.* **7**, 279 (1968).

10. Ammon, R., and Meyer, H., Zur stereochemischen Spezifität der Acetylcholinesterase. *Hoppe-Seyler's Z. Physiol. Chem.* **314**, 198 (1959).

11. Ang. M., Fell, S., and Foldes, F. F., unpublished data (1974).

12. Angel, C. R., Mahin, D. T., Farris, R. D., Woodward, K. T., Yahas, J. M., and Storer, J. B., Heritability of plasma cholinesterase activity in inbred mouse strains. *Science* **156**, 529 (1967).

13. Ansell, G. B., and Spanner, S., The origin and turnover of choline in the brain. *In* "Drugs and Cholinergic Mechanisms in the CSN" (E. Heilbronn and A. Winter, eds.), p. 143. Försvarets Forskningsanstalt, Stockholm, 1970.

14. Antopol, W., Schifrin, A., and Tuchman, L., Decreased cholinesterase activity of serum in jaundice and in biliary disease. *Proc. Soc. Exp. Biol. Med.* **38**, 363 (1938).

15. Argent, D. S., Dennick, O. P., and Hofbiger, F., Prolonged apnoea after suxamethonium in man. *Br. J. Anaesth.* **27**, 24 (1955).

16. Artusio, J. F., Jr., Riker, W. F., Jr., and Wescoe, W. C., Studies on the inter-relationship of certain cholinergic compounds. IV. Anticurare action in anesthetized man. *J. Pharmacol. Exp. Ther.* **100**, 227 (1950).

17. Auditore, J. V., and Sastry, B. V. R., Stereospecificity of erythrocyte acetylcholinesterase. *Arch. Biochem. Biophys.* **105**, 506 (1964).

18. Auerbach, M. E., Davis, D. L., and Foldes, F. F., Micromethod for the colorimetric determination of tetracaine. *Fed. Proc., Fed. Am. Soc. Exp. Biol.* **11**, 319 (1952).

19. Augustinsson, K. B., Cholinesterases; a study in comparative enzymology. *Acta Physiol. Scand.* **15**, Suppl. 52, 1 (1948).

20. Augustinsson, K. B., Substrate concentration and specificity of choline ester-splitting enzymes. *Arch. Biochem.* **23**, 111 (1949).

21. Augustinsson, K. B., The normal variation of human blood cholinesterase activity. *Acta Physiol. Scand.* **35**, 40 (1955).

22. Augustinsson, K. B., Electrophoresis studies on blood plasma esterases. I. Mammalian plasmata. *Acta Chem. Scand.* **13**, 571 (1959).

23. Augustinsson, K. B., Electrophoresis studies on blood plasma esterases. II. Avian, reptilian, amphibian and piscine plasmata. *Acta Chem. Scand.* **13**, 1081 (1959).

24. Augustinsson, K. B., Multiple forms of esterase in vertebrate blood plasma. *Ann. N.Y. Acad. Sci.* **94**, 844 (1961).

25. Augustinsson, K. B., Classification and comparative enzymology of the cholinesterases and methods for their determination. *In* "Cholinesterases and Anti-Cholinesterase Agents" (G. B. Koelle, ed.), p. 89. Springer-Verlag, Heidelberg, Berlin, New York, 1963.

26. Augustinsson, K. B., The nature of an "anionic" site in butyrylcholinesterase compared with that of a similar site in acetylcholinesterase. *Biochim. Biophys. Acta* **128**, 351 (1966).

27. Augustinsson, K. B., and Isacshen, T., The enzymatic hydrolysis of the β-methyl derivatives of acetylcholine and acetylthiocholine. *Acta Chem. Scand.* **11**, 750 (1957).

28. Austin, L., and Berry, W. K., Two selective inhibitors of cholinesterase. *Biochem. J.* **54**, 695 (1953).

29. Aven, M. H., Light, A., and Foldes, F. F., Hydrolysis of procaine in various mammalian plasmas. *Fed. Proc., Fed. Am. Soc. Exp. Biol.* **12**, 299 (1953).

30. Baart, N., Shanor, S. P., Van Hees, G. R., Erdos, E. G., and Foldes, F. F., Inhibitory effect of local anesthetics and their halogenated analogs on human plasma and red cell cholinesterase. *Fed. Proc., Fed. Am. Soc. Exp. Biol.* **15**, 395 (1956).

31. Ballantyne, B., Histochemical and biochemical aspects of cholinesterase activity of adipose tissue. *Arch. Int. Pharmacodyn. Ther.* **173**, 343 (1968).

32. Barker, L. A., Dowdall, M. J., Essman, W. B., and Whittaker, V. P., The compartmentation of acetylcholine in cholinergic nerve terminals. *In* "Drugs and Cholinergic Mechanisms in the CNS" (E. Heilbronn and A. Winter, eds.), p. 193. Försvarets Forskningsanstalt, Stockholm, 1970.

33. Barlow, R. B., "Introduction to Chemical Pharmacology," 2nd ed. Methuen, London, 1964.

34. Barlow, R. B., and Dixon, O. D., Choline acetyltransferase in the nettle urtica dioica L. *Biochem. J.* **132**, 15 (1973).

35. Barrnett, R. J., and Ball, E. G., Metabolic and ultrastructural changes induced in adipose tissue by insulin. *J. Biophys. Biochem. Cytol.* **8**, 83 (1960).

36. Beckett, A. H., Vaughan, C. L., and Mitchard, M., Inhibition of the pseudocholinesterase in horse serum by some choline analogues. *Biochem. Pharmacol.* **17**, 1595 (1968).

37. Beleslin, D., and Polak, R. L., Depression by morphine and chloralose of acetylcholine release from the cat's brain. *J. Physiol. (London)* **177**, 411 (1965).

38. Berends, F., Reactiveringen "Veroudering" van Esterasen Geremd met Organische Fosforverbindengen. Thesis, University of Leiden (1964).

39. Berends, F., Stereospecificity in the reactivation and aging of butyryl-cholinesterase inhibited by organophosphates with an asymmetrical P-atom. *Biochim. Biophys. Acta* **81**, 190 (1964).

40. Bergmann, F., Fine structure of the active surface of cholinesterase and the mechanism of enzymatic ester hydrolysis. *Discuss. Faraday Soc.* **20**, 126 (1955).

40a. Bergmann, F., The structure of the active surface of cholinesterases and the mechanism of their catalytic action in ester hydrolysis. *Adv. Catal.* **10**, 131 (1958).

41. Bergmann, F., and Segal, R., The relationship of quaternary ammonium salts to the anionic sites of true and pseudo cholinesterase. *Biochem. J.* **58**, 692 (1954).

42. Bergmann, F., and Segal, R., The characterization of tissue cholinesterases. *Biochim. Biophys. Acta* **16**, 513 (1955).

43. Bergmann, F., Wurzel, M., and Shimoni, E., The enzymatic hydrolysis of acid anhydrides. *Biochem. J.* **55**, 888 (1953).

43a. Bergmann, F., Rimon, S., and Segal, R., Effect of pH on the activity of eel esterase towards different substrates. *Biochem. J.* **68**, 493 (1958).

44. Berman, R., Wilson, I. B., and Nachmansohn, D., Choline acetylase specificity in relation to biological function. *Biochim. Biophys. Acta* **12**, 315 (1953).

45. Berry, J. F., and Whittaker, V. P., The acyl-group specificity of choline acetylase. *Biochem. J.* **73**, 447 (1959).

46. Berry, W. K., and Davis, D. R., Factors influencing the rate of "aging" of a series of alkyl methylphosphorylacetylcholinesterases. *Biochem. J.* **100**, 572 (1966).

47. Beznak, A. B. L., Effect of acetylcholinesterase on the surviving frog heart. *Nature (London)* **181**, 1190 (1958).

48. Bidstrup, P. L., Bonnell, J. A., and Beckett, A. G., Paralysis following poisoning by a new organic phosphorus insecticide (Mipafox). *Br. Med. J.* **1**, 1068 (1953).

49. Blaber, L. C., Studies on the mode of action of drugs which facilitate neuro-muscular transmission. Thesis, University of London (1962).

50. Borders, R. W., Stephen, C. R., Nowill, W. K., and Martin, R., The interrelationship of succinylcholine and the blood cholinesterases during anesthesia. *Anesthesiology* **16**, 401 (1955).

51. Bourne, J. G., Collier, H. O. J., and Somers, G. F., Succinylcholine (Succinoylcholine). Muscle relaxant of short duration. *Lancet* **1**, 1225 (1952).

51a. Brestkin, A. P., Ivanova, L. A., and Svechnikova, V. V., On the influence of choline on the rate of hydrolysis of acetylcholine under the action of bovine erythrocyte cholinesterase. *Biokhimiya* **31**, 416 (1966).

52. Brodie, B. B., Distribution and fate of drugs. *In* "Absorption and Distribution of Drugs" (T. B. Binns, ed.), p. 199. Williams & Wilkins, Baltimore, Maryland, 1964.

53. Brodie, B. B., and Maickel, R. P., Comparative biochemistry of drug metabolism. *In* "Metabolic Factors Controlling Duration of Drug Action" (B. B. Brodie and E. G. Erdos, eds.), p. 299. Macmillan, New York, 1962.

54. Brown, B. B., and Bailey, L., Neuromuscular blocking action of gabacholine and congeners. *Pharmacologist* **2**, 77 (1960).

55. Brown, B. B., and Bailey, L., personal communication (1961).

56. Browning, E. T., and Schulman, M. P., Acetylcholine synthesis by cortex slices of rat brain. *J. Neurochem.* **15**, 1391 (1968).

57. Bull, G. B., Hebb, C. O., and Ratković, D., Choline acetylase in the human placenta at different stages of development. *Nature (London)* **190**, 1202 (1961).

58. Bullock, T. H., Grundfest, H., Nachmansohn, D., and Rothenberg, M. A., Generality of the role of acetylcholine in nerve and muscle conduction. *J. Neurophysiol.* **10**, 11 (1947).

59. Burgen, A. S., and Chipman, L. M., The location of cholinesterase in the central nervous system. *Q. J. Exp. Physiol. Cogn. Med. Sci.* **37**, 61 (1952).

59a. Burnett, W., and Cohen, Y., Liver function after surgery: A study of 50 cases with particular reference to serum cholinesterase. *Br. J. Anaesth.* **72**, 66 (1955).

60. Bush, G. H., The use of muscle relaxants in burnt children. *Anaesthesia* **19**, 231 (1964).

60a. Callahan, J. F., and Kruckenberg, S. M., Erythrocyte cholinesterase activity of domestic and laboratory animals: Normal levels for nine species. *Am. J. Vet. Res.* **28**, 1509 (1967).

61. Cannavà, A., Azione in vitro degli alcaloidi fenantrenici dell'oppio e di alcuni loro derivati sulla attività procainoesterasica del sangue. *Arch. Ital. Sci. Farmacol.* **3**, 137 (1953).

62. Cavallito, C. J., Some speculations on the chemical nature of postjunctional membrane receptors. *Fed. Proc., Fed. Am. Soc. Exp. Biol.* **26**, 1647 (1967).

63. Cavallito, C. J., Bonding characteristics of acetylcholine stimulants and antagonists and cholinergic receptors. *Ann. N.Y. Acad. Sci.* **144**, 900 (1967).

63a. Cavallito, C. J., Napoli, M. D., and O'Dell, T. B., Neuromuscular blocking activities of two short chain linked bisquaternary ammonium compounds. *Arch. Int. Pharmacodyn. Ther.* **149**, 188 (1964).

64. Cavallito, C. J., White, H. L., and Yun, H. S., Inhibitors of choline acetyltransferase. *In* "Drugs and Cholinergic Mechanisms in the CNS" (E. Heilbronn and A. Winter, eds.), p. 97. Försvarets Forskningsanstalt, Stockholm, 1970.

65. Cavallito, C. J., Yun, H. S., Smith, J. C., and Foldes, F. F., Choline acetyltransferase inhibitors. Configurational and electronic features of styrylpyridine analogs. *J. Med. Chem.* **12**, 134 (1969).

66. Cavallito, C. J., Yun, H. S., Kaplan, T., Smith, J. C., and Foldes, F. F., Choline acetyltransferase inhibitors. Dimensional and substituent effects among styrylpyridine analogs. *J. Med. Chem.* **13**, 221 (1970).

67. Cavallito, C. J., Yun, H. S., Edwards, M. L., and Foldes, F. F., Choline acetyltransferase inhibitors. Styrylpyridine analogs with nitrogen-atom modifications. *J. Med. Chem.* **14**, 130 (1971).

68. Celesia, G. G., and Jasper, J. J., Acetylcholine released from cerebral cortex in relation to state of activation. *Neurology* **16**, 1053 (1966).

69. Chang, C. H., and Gaddum, J. H., Choline esters in tissue extracts. *J. Physiol. (London)* **79**, 255 (1933).

70. Changeux, J. P., Ryter, A., and Leuzinger, W., On the association of tyrocidine with acetylcholinesterase. *Proc. Natl. Acad. Sci. U.S.A.* **62**, 986 (1969).

71. Chao, L. P., and Wolfgram, F., Purification and some properties of choline acetyltransferase (E.C.2.3.1.6) from bovine brain. *J. Neurochem.* **20**, 1075 (1973).

72. Chase, J. F. A., and Tubbs, P. K., Specific inhibitors of carnitine acetyltransferase and other acetyltransferases. *Biochem. J.* **100**, 47P (1966).

72a. Childs, A. F., Davies, D. R., Green, A. L., and Rutland, J. P., The reactivation by oximes and hydroxamic acids of cholinesterases inhibited by organo-phosphorus compounds. *Br. J. Pharmacol. Chemother.* **10**, 462 (1955).

73. Cimasoni, G., Inhibition of cholinesterases by fluoride in vitro. *Biochem. J.* **99**, 133 (1966).

74. Clark, S. W., Glaubiger, G. A., and LaDu, B. N., Properties of plasma cholinesterase variants. *Ann. N.Y. Acad. Sci.* **151**, 710 (1968).

75. Clitherow, J. W., Mitchard, M., and Harper, N. J., The possible biological function of pseudocholinesterase. *Nature (London)* **199**, 1000 (1963).

75a. Coates, P. M., and Simpson, N. E., Genetic variations in human erythrocyte acetylcholinesterase. *Science* **175**, 1466 (1972).

75b. Cohen, J. A., Oosterbaan, R. A., and Warringa, M. G. P. J., The turnover number of aliesterase, pseudo- and true cholinesterase and the combination of these enzymes with diisopropylfluorophosphonate. *Biochim. Biophys. Acta* **18**, 228 (1955).

75c. Cohen, J. A., Oosterbaan, R. A., and Jansz, H. S., The chemical structure of the reactive group of esterases. *Discuss. Faraday Soc.* **20**, 114 (1955).

76. Cohen, J. A., Oosterbaan, R. A., Jansz, H. S., and Berends, F., The active site of esterases. *J. Cell. Physiol.* **54**, 231 (1959).

77. Cohen, J. A., and Oosterbaan, R. A., The active site of acetylcholinesterase and related esterases and its reactivity towards substrates and inhibitors. *In* "Cholinesterases and Anticholinesterase Agents" (G. B. Koelle, ed.), p. 299. Springer-Verlag, Heidelberg, Berlin, New York, 1963.

77a. Cole, B. R., and Leadbeater, L., Estimation of the stability of dry horse serum cholinesterase by means of an accelerated storage test. *J. Pharm. Pharmacol.* **20**, 48 (1968).

78. Collier, B., Discussion. *In* "Drugs and Cholinergic Mechanisms in the CNS" (E. Heilbronn and A. Winter, eds.), p. 173. Försvarets Forskningsanstalt, Stockholm, 1970.

79. Couteaux, R., Morphological and cytochemical observations on the postsynaptic membrane at motor end-plates and ganglionic synapses. *Exp. Cell Res. Suppl.* **5**, 294 (1958).

80. Couteaux, R., and Taxi, J., Recherches histochimiques sur la distribution des activités cholinestérasiques au niveau de la synapse myoneurale. *Arch. Anat. Microsc. Morphol. Exp.* **41**, 352 (1952).

81. Crossland, J., The significance of brain acetylcholine. *J. Ment. Sci.* **99**, 247 (1953).

82. Crossland, J., Acetylcholine and the morphine abstinence syndrome. *In* "Drugs and Cholinergic Mechanisms in the CNS" (E. Heilbronn and A. Winter, eds.), p. 355. Försvarets Forskningsanstalt, Stockholm, 1970.

83. Curtis, D. R., Acetylcholine as a central transmitter. *Can. J. Biochem. Physiol.* **41**, 2611 (1963).

84. Curtis, D. R., and Ryall, R. W., The excitation of Renshaw cells by cholinomimetics. *Exp. Brain Res.* **2**, 49 (1966).

85. Dale, H. H., The action of certain esters and ethers of choline and their relation to muscarine. *J. Pharmacol. Exp. Ther.* **6**, 147 (1914).

85a. Daly, M. deB., and Wright, P. G., The effects of anticholinesterases upon peripheral vascular resistance in the dog. *J. Physiol. (London)* **133**, 475 (1956).

86. Davies, R. O., Marton, A. V., and Kalow, W., The action of normal and atypical cholinesterase of human serum upon a series of esters of choline. *Can. J. Biochem.* **38**, 545 (1960).

87. Davis, R., and Koelle, G. B., Electron microscopic localization of acetylcholinesterase and nonspecific cholinesterase at the neuromuscular junction by the gold-thiocholine and gold-thiolacetic acid methods. *J. Cell Biol.* **34**, 157 (1967).

88. Denz, F. A., On the histochemistry of the myoneural junction. *Br. J. Exp. Pathol.* **34**, 329 (1953).

89. De Robertis, E., Morphological bases of synaptic processes and neurosecretion. *In* "Regional Neurochemistry" (S. S. Kety and J. Elkes, eds.), p. 248. Pergamon, Oxford, 1961.

90. De Robertis, E., Contribution of electronmicroscopy to some neuropharmacological problems. *Biochem. Pharmacol.* **9**, 49 (1962).

91. De Robertis, E., and Bennett, H. S., Some features of the submicroscopic morphology of synapses in frog and earthworm. *J. Biophys. Biochem. Cytol.* **1**, 47 (1955).

92. De Robertis, E., Pellegrino de Traldi, A., Rodrigues de Lores Arnaiz, G., and Salganicoff, L., Cholinergic and non-cholinergic nerve endings in rat brain. I. Isolation and subcellular distribution of acetylcholine and acetylcholinesterase. *J. Neurochem.* **9**, 23 (1962).

93. Dettbarn, W. D., The acetylcholine system in peripheral nerve. *Ann. N.Y. Acad. Sci.* **144**, 483 (1967).

94. Dixon, E. M., Variation in human cholinesterase activity. *Diss. Abstr.* **17**, 2567 (1957).

95. Dixon, M., and Webb, E. C., "Enzymes," 2nd ed. Academic Press, New York, 1964.

96. Dixon, W. E., On the mode of action of drugs. *Med. Mag. (London)* **16**, 454 (1907).

97. Doenicke, A., Klinische Bedeutung der Pseudocholinesterase. *In* "Pseudocholinesterasen" (H. W. Goedde, A. Doenicke, and K. Altland, eds.), p. 128. Springer-Verlag, Heidelberg, Berlin, New York, 1967.

98. Doenicke, A., and Holle, F., Das Verhalten der Leberfunktion im postoperativen Schock. *Fortschr. Med.* **80**, 253 (1962).

99. Doenicke, A., Gürtner, T., Kreutzberg, G., Remes, J., Spiess, W., and Steinbereithner, K., Kritische Betrachtungen über einen Fall ohne Serum-cholinesterase. *Proc. Eur. Congr. Anaesthesiol., 1st, Vienna, Sept. 3–9, 1962* Vol. 2, p. 187 (1962).

100. Doenicke, A., Gürtner, T., Kreutzberg, G., Remes, J., Spiess, W., and Steinbereithner, K., Serum cholinesterase anenzymia: Report of a case confirmed by enzyme-histochemical examination of liver-biopsy specimen. *Acta Anaesthesiol. Scand.* **7**, 59 (1963).

101. Donhoffer, A., Feinere lokalization verschiedener cholinesterasen der nervösen Darmgeflechte. *Acta Morphol. Acad. Sci. Hung.* **8**, 375 (1959).

102. Eccles, J. C., Fatt, P., and Koketsu, K., Cholinergic and inhibitory synapses in a pathway from motor-axon collaterals to motoneurones. *J. Physiol. (London)* **216**, 524 (1954).

103. Ecobichon, D. J., and Israel, Y., Characterization of the esterases from electric tissue of electrophorus by starch-gel electrophoresis. *Can. J. Biochem.* **45**, 1099 (1967).

104. Ecobichon, D. J., and Kalow, W., Some properties of the soluble esterases of liver. *Can. J. Physiol. Pharmacol.* **39**, 1329 (1961).

105. Eldefrawi, M. E., Tripathi, R. K., and O'Brien, R. D., Acetylcholinesterase isoenzymes from the housefly brain. *Biochim. Biophys. Acta* **212**, 308 (1970).

106. Ehrenpreis, S., Acetylcholine and nerve activity. *Nature (London)* **201**, 887 (1964).

107. Ehrenpreis, S., Possible nature of the cholinergic receptor. *Ann. N.Y. Acad. Sci.* **144**, 720 (1967).

108. Ehrenpreis, S., Molecular aspects of cholinergic mechanisms. *In* "Drugs Affecting the Peripheral Nervous System" (A. Burger, ed.), Vol. 1, p. 1. Dekker, New York, 1968.

109. Ehrenpreis, S., Fleish, J. H., and Mittag, T. W., Approaches to the molecular nature of pharmacological receptors. *Pharmacol. Rev.* **21**, 131 (1969).

110. Ehrenpreis, S., Hehir, R. M., and Mittag, T. W., Assay and properties of essential (junctional) cholinesterases of the rat diaphragm. *In* "Cholinergic Ligand Interactions" (D. J. Triggle, J. F. Moran, and E. A. Barnard, eds.), p. 67. Academic Press, New York, 1971.

111. Engelhard, N., and Erdmann, W. D., Ein neuer Reaktivator für durch alkyl-phosphatgehemmte Acetylcholinesterase. *Klin. Wochenschr.* **41**, 525 (1963).

112. Engelhard, N., and Erdmann, W. D., Beziehungen zwischen chemischer Struktur und Cholinesterase reaktivierender Wirksamkeit bei einer Reihe neuer bis-quartärer Pyridin-4-aldoxime. *Arzneim.-Forsch.* **14**, 870 (1964).

113. Engelhard, N., Prchal, K., and Nenner, M., Acetylcholinesterase. *Angew. Chem.* **6**, 615 (1967).

114. Eränkö, O., and Teräväinen, H., Distribution of esterases in the myoneural junction of the striated muscle of the rat. *J. Histochem. Cytochem.* **15**, 399 (1967).

115. Erdmann, W. D., Vergleichende Untersuchungen über das Penetrations-vermogen einiger Esterase-reaktivierender Oxime in das zentrale Nervensystem. *Arzneim.-Forsch.* **15**, 135 (1965).

116. Erdmann, W. D., and Clarmann, M. V., Ein neuer Esterase-Reaktivator für die Behandlung von Vergiftungen mit Alkylphosphaten. *Dtsch. Med. Wochenschr.* **88**, 2201 (1963).

117. Erdos, E. G., Baart, N., Foldes, F. F., and Zsigmond, E. K., Activation of enzymatic hydrolysis of benzolcholine by tryptamine. *Science* **126**, 1176 (1957).

118. Erdos, E. G., Foldes, F. F., Baart, N., and Shanor, S. P., Activating effect of tryptamine, benzylamine and histamine on plasma cholinesterase. *Fed. Proc., Fed. Am. Soc. Exp. Biol.* **16**, 294 (1957).

119. Erdos, E. G., Baart, N., Shanor, S. P., and Foldes, F. F., The inhibitory effect of chlorpromazine and chlorpromazine sulfoxide on human cholinesterases. *Arch. Int. Pharmacodyn. Ther.* **117**, 163 (1958).

120. Erdos, E. G., Foldes, F. F., Zsigmond, E. K., Baart, N., and Zwartz, J. A., Acceleration of plasma cholinesterase activity by quaternary ammonium salts. *Science* **128**, 92 (1958).

121. Erdos, E. G., Foldes, F. F., Baart, N., Zsigmond, E. K., and Zwartz, J. A., The acceleratring effect of narcotic analgesics on the hydrolysis of aromatic substrates by human plasma cholinesterase. *Biochem. Pharmacol.* **2** 97 (1959).

122. Evans, F. T., Gray, P. W. S., Lehmann, H., and Silke, E., Sensitivity to succinylcholine in relation to serum-cholinesterase. *Lancet* **1**, 1229 (1952).

123. Everett, J. W., and Sawyer, C. H., Effects of castration and treatment with sex steroids on the synthesis of serum cholinesterase in the rat. *Endocrinology* **39**, 323 (1946).

124. Faber, M., Relationship between serum choline esterase and serum albumin. *Acta Med. Scand.* **114**, 72 (1943).

125. Fatt, P., and Katz, B., Spontaneous subthreshold activity at motor nerve endings. *J. Physiol. (London)* **117**, 109 (1952).

126. Feldberg, W., Present views on mode of action of acetylcholine in central nervous system. *Physiol. Rev.* **25**, 596 (1945).

127. Feldberg, W., and Vogt, M., Acetylcholine synthesis in different regions of the central nervous system. *J. Physiol. (London)* **107**, 372 (1948).

128. Finster, M., personal communication (1974).

129. Foldes, F. F., Succinylmonocholine iodide: Its enzymatic hydrolysis and neuromuscular activity. *Proc. Soc. Exp. Biol. Med.* **83**, 187 (1953).

130. Foldes, F. F., "Muscle Relaxants in Anesthesiology," p. 80. Thomas, Springfield, Illinois, 1957.

131. Foldes, F. F., The pharmacology of neuromuscular blocking agents in man. *Clin. Pharmacol. Ther.* **1**, 345 (1960).

132. Foldes, F. F., The choice and mode of administration of relaxants. *In* "Muscle Relaxants" (F. F. Foldes, ed.), p. 1. Davis, Philadelphia, Pennsylvania, 1966.

133. Foldes, F. F., Adatok a Humán Cholinesterasek és Choline Acetylase Biokémiájához. *Orvostudomany* **19**, 15 (1968).

134. Foldes, F. F., Present concepts of the clinical use of muscle relaxants. *Proc. Int. Anesth. Symp. 4th, Varna, Sept. 15–20, 1969* Vol. 4, p. 903 (1969).

135. Foldes, F. F., Presynaptic aspects of neuromuscular transmission and block. *Anaesthesist* **20**, 6 (1971).

136. Foldes, F. F., and Aven, M. H., The hydrolysis of procaine and 2-chloroprocaine in spinal fluid. *J. Pharmacol. Exp. Ther.* **105**, 259 (1952).

137. Foldes, F. F., and Deery, A., unpublished data (1972).

137a. Foldes, F. F., and Deery, A., unpublished data (1974).

138. Foldes, F. F., and Foldes, V. M., α-amino fatty acid esters of choline: Interaction with ChE and neuromuscular activity in man. *J. Pharmacol. Exp. Ther.* **150**, 220 (1965).

139. Foldes, F. F., and McNall, P. G., 2-chloroprocaine: A new local anesthetic agent. *Anesthesiology* **13**, 287 (1952).

140. Foldes, F. F., and McNall, P. G., Toxicity of local anesthetics in man. *Dent. Clin. North Am.* July 257 (1961).

140a. Foldes, F. F., and McNall, P. G., Myasthenia Gravis: A guide for anesthesiologists. *Anesthesiology* **23**, 837 (1962).

141. Foldes, F. F., and Rhodes, D. H., Jr., The role of plasma cholinesterase in anesthesiology. *Anesth. Analg. (Cleveland)* **32**, 305 (1953).

142. Foldes, F. F., and Smith, J. C., The interaction of human cholinesterases with anticholinesterases used in the therapy of myasthenia gravis. *Ann. N.Y. Acad. Sci.* **135**, 287 (1966).

143. Foldes, F. F., Abernethy, G. S., Cavallito, C. J., Hollinger, I., and Pan, T., Tetraquaternary neuromuscular blocking agents. *Abstr. Sci. Pap. Annu. Meet. Am. Soc. Anesthesiol.* p. 159 (1972).

144. Foldes, F. F., Ambrus, J. L., Back, N., Bardos, T. J., and Foldes, V. M., Relationship between the anticholinesterase activity and side effect liability of AB 132 in man. *Fed. Proc., Fed. Am. Soc. Exp. Biol.* **21**, 335 (1962).

145. Foldes, F. F., Arai, T., Gentsch, H. H., and Zarday, Z., The influence of glucocorticoids on plasma cholinesterase. *Proc. Soc. Exp. Biol. Med.* **146**, 918 (1974).

146. Foldes, F. F., Aven, N. H., and Davis, D. L., The effect of preincubation on the inhibitory effect of neostigmine and dimethylcarbamate of (2-hydroxy-5-phenylbenzyl) trimethylammonium bromide (RO2-683) on the enzymatic hydrolysis of procaine. *J. Pharmacol. Exp. Ther.* **108**, 330 (1953).

147. Foldes, F. F., Baart, N., Shanor, S. P., and Erdos, E. G., The inhibitory effect of neuromuscular blocking agents and their antagonists on human cholinesterases. *Anesthesiology* **18**, 163 (1957).

148. Foldes, F. F., Baart, N., Shanor, S. P., and Erdos, E. G., The inhibitory effect of neuromuscular blocking agents and their antagonists on human cholinesterases. *Atti Congr. Soc. Ital. Anestesiol. 11th, Venice, Sept. 12–15, 1958* p. 511 (1958).

148a. Foldes, F. F., Colavincenzo, J. W., and Birch, J. H., Epidural anesthesia: A reappraisal (conclusions). *Anesth. Analg. (Cleveland)* **35**, 89 (1956).

149. Foldes, F. F., Davidson, G. M., Duncalf, D., and Kuwabara, S., The intravenous toxicity of local anesthetic agents in man. *Clin. Pharmacol. Ther.* **6**, 328 (1965).

150. Foldes, F. F., Davis, D. L., and Shanor, S. P., Comparison of hydrolysis of acetylcholine, benzoylcholine, succinylcholine and procaine in human plasma. *Fed. Proc., Fed. Am. Soc. Exp. Biol.* **13**, 354 (1954).

151. Foldes, F. F., Davis, D. L., Shanor, S. P., and van Hees, G. R., Hydrolysis of ester-type local anesthetics and their halogenated analogs by purified plasma cholinesterase. *J. Am. Chem. Soc.* **77**, 5149 (1955).

152. Foldes, F. F., Erdos, E. G., Baart, N., Zwartz, J., and Zsigmond, E. K., Inhibition of

human cholinesterases by narcotic analgesics and their antagonists. *Arch Int. Pharmacodyn. Ther.* **120**, 286 (1959).

153. Foldes, F. F., Erdos, E. G., Zsigmond, E. K., and Zwartz, J. A., Reactivation of neostigmine inhibited human plasma cholinesterase. *J. Pharmacol. Exp. Ther.* **129**, 394 (1960).

154. Foldes, F. F., Foldes, V. M., Smith, J. C., and Zsigmond, E. K., The relation between plasma cholinesterase and prolonged apnea caused by succinylcholine. *Anesthesiology* **24**, 208 (1963).

155. Foldes, F. F., Foldes, V. M., and McNall, P. G., The use of echothiophate in myasthenia gravis. *Clin. Pharmacol. Ther.* **7**, 620 (1966).

156. Foldes, F. F., Hillmer, N. R., Molloy, R. E., and Monte, A. B., Potentiation of the neuromuscular effect of succinylcholine by hexafluorenium. *Anesthesiology* **21**, 50 (1960).

157. Foldes, F. F., Kepes, E. R., Kronfeld, P. P., and Shiffman, H. P., A rational approach to neuroleptanesthesia. *Anesth. Analg. (Cleveland)* **45**, 642 (1966).

158. Foldes, F. F., McNall, P. G., Davis, D. L., Ellis, C. H., and Wnuck, A. L., Substrate competition between procaine and succinylcholine diiodide for plasma cholinesterase. *Science* **117**, 383 (1953).

159. Foldes, F. F., Molloy, R. E., Zsigmond, E. K., and Zwartz, J. A., Hexafluorenium: Its anticholinesterase and neuromuscular activity. *J. Pharmacol. Exp. Ther.* **129**, 400 (1960).

159a. Foldes, F. F., Molloy, R. E., McNall, P. G., and Koukal, L. R., The comparison of the intravenous toxicity of local anesthetic agents in man. *J. Am. Med. Assoc.* **172**, 1493 (1960).

160. Foldes, F. F., Swerdlow, M., Lipschitz, E., and van Hees, G. R., Comparison of enzymatic hydrolysis of suxamethonium with their respiratory effects. *Fed. Proc., Fed. Am. Soc. Exp. Biol.* **14**, 339 (1955).

161. Foldes, F. F., Swerdlow, M., Lipschitz, E., van Hees, G. R., and Shanor, S. P., Comparison of the respiratory effects of suxamethonium and suxethonium in man. *Anesthesiology* **17**, 559 (1956).

162. Foldes, F. F., van Hees, G. R., Shanor, S. P., and Baart, N., Interrelationship of suxamethonium, suxethonium and succinylmonocholine and human cholinesterase. *Fed. Proc., Fed. Am. Soc. Exp. Biol.* **15**, 422 (1956).

163. Foldes, F. F., van Hees, G. R., Davis, D. L., and Shnor, S. P., The structure-action relationship of urethane type cholinesterase inhibitors. *J. Pharmacol. Exp. Ther.* **122**, 457 (1958).

164. Foldes, F. F., Zsigmond, E. K., Foldes, V. M., and Erdos, E. G., The distribution of acetylcholinesterase and butyrylcholinesterase in the human brain. *J. Neurochem.* **9**, 559 (1962).

165. Foldes, V. M., Foldes, F. F., and Zsigmond, E. K., The in vivo and in vitro anticholinesterase effect of phospholine iodide in man. *Fed. Proc., Fed. Am. Soc. Exp. Biol.* **20**, 2167 (1961).

166. Fonnum, F., The "compartmentation" of choline acetyltransferase within the synaptosome. *Biochem. J.* **103**, 262 (1967).

167. Fonnum, F., Subcellular localization of choline acetyltransferase in brain. *In* "Drugs and Cholinergic Mechanisms in the CNS" (E. Heilbronn and A. Winter, eds.), p. 83. Försvarets Forskningsanstalt, Stockholm, 1970.

168. Forbath, A., Lehmann, H., and Silk, E., Prolonged apnea following succinylcholine. *Lancet* **2**, 1067 (1953).

169. Fouts, J. R., Physiological impairment of drug metabolism. *In* "Metabolic Factors Controlling Duration of Drug Action" (B. B. Brodie and E. G. Erdos, eds.), p. 257. Macmillan, New York, 1962.

170. Fraser, P. J., Acceleration of the enzymatic hydrolysis of benzoylcholine. *Br. J. Pharmacol. Chemother.* **11**, 7 (1956).

171. Friede, R. L., and Fleming, L. M., A comparison of cholinesterase distribution in the cerebellum of several species. *J. Neurochem.* **11**, 1 (1964).

171a. Friess, S. L., and Baldridge, H. D., The acetylcholinesterase surface. VI. Further studies with cyclic isomers as inhibitors and substrates. *J. Am. Chem. Soc.* **78**, 2482 (1956).

171b. Friess, S. L., and McCarville, W. J., Nature of the acetyl cholinesterase surface. II. The ring effect in enzymatic inhibition of the substituted ethylenediamine type. *J. Am. Chem. Soc.* **76**, 2260 (1954).

172. Fuhner, H., Untersuchungen über die periphere Wirkung des Physostigmins. *Arch. Exp. Pathol. Pharmakol.* **82**, 205 (1917-1918).

173. Fuhner, H., Untersuchungen über den Synergismus von giften. IV. Die Chemische Erregbarkeits-steigerung glatter Muskulatur. *Arch. Exp. Pathol. Pharmakol.* **82**, 51 (1917-1918).

174. Funke, A., Bagot, J., and Depierre, F., Anticholinestérasiques. I. Synthèse de di-phénoxyalcanes porteurs d'une ou deux fonctions phénoliques libres. *C. R. Hebd. Seances Acad. Sci. (Paris)* **239**, 329 (1954).

175. Funnel, H., and Oliver, W. T., Proposed physiological function for plasma cholinesterase. *Nature (London)* **208**, 689 (1965).

176. Gerebtzoff, M. A., "Cholinesterase. A Histochemical Contribution to the Solution of Some Functional Problems." Pergamon, Oxford, 1959.

177. Giacobini, E., and Holmstedt, B., Cholinesterase content of certent of certain regions of the spinal cord as judged by histochemical and Cartesian diver technique. *Acta Physiol. Scand.* **42**, 12 (1958).

178. Giarman, N. J., and Pepen, G., Drug-induced changes in brain acetylcholine. *Br. J. Pharmacol. Chemother.* **19**, 226 (1962).

179. Glick, D., Studies on the specificity of choline esterase. *J. Biol. Chem.* **125**, 729 (1938).

180. Glick, D., Some additional observations on the specificity of choline esterase. *J. Biol. Chem.* **137**, 357 (1941).

181. Glover, V. A. S., and Potter, L. T., Purification and properties of choline acetyltransferase from ox brain striate nuclei. *J. Neurochem.* **18**, 571 (1971).

182. Goedde, H. W., and Altland, K., Biochemic und Genetike der Pseudocholinesterasen. *In* "Cholinesterasen: Pharmakogenetik, Biochemic Klinik" (H. W. Goedde, A. Doenicke, and K. Altland, eds.), p. 1. Springer-Verlag, Berlin, Heidelberg, and New York, 1967.

183. Goedde, H. W., and Baitsch, H., Nomenclature of pseudocholinesterase polymorphism. *Br. Med. J.* **2**, 310 (1964).

184. Goedde, H. W., and Baitsch, H., On nomenclature of pseudocholinesterse polymorphism. *Acta Genet. Stat. Med.* **14**, 366 (1964).

185. Goedde, H. W., and Fuss, W., Differenzierung von Pseudocholinesterase-Varienten im Diffusionstest. *Klin. Wochenschr.* **42**, 286 (1964).

186. Goedde, H. W., and Fuss, W., Untersuchungen zur Phylogenetic der Pseudo-cholinesterasen. *Humangenetik* **1**, 126 (1964).

187. Goedde, H. W., Altland, K., and Scholler, L., Pharmakogenetische Reaktion auf Succinyldicholin: Therapie der verlängerten Apnoe. *Med. Klin. (Munick)* **62**, 1631 (1967).

188. Goedde, H. W., Altland, K., and Schloot, W., Therapy of prolonged apnea after suxamethonium with purified pseudocholinesterase: New data on kinetics of the hydroly-sis of succinyldicholine and succinylmonocholine and further data on *N*-acetyltransferase-polymorphism. *Ann. N.Y. Acad. Sci.* **151**, 742 (1968).

189. Goedde, H. W., Gehring, D., and Hofmann, R. A., On the problem of a "silent gene" in pseudocholinesterase polymorphism. *Biocheim. Biophys. Acta* **107**, 391 (1965).

190. Goedde, H. W., Gehring, D., and Hofman, R. A., Biochemische Untersuchungen zur Frage der Existenz eines "siltent gene" in Polymorphismus der Pseudo-cholinesterasen. *Humangenetik* **1**, 607 (1965).

191. Goedde, H. W., Held, K. R., and Altland, K., Hydrolysis of succinyldicholine and succinylmonocholine in human serum. *Mol. Pharmacol.* **4**, 274 (1968).

192. Gold, A. J., Weller, J. M., and Freeman, G., Metabolic and acid-base changes following acute cholinesterase inhibition. *Am. J. Physiol.* **188**, 321 (1957).

193. Goldberg, A. M., and McCaman, R. E., A quantitative microchemical study of choline

acetyltransferase in the cerebellum of several species. *Life Sci.* **6**, 1493 (1967).

194. Goutier-Pirotte, M., and Gerebtzoff, M. A., Acetylcholinesterase in the guinea pig placenta; initial results of histochemical and biochemical research. *Arch. Int. Physiol. Biochim.* **63**, 445 (1955).

195. Graig, F. A., and Foldes, F. F., unpublished data (1969).

196. Gray, E. G., and Whittaker, V. P., The isolation of nerve endings from brain: an electron-microscopic study of cell fragments derived by homogenization and centrifugation. *J. Anat.* **96**, 79 (1962).

197. Green, A. L., and Nicholls, J. D., The reactivation of phosphorylated chymotrypsin. *Biochem. J.* **72**, 70 (1959).

198. Grob, D., and Johns, R. J., Use of oximes in the treatment of intoxication by anticholinesterase compounds in normal subjects. *Am. J. Med.* **24**, 497 (1958).

199. Grob, D., and Johns, R. J., Use of oximes in the treatment of intoxication by anticholinesterase compounds in patients with myasthenia gravis. *Am. J. Med.* **24**, 512 (1958).

200. Gronert, G. A., Dotin, L. N., Ritchey, C. R., and Mason, A. D., Jr., Succinyl-choline-induced hyperkalemia in burned patients. II. *Anesth. Analg. (Cleveland)* **48**, 958 (1969).

201. Gurtner, T., Kreutzberg, G., and Doenicke, A., Comparative studies on cholinesterase activity in serum and liver cells. *Acta Anaesthesiol. Scand.* **7**, 69 (1963).

202. Hardegg, W., Rieken, E., and Schmalz, H., Über die Kinetik der Benzoyl-cholinspaltung durch die Serum-Cholinesterase und ihre Beeinflussung durch Cholin. *Biochem. Z.* **324**, 115 (1953).

203. Harris, H., and Whittaker, M., Differential inhibition of human serum cholinesterase with fluoride: Recognition of two new phenotypes. *Nature (London)* **191**, 496 (1961).

204. Harris, H., Hopkinson, D. A., and Robson, E. B., Two-dimensional electrophoresis of pseudocholinesterase components in human serum. *Nature (London)* **196**, 1296 (1962).

205. Harris, H., Hopkinson, D. A., Robson, E. B., and Whittaker, M., Genetical studies on a new variant of serum cholinesterase detected by electrophoresis. *Ann. Hum. Genet.* **26**, 359 (1963).

205a. Harris, H., Hopkinson, D. A., and Robson, E. B., Two-dimensional electrophoresis of pseudocholinesterase components in human serum. *Nature (London)* **196**, 1296 (1962).

206. Harris, L. W., Fleisher, J. H., Clark, J., and Cliff, W. J., Dealkylation and loss of capacity for reactivation of cholinesterase inhibited by sarin. *Science* **154**, 404 (1966).

207. Haupt, H., Heide, K., Zwisler, O., and Schwick, H. G., Isolierung und physikalisch-chemische charakterisierung der Cholinesterase aus Human serum. *Blut* **14**, 65 (1966).

208. Hazard, R., Cornec, A., and Pignard, P., La dihydrooxycodéinone activateur de la procainestérase. *C. R. Seances Soc. Biol. Ses Fil.* **144**, 356 (1950).

209. Hazard, R., Uriel, J., and Larno, S., Apparente identité de la cholinesterase et de la procainesterase sériques d'origine humaine. *J. Physiol. (Paris)* **59**, 5 (1967).

210. Heath, D. F., "Organophosphorus Poisons—Anticholinesterases and Related Compounds," 1st ed. Pergamon, Oxford, 1961.

210a. Heath, D. F., The toxic action of some phosphorus anticholinesterases with cationic groups. *Biochem. Pharmacol.* **6**, 244 (1961).

211. Hebb, C., Formation, storage, and liberation of acetylcholine. *In* "Cholinesterases and Anticholinesterase Agents" (G. B. Koelle, ed.), p. 55. Springer-Verlag, Heidelberg, Berlin, New York, 1963.

212. Hebb, C., Biosynthesis of acetylcholine in nervous tissue. *Physiol. Rev.* **52**, 918 (1972).

213. Hebb, C. O., and Smallman, B. N., Intracellular distribution of choline acetylase. *J. Physiol. (London)* **134**, 385 (1956).

214. Hebb, C. O., and Waites, G. M., Choline acetylase in antero- and retrograde degeneration of a cholinergic neuron. *J. Physiol. (London)* **132**, 667 (1956).

215. Hebb, C. O., and Whittaker, V. P., Cellular distributions of acetylcholine and choline acetylase. *J. Physiol. (London)* **142**, 187 (1958).

216. Heilbronn, E., Purification of cholinesterase from horse serum. *Biochim. Biophys. Acta* **58**, 222 (1962).

217. Heilbronn, E., In vitro reactivation and "ageing" of tabun-inhibited blood cholinesterases. Studies with N-methyl pyridinium-2-aldoxime methane sulphonate and N,N'-trimethylene bis (pyridinium-4-aldoxime) dibromide. *Biochem. Pharmacol.* **12**, 25 (1963).

218. Heilbronn, E., Phosphorylated cholinesterases, their formation, reactions and induced hydrolysis. *Sven. Kem. Tidskr.* **77**, 11 (1965).

219. Heilbronn, E., and Tolagen, B., Toxogenin in sarin, soman and tabun poisoning. *Biochem. Pharmacol.* **14**, 73 (1965).

220. Heinecker, R., and Mayer, I., Das Verhalten der Serumcholinesterase nach Myokard-infarkt. *Klin. Wochenschr.* **35**, 340 (1957).

221. Hemsworth, B. A., and Foldes, F. F., Preliminary pharmacological screening of styryl-pyridine choline acetyltransferase inhibitors. *Eur. J. Pharmacol.* **11**, 187 (1970).

222. Herschberg, A. D., and Frommel, F., Cholinestérase sérique et glandes sexuelles. *Annea Endocrinol.* **9**, 117 (1948).

223. Hirano, A., and Dembitzer, H. M., Cerebellar alterations in the Weaver mouse. *J. Cell Biol.* **56**, 478 (1973).

224. Hobbiger, F., Effect of nicotinhydroxamic acid methiodide on human plasma cholinesterase inhibited by organophosphate containing a dialkylphosphate group. *Br. J. Pharmacol. Chemother.* **10**, 356 (1955).

225. Hobbiger, F., Chemical reactivation of phosphorylated human and bovine true cholinesterase. *Br. J. Pharmacol. Chemother.* **11**, 295 (1956).

226. Hobbiger, F., The inhibition of organophosphorus compounds and its reversal. *Proc. R. Soc. Med.* **54**, 403 (1961).

227. Hobbiger, F., and Vojvodíc, V., The reactivating and antidotal actions of N,N'-trimethyl-enebis (pyridinium-4-aldoxime)(TMB-4) and N,N'-oxydimethyl-enebis (pyridinium-4-aldoxime) (toxogenin) with particular reference to their effect on phosphorylated acetylcholinesterase in brain. *Biochem. Pharmacol.* **15**, 1677 (1966).

228. Hobbiger, F., O'Sullivan, D. G., and Sadler, P. W., New potent reactivators of acetocholinesterase inhibited by tetraethyl pyrophosphate. *Nature (London)* **182**, 1672 (1958).

229. Hofstee, B. H. J., Spectrophotometric determinations of esterases. *Science* **114**, 128 (1951).

230. Hokin, L. E., and Hokin, M. R., The role of phosphatidic acid and phosphoionositide in transmembrane transport elicited by acetylcholine and other humoral agents. *Int. Rev. Neurobiol.* **2**, 99 (1960).

231. Hokin, M. R., Hokin, L. E., and Shelp, W. D., The effects of acetylcholine on the turnover of phosphatidic acid and phosphoinositide in sympathetic ganglia, and in various parts of the central nervous system in vitro. *J. Gen. Physiol.* **44**, 217 (1960).

232. Holle, F., and Doenicke, A., Cholinesterase in der Chirurgie. *Ergeb. Chir. Orthop.* **43**, 77 (1961).

232a. Holmstedt, B., A modification of the thiocholine method for the determination of cholinesterase. I. Biochemical evaluations of selective inhibitors. *Acta Physiol. Scand.* **40**, 322 (1957).

233. Hoskin, F. C. G., Sterospecificity in the reactions of acetylcholinesterase. *Proc. Soc. Exp. Biol. Med.* **113**, 320 (1963).

234. Hoskin, F. C. G., and Trick, G. S., Stereospecificity in the enzymatic hydrolysis of tabun and acetyl-methylcholine chloride. *Can. J. Biochem. Physiol.* **33**, 963 (1955).

235. Hunt, R., Note on a blood pressure lowering body in the suprarenal gland. *Am. J. Physiol.* **3**, 18 (1900).

236. Hunt, R., Further observations on the blood-pressure-lowering bodies in extracts of the suprarenal glands. *Am. J. Physiol.* **5**, 6 (1901).

237. Hunt, R., Some physiological actions of the homocholins and of some of their derivatives. *J. Pharmacol. Exp. Ther.* **6**, 477 (1915).

238. Hunt, R., and Taveau, R. DeM., On the physiological action of certain cholin derivatives and new methods for detecting cholin. *Br. Med. J.* **2**, 1788 (1906).

239. Hunt, R., and Taveau, R. DeM., The effects of a number of derivatives of choline and analogous compounds on the blood pressure. *U. S. Public Health Mar. Hosp. Serv. Hyg. Lab. Bull.* **73**, 1 (1911).

240. Hunter, A. R., Tensilon: A new anticurare agent. *Br. J. Anaesth.* **24**, 175 (1952).

241. Irwin, R. L., and Hein, M. M., The substrate specificity of atypical cholinesterase in relation to phenotypes. *Biochem. Pharmacol.* **15**, 145 (1966).

242. Jaffe, M. J., Evidence for the regulation of phytochrome-mediated processes in bean roots by the neurohumor, acetylcholine. *Plant Physiol.* **46**, 768 (1970).

243. Jandorf, B. J., Michel, H. O., Schaffer, N. K., Egan, R., and Summerson, W. H., The mechanism of reaction between esterases and phosphorus-containing anti-esterases. *Discuss. Faraday Soc.* **20**, 134 (1955).

243a. Jandorf, B. J., Crowell, E. A., and Levin, A. P., Role of hydroxamic acids in prevention and reversal of cholinesterase inactivation by DFP and sarin. *Fed. Proc., Fed. Am. Soc. Exp. Biol.* **14**, 231 (1955).

244. Jansz, H. S., and Cohen, J. A., Pseudocholinesterase from horse serum. I. Purification and properties of the enzyme. *Biochim. Biophys. Acta* **56**, 531 (1962).

245. Jansz, H. S., Oosterbaan, R. A., Berends, F., and Cohen, J. A., Studies on the active site of esterases. *In* "Molecular Basis of Enzyme Action and Inhibition" (P. A. E. Desnuelle, ed.), 1st ed., p. 45. Pergamon, Oxford, 1963.

246. Jenkins, T., Balinsky, D., and Patient, D. W., Cholinesterase in plasma: First reported absence in the Bantu: Half-life determination. *Science* **156**, 1748 (1967).

247. Jondorf, W. R., Maickel, R. P., and Brodie, B. B., Inability of newborn mice and guinea pigs to metabolize drugs. *Biochem. Pharmacol.* **1**, 352 (1959).

248. Jung, M. J., and Belleau, B., Purification and fractionation of acetyl-cholinesterase into subspecies by affinity chromatography on a *d*-tubocurarine-sepharose column. *Mol. Pharmacol.* **6**, 589 (1972).

249. Juul, P., Human plasma cholinesterase isoenzymes. *Clin. Chim. Acta* **19**, 205 (1968).

250. Kabachnik, M. I., Brestkin, A. P., Godovikov, N. N., Michelson, M. J., Rozengart, E. V., and Rozengart, V. I., Hydrophobic areas on the active surface of cholinesterases. *Pharmacol. Rev.* **22**, 355 (1970).

251. Kaita, A. A., and Goldberg, A. M., Control of acetylcholine synthesis—the inhibition of choline acetyltransferase by acetylcholine. *J. Neurochem.* **16**, 1185 (1969).

252. Kalow, W., Hydrolysis of local anesthetics by human serum cholinesterase. *J. Pharmacol. Exp. Ther.* **104**, 122 (1952).

253. Kalow, W., Familial incidence of low pseudocholinesterase level. *Lancet* **2**, 576 (1956).

254. Kalow, W., Cholinesterase types. *Biochem. Hum. Genet., Ciba Found. Symp. 1959* p. 39 (1959).

255. Kalow, W., and Davies, R. O., The activity of various esterase inhibitors towards atypical human serum cholinesterase. *Biochem. Pharmacol.* **1**, 183 (1959).

256. Kalow, W., and Genest, K., Method for detection of atypical forms of human serum cholinesterases: Determination of dibucaine numbers. *Can. J. Biochem. Physiol.* **35**, 339 (1957).

257. Kalow, W., and Gunn, D. R., The relationship between dose of succinylcholine and duration of apnea in man. *J. Pharmacol. Exp. Ther.* **120**, 203 (1957).

257a. Kalow, W., and Gunn, D. R., Some statistical data on atypical cholinesterase of human serum. *Ann. Hum. Genet.* **23**, 239 (1959).

258. Kalow, W., and Lindsay, H., Abnormal behavior of human serum cholinesterase. *J. Pharmacol. Exp. Ther.* **116**, 34 (1956).

259. Kalow, W., and Staron, N., On distribution and inheritance of atypical forms of human serum cholinesterase as indicated by dibucaine numbers. *Can. J. Biochem. Physiol.* **35**, 1305 (1957).

260. Kalow, W., Genest, K., and Staron, N., Qualitative variation of serum cholinesterase activity in man as defined by dibucaine numbers. *Fed. Proc., Fed. Am. Soc. Exp. Biol.* **15**, 444 (1956).

260a. Kamijo, K., and Koelle, G. B., The histochemical localization of specific cholinesterase in the conduction system of beef heart. *J. Pharmacol. Exp. Ther.* **113**, 30 (1955).
261. Kaplan, E., Herz, F., and Hsu, K. S., Erythrocyte acetylcholinesterase activity in ABO hemolytic disease of the newborn. *Pediatrics* **33**, 205 (1964).
262. Karczmar, A. G., Ontogenesis of cholinesterases. *In* "Cholinesterases and Anticholinesterase Agents" (G. B. Koelle, ed.), p. 129. Springer-Verlag, Berlin, 1963.
263. Karczmar, A. G., Discussion of paper by Nachmansohn, D. Molecular forces controlling bioelectric currents in membranes. *In* "Nerve as a Tissue" (K. Rodahl, ed.), p. 273. Harper (Hoeber), New York, 1966.
264. Karczmar, A. G., Central cholinergic pathways and their behavioral implications. *In* "Principles of Psychopharmacology" (W. G. Clark *et al.*, eds.), p. 57. Academic Press, New York, 1970.
265. Karczmar, A. G., History of the research with anticholinesterase agents. *In* "Anticholinesterase Agents" (A. G. Karczmar, ed.), Int. Encycl. Pharmacol. Ther., Sect. 13, Vol. I, p. 1. Pergamon, Oxford, New York, 1970.
266. Karlin, A., The association of acetylcholinesterase and membrane in subcellular fractions of the electric tissue of electrophorus. *J. Cell Biol.* **25**, 159 (1965).
267. Karlin, A., Chemical distinctions between acetylcholinesterase and the acetylcholine receptor. *Biochim. Biophys. Acta* **139**, 358 (1967).
268. Karlin, A., and Winnik, M., Reduction and specific alkylation of the receptor for acetylcholine. *Proc. Natl. Acad. Sci. U.S.A.* **60**, 668 (1968).
269. Kasa, P., and Csernovsky, E., Electron microscopic localization of acetylcholinesterase in the superior cervical ganglion of the rat. *Acta Histochem.* **28**, 274 (1967).
270. Kasa, P., and Csillik, B., Electron microscopic localization of cholinesterase by a copper-lead-thiocholine technique. *J. Neurochem.* **13**, 1345 (1966).
271. Kasa, P., Mann, S. P., and Hebb, C., Localization of choline acetyltransferase. *Nature (London)* **226**, 812 (1970).
272. Kato, G., Acetylcholinesterase I. A study by nuclear magnetic resonance of the binding of inhibitors to the enzyme. *Mol. Pharmacol.* **8**, 575 (1972).
273. Kato, G., Acetylcholinesterase II. A study by nuclear magnetic resonance of the acceleration of acetylcholinesterase by atropine and inhibition by eserine. *Mol. Pharmacol.* **8**, 582 (1972).
274. Katz, B., "Nerve, Muscle and Synapse." McGraw-Hill, New York, 1966.
275. Kaulla, von K., and Holmes, J. H., Changes following anticholinesterase exposures. Blood coagulation studies. *Arch. Environ. Health* **2**, 168 (1961).
276. Klein, H., Gärtner, K., and Günther, R., Die Variante C$_5$ der Cholinesterasen des Serums. *Dtsch. Z. Gesamte Gerichtl. Med.* **61**, 137 (1967).
277. Kloot, W. G. Van der, The effect of enzyme inhibitors on the resting potential and on the ion distribution of the sartorius muscle of the frog. *J. Gen. Physiol.* **41**, 879 (1958).
278. Klupp, H., and Kraupp, O., Über die freisetzung von Kalium aus der Muskalatus unter der Einwirkung einiger Muskelrelaxantien. *Arch. Int. Pharmacodyn. Ther.* **98**, 340 (1954).
279. Koelle, G. B., The histochemical differentiation of types of cholinesterases and their localizations in tissues of the cat. *J. Pharmacol. Exp. Ther.* **100**, 158 (1950).
280. Koelle, G. B., The elimination of enzymatic diffusion artifacts in the histochemical localization of cholinesterases and a survey of their distributions. *J. Pharmacol. Exp. Ther.* **103**, 153 (1951).
281. Koelle, G. B., ed., "Cholinesterases and Anticholinesterase Agents." Springer-Verlag, Heidelberg, Berlin, New York, 1963.
282. Koelle, G. B., Cytological distributions and physiological functions of cholinesterases. *In* "Cholinesterases and Anticholinesterase Agents" (G. B. Koelle, ed.), p. 187. Springer-Verlag, Heidelberg, Berlin, New York, 1963.
283. Koelle, G. B., The neurohumoral theory. *In* "Nerve as a Tissue" (K. Rodahl, ed.), p. 287. Harper (Hoeber), New York, 1966.
284. Koelle, G. B., Part I. Correlations between ultrastructures, cytochemistry and function at

the synapse. Current concepts of synaptic structure and function. *Ann. N. Y. Acad. Sci.* **183**, 5 (1971).

285. Koelle, G. B., Acetylcholine—current status in physiology, pharmacology and medicine. *N. Engl. J. Med.* **286**, 1086 (1972).

286. Koelle, G. B., Davis, R., and Smyrl, E. G., New findings concerning the localization by electron microscopy of acetylcholinesterase in autonomic ganglia. *Prog. Brain Res.* **34**, 371 (1971).

287. Koelle, W. A., and Koelle, G. B., The localization of external or functional acetylcholinesterase at the synapses of autonomic ganglia. *J. Pharmacol. Exp. Ther.* **126**, 1 (1959).

288. Koenig, E., and Koelle, G. B., Acetylcholinesterase in cholinergic neurons following irreversible inactivation. *J. Neurochem.* **8**, 169 (1961).

289. Kondritzer, A. A., Zvirblis, P., Goodman, A., and Paplanus, S. H., Blood plasma levels and elimination of salts of 2-PAM in man after oral administration. *J. Pharm. Sci.* **57**, 1142 (1968).

290. Korolkovas, A., "Essentials of Molecular Pharmacology. Background for Drug Design" p. 205. Wiley (Interscience), New York, 1970.

291. Köver, A., Kónya, L., Kovács, L., and Szöör, Á., Positive inotropic action of cholinesterase on the hypodynamic frog heart. *Acta Physiol. Acad. Sci. Hung.* **22**, 145 (1962).

292. Kramer, T., Preliminary report on the innervation of the embryonic chick heart. *Anat. Rec.* **106**, 210 (1950).

293. Krupka, R. M., Fluoride inhibition of acetylcholinesterase. *Mol. Pharmacol.* **2**, 558 (1966).

294. Krupka, R. M., and Laidler, K. J., Molecular mechanisms for hydrolytic enzyme action. I. Apparent non-competitive inhibition, with special reference to acetylcholinesterase. II. Inhibition of acetylcholinesterase by excess substrate. III. A general mechanism for the inhibition of acetylcholinesterase. IV. The structure of the active center and the reaction mechanism. *J. Am. Chem. Soc.* **83**, 1445 (1961).

295. Kupfer, C., and Koelle, G. B., A histochemical study of cholinesterase during formation of the motor and plate of the albino rat. *J. Exp. Zool.* **116**, 397 (1951).

296. Kurokawa, M., Machijama, Y., and Kato, M., Distribution of acetylcholine in the brain during various states of activity. *J. Neurochem.* **10**, 341 (1963).

297. Lakos, T., Csinady, L., and Kovacs, T., Die Rolle der Cholinesterase im Kationentransport des muskels. *Acta Physiol. Acad. Sci. Hung.* **16**, Suppl. 44 (1959).

298. La Motta, R. V., McComb, R. B., Noll, C. R., Jr., Wetstone, H. J., and Reinfrank, R. F., Multiple forms of serum cholinesterase. *Arch. Biochem. Biophys.* **124**, 299 (1968).

299. Lawler, H. C., Turnover time of acetylcholinesterase. *J. Biol. Chem.* **236**, 2296 (1961).

300. Lehmann, H., and Liddell, J., Genetical variants of human serum pseudocholinesterase. *Prog. Med. Genet.* **3**, 75 (1964).

301. Lehmann, H., and Ryan, E., The familial incidence of low pseudocholinesterase level. *Lancet* **2**, 124 (1956).

302. Lehmann, H., Liddell, J., Blackwell, B., O'Connor, D. C., and Daws, A. V., Two further serum pseudocholinesterase phenotypes as causes of suxamethonium apnoea. *Br. Med. J.* **1**, 1116 (1963).

303. Lehmann, H., Patson, V., and Ryan, E., The inheritance of an idiopathic low plasma pseudocholinesterase level. *J. Clin. Pathol.* **11**, 554 (1958).

304. Lehmann, H., Silk, E., and Liddell, J., Pseudo-cholinesterase. *Br. Med. Bull.* **17**, 230 (1961).

305. Leopold, I. H., and Krishna, N., Local use of anticholinesterase agents in ocular therapy. *In* "Cholinesterases and Anticholinesterase Agents" (G. B. Koelle, ed.), p. 1051. Springer-Verlag, Berlin, 1963.

306. Leuzinger, W., Structure and function of acetylcholinesterases. *Prog. Brain Res.* **31**, 241 (1969).

307. Leuzinger, W., The number of catalytic sites in acetylcholinesterase. *Biochem. J.* **123**, 139 (1971).

308. Leuzinger, W., and Baker, A. L., Acetylcholinesterase, I. Large scale purification, homogeneity and amino acid analysis. *Proc. Natl. Acad. Sci. U. S. A.* **57**, 446 (1967).

309. Leuzinger, W., Baker, A. L., and Cauvin, E., Acetylcholinesterase. II. Crystallization, absorption spectra, isoionic point. *Proc. Natl. Acad. Sci. U. S. A.* **59**, 620 (1968).

310. Leuzinger, W., Goldberg, M., and Cauvin, E., Molecular properties of acetylcholinesterases. *J. Mol. Biol.* **40**, 217 (1969).

310a. Leuzinger, W., Goldberg, M., and Cauvin, E., Molecular properties of acetylcholinesterase. *J. Mol. Biol.* **40**, 365 (1969).

311. Leuzinger, W., and Schneider, M., Acetylcholine-induced excitation on bilayers. *Experientia* **28**, 256 (1972).

312. Levine, M. G., and Hoyt, R. E., Serum cholinesterase in some pathological conditions. *Proc. Soc. Exp. Biol. Med.* **70**, 50 (1949).

313. Levine, M. G., and Hoyt, R. E., The relationship between human serum cholinesterase and serum albumin. *Science* **111**, 286 (1950).

314. Lewis, P. R., and Shute, C. C. D., The distribution of cholinesterase in cholinergic neurons demonstrated with the electron microscope. *J. Cell Sci.* **1**, 381 (1966).

315. Liddell, J., Lehmann, H., and Silk, E., A "silent" pseudocholinesterase gene. *Nature (London)* **193**, 561 (1962).

316. Liddell, J., Lehmann, H., Davies, D., and Sharih, A., Physical separation of pseudocholinesterase variants in human serum. *Lancet* **1**, 463 (1962).

317. Liddell, J., Newman, G. E., and Brown, D. F., A pseudocholinesterase variant in human tissues. *Nature (London)* **198**, 1090 (1963).

318. Loewi, O., Über humorale übertragbarkeit der Herzenervenwirkung. I. Mitteilung. *Pfluegers Arch. Gesamte Physiol. Menschen Tiere* **189**, 239 (1921).

318a. Loewi, O., and Mansfield, G., Über den Wirkungsmodus des Physostigmins. *Arch. Exp. Pathol. Pharmakol.* **62**, 180 (1910).

319. Loewi, O., and Navratil, E., Über humorale Übertragbarkeit der Herzenervenwirkung. X. Über das Schicksal des Vagusstoffes. *Pfluegers Arch. Gesamte Physiol. Menschen Tiere* **214**, 678 (1926).

320. Long, J. P., Structure-activity relationships of the reversible anti-cholinesterase agents. *In* "Cholinesterases and Anti-cholinesterase Agents" (G. B. Koelle, ed.), p. 374. Springer-Verlag, Heidelberg, Berlin, New York, 1963.

321. Lüttringhaus, A., and Hagedorn, I., Quartäre Hydroxyiminomethylpyridiniumsalze. Das Dichlorid des Bis-[4-hydroxylminomethylpyridinium-(1)-methyl]-ather ("Lüth 6") ein neuer Reaktivator der durch organische Phosphorsaureester gehemten Acetylcholinesterase. *Arzneim-Forsch* **14**, 1 (1961).

322. Macfarlane, D. W., Pelikan, E. W., and Unna, K. R., Evaluation of curarizing drugs in man. V. Antagonism to curarizing effects of d-tubocurarine by neostigmine m-hydroxy phenyltrimethylammonium and m-hydroxy phenylethyldimethylammonium. *J. Pharmacol. Exp. Ther.* **100**, 382 (1950).

323. McIsaac, R. J., and Koelle, G. B., Comparison of the effects of inhibition of external, internal, and total acetylcholinesterase upon ganglionic transmission. *J. Pharmacol. Exp. Ther.* **126**, 9 (1959).

324. Maier, E. H., and Fischer, R., Serum-cholinesterase-aktivität und Lebermorphologie. *Klin. Wochenschr.* **32**, 566 (1954).

325. Main, A. R., Affinity and phosphorylation constants for the inhibition of esterases by organophosphates. *Science* **144**, 992 (1964).

326. Malmström, B. G., Levin, O., and Boman, H. G.; Chromatography of human serum cholinesterase. *Acta Chem. Scand.* **10**, 1077 (1956).

327. Malthe-Sørensen, D., and Fonnum, F., Multiple forms of choline acetyltransferase from rat brain. *Nature (London) New Biol.* **229**, 127 (1971).

328. Malthe-Sørenson, D., and Fonnum, F., Multiple forms of choline acetyltransferase in several species demonstrated by isoelectric focusing. *Biochem. J.* **127**, 229 (1972).

329. Manner, G., Foldes, F. F., Kuleba, M., and Deery, A. M., Morphine tolerance in a human neuroblastoma line: Changes in choline acetylase and cholinesterase activities. *Experientia* **30**, 137 (1974).

330. Manner, G., Hirano, A., Dembitzer, H. M., and Foldes, F. F., unpublished data.

331. Markert, C. L., and Møller, F., Multiple forms of enzymes: Tissue, ontogenetic and species specific patterns. *Proc. Natl. Acad. Sci. U. S. A.* **45**, 753 (1959).

332. Massoulié, J., and Rieger, F., L'acétylcholinéstérase des organes électriques de poissons (torpille et gymnote); complexes membranaires. *Eur. J. Biochem.* **11**, 441 (1969).

333. Mayrhofer, O., Die Nebenwirkungen des Succinylcholins und ihre Verhuetung. *Curare Curare-like Agents, Proc. Int. Symp.* **1958** p. 376.

334. McCance, R. A., Hutchinson, A. O., Dean, R. F. A., and Jones, P. E. H., The cholinesterase activity of the serum of newborn animals and of colostrum. *Biochem. J.* **45**, 493 (1949).

335. Mendel, B., and Rudney, H., Studies on cholinesterase; cholinesterase and pseudo-cholinesterase. *Biochem. J.* **37**, 59 (1943).

335a. Mendel, B., Mundell, D. B., and Rudney, H., Studies on cholinesterase. III. Specific tests for true cholinesterase and pseudo-cholinesterase. *Biochem. J.* **37**, 473 (1943).

336. Mengle, D. C., and O'Brien, R. D., The spontaneous and induced recovery of fly-brain cholinesterase after inhibition by organophosphates. *Biochem. J.* **75**, 201 (1960).

337. Merrill, G. G., Neostigmine toxicity. Report of fatality following diagnostic test for myasthenia. *J. Am. Med. Assoc.* **137**, 362 (1948).

338. Metzger, H. P., and Wilson, I. B., The acceleration of the acetylcholinesterase catalyzed hydrolysis of acetyl fluoride. *Biochem. Biophys. Res. Commun.* **28**, 263 (1967).

339. Michaelson, I. A., The subcellular distribution of acetylcholine, choline acetyltransferase and acetylcholinesterase in nerve tissue. *Ann. N. Y. Acad. Sci.* **144**, 387 (1967).

340. Michel, H. O., Development of resistance of alkylphosphorylated cholinesterase to reactivation by oximes. *Fed. Proc.*, **17**, 275 (1958).

341. Mitchell, J. F., The spontaneous and evoked release of acetylcholine from the cerebral cortex. *J. Physiol. (London)* **165**, 98 (1963).

341a. Mitchell, J. F., Acetylcholine release from the brain. *In* "Mechanisms of Release of Biogenic Amines" (U. S. von Euler, S. Rosell, and B. Uvnäs, eds.), p. 425. Pergamon, Oxford, 1966.

342. Morishima, H. O., Daniel, S. S., Finster, M., Poppers, P. J., and James, L. S., Transmission of mepivacaine hydrochloride (carbocaine) across the human placenta. *Anesthesiology* **27**, 147 (1966

343. Morris, D., and Grewaal, D. S., Isotopic exchange between acetylcholine and [Me-^{14}C] choline catalyzed by human placental choline acetyltransferase. *Biochem. J.* **114**, 85P (1969).

344. Morris, D., and Grewaal, D. S., Human placental choline acetyltransferase. *Eur. J. Biochem.* **22**, 563 (1971).

345. Motulsky, A. G., Pharmacogenetics. *Prog. Med. Genet.* **3**, 49 (1964).

346. Mounter, L. A., Alexander, H. C., and Tuck, K. D., The pH dependence and dissociation constants of esterases and proteases treated with diisopropyl fluorophosphate. *J. Biol. Chem.* **226**, 867 (1957).

347. Mundell, D. B., Plasma cholinesterase in male and female rats. *Nature (London)* **153**, 557 (1944).

348. Muralt, A. von, Observations on chemical wave transmission in excited nerve. *Proc. R. Soc. London* **123**, 397 (1937).

349. Muralt, A. von, Lotmar, W., and Wildbrandt, W., Physikalischchemische Messungen an Nervenextrakten. Nachweis einer Aktionssubstanz der Nervenerregung. *Proc. Int. Physiol. Congr., 16th* **2**, 924 (1938).

350. Myers, D. K., Differentiation of three types of competitive cholinesterase inhibitors. *Arch. Biochem.* **31**, 29 (1951).

351. Myers, D. K., Effect of salt on the hydrolysis of acetylcholine by cholinesterases. *Arch. Biochem.* **37**, 469 (1952).

352. Nachmansohn, D., Action of ions on choline esterase. *Nature (London)* **145**, 513 (1940).

353. Nachmansohn, D., "Chemical and Molecular Basis of Nerve Activity." Academic Press, New York, 1959.

354. Nachmansohn, D., Choline acetylase. *In* "Cholinesterases and Anticholinesterase Agents" (G. B. Koelle, ed.), p. 40. Springer-Verlag, Heidelberg, Berlin, New York, 1963.

355. Nachmansohn, D., Chemical control of the permeability cycle in excitable membranes during activity. *Is. J. Med. Sci.* **1**, 1201 (1965).

356. Nachmansohn, D., Chemical control of the permeability cycle in excitable membranes during electrical activity. *Ann. N. Y. Acad. Sci.* **137**, 877 (1966).

357. Nachmansohn, D., Chemical forces controlling permeability changes of excitable membranes during electrical activity. *In* "Nerve as a Tissue" (K. Rodahl, ed.), p. 141. Harper (Hoeber), New York, 1966.

358. Nachmansohn, D., Proteins of excitable membranes. *J. Gen. Physiol.* **54**, 187S (1969).

359. Nachmansohn, D., and Berman, M., Studies on choline acetylase; on preparation of the coenzyme and its effect on the enzyme. *J. Biol. Chem.* **165**, 551 (1946).

360. Nachmansohn, D., and Lederer, E., Sur la biochimie de la cholinestérase; préparation de l'enzyme, rôle des groupements-SH. *Bull. Soc. Chim. Biol.* **21**, 797 (1939).

360a. Nachmansohn, D., and Lederer, E., Sur quelques propriétés chimiques de la cholinestérase. *C. Seances R. Soc. Biol. Ses. Fil.* **130**, 321 (1939).

361. Nachmansohn, D., and Machado, A. L., The formation of acetylcholine. A new enzyme "choline acetylase." *J. Neurophysiol.* **6**, 397 (1943).

362. Nachmansohn, D., and Rothenberg, M. A., Studies on cholinesterase. I. On the specificity of the enzyme in nerve tissue. *J. Biol. Chem.* **158**, 653 (1945).

363. Nachmansohn, D., and Weiss, M. S., Studies on choline acetylase; effect of citric acid. *J. Biol. Chem.* **172**, 677 (1948).

364. Neitlich, H. W., Increased plasma cholinesterase activity and succinylcholine resistance: A genetic variant. *J. Clin. Invest.* **45**, 380 (1966).

365. Nishi, S., Soeda, H., and Koketsu, K., Release of acetylcholine from sympathetic preganglionic nerve terminals. *J. Neurophysiol.* **30**, 114 (1967).

365a. Nyhan, W. L., Toxicity of drugs in the neonatal period. *J. Pediatr.* **59**, 1 (1961).

366. Oosterbaan, R. A., Constitution of DFP enzymes. *In* "Proceedings of the Conference on Structure and Reactions of DFP Sensitive Enzymes" (E. Heilbronn, ed.), p.25. Swedish Research Institute of National Defence, Stockholm, 1967.

367. Oosterbaan, R. A., and Jansz, H. S., Cholinesterases, esterases and lipases. *Compr. Biochem.* **16**, 1 (1965).

368. Oosterbaan, R. A., Jansz, H. S., and Cohen, J. A., Studies with [18]O on the mechanism of hydrolytic enzyme reactions. *Abstr., Int. Congr. Biochem., 5th, Moscow, Aug. 10–16, 1961* p. 119 (1963).

369. Ord, M. G., and Thompson, R. H. S., The distribution of cholinesterase types in mammalian tissues *Biochem. J.* **46**, 346 (1950).

370. Pantuck, E. J., Ecothiopate iodide eye drops and prolonged response to suxamethonium. *Br. J. Anaesth.* **38**, 406 (1966).

371. Persson, B. O., Larsson, L., Schuberth, J., and Sörbo, B., 3-bromo-acetonyltrimethylammonium bromide, a choline acetylase inhibitor. *Acta Chem. Scand.* **21**, 2283 (1967).

372. Petty, C. S., Organic phosphate insecticide poisoning. Residual effects in two cases. *Am. J. Med.* **24**, 467 (1958).

373. Plattner, F., Der Nachweis der Vagusstoffes bein Saugetier. *Pfluegers Arch. Gesamte Physiol. Menschen Tiere.* **214**, 112 (1926).

374. Polak, R. L., An analysis of the stimulating action of atropine on release synthesis of

acetyl choline in cortical slices from rat brain. *In* "Drugs and Cholinergic Mechanisms in the CNS" (E. Heilbronn, and A. Winter, eds.), p. 323. Försvarets Forskningsanstalt, Stockholm, 1970.

375. Poppers, P. J., and Finster, M., The use of prilocaine hydrochloride (Citanest) for epidural analgesia in obstetrics. *Anesthesiology* **29**, 1134 (1968).

376. Potter, L. T., and Glover, V. A. S., Choline acetyltransferase from mammalian brains. *In* "Drugs and Cholinergic Mechanisms in the CNS" (E. Heilbronn and A. Winter, eds.), p.75. Försvarets Forskningsanstalt, Stockholm, 1970.

377. Poziomek, E. J., Hackley, B. E., Jr., and Steinberg, G. M., Pyridinium aldoximes. *J. Org. Chem.* **23**, 714 (1958).

378. Richter, D., and Crossland, J., Variations in acetylcholine content of the brain with physiological state. *Am. J. Physiol.* **159**, 247 (1949).

379. Robertson, G. S., Serum cholinesterase deficiency. I. Disease and inheritance. *Br. J. Anaesth.* **38**, 355 (1966).

380. Robins, G. S., Zsigmond, E. K., and Shanor, S. P., *Plasma cholinesterase activity and the safe use of succinylcholine in cancer patients.* Presented at the *6th Annu. Meet. Am. Soc. Clin. Pharmacol. Chemother.*, Atlantic City, May 1-3 (1969).

381. Rosenberg, P., Kremzner, L. T., McCreery, D., and Willette, R. E., Inhibition of choline acetyltransferase activity in squid giant axon. *Biochim. Biophys. Acta* **268**, 49 (1972).

382. Rosner, V., Kepes, E. R., and Foldes, F. F., The effects of atropine and neostigmine on heart rate and rhythm. *Br. J. Anaesth.* **43**, 1066 (1971).

383. Rothenberg, M. A., and Nachmansohn, D., Studies on cholinesterase. III. Purification of the enzyme from electric tissue by fractional ammonium sulfate precipitation. *J. Biol. Chem.* **168**, 223 (1947).

384. Salpeter, M. M., and Eldefrawi, M. E., Sizes of end plate compartments, densities of acetylcholine receptor and other quantitative aspects of neuromuscular transmission. *J. Histochem. Cytochem.* **21**, 769 (1973).

385. Sastry, B. V. R., and Henderson, G. I., Kinetic mechanisms of human placental choline acetyltransferase. *Biochem. Pharmacol.* **21**, 787 (1972).

386. Sawyer, C. H., and Everett, J. W., Effects of various hormonal conditions in intact rat on synthesis of serum cholinesterase. *Endocrinology* **39**, 307 (1946).

387. Sawyer, C. H., and Everett, J. W., Cholinesterases in rat tissues and the site of serum non-specific cholinesterase production. *Am. J. Physiol.* **148**, 675 (1947).

388. Sawyer, C. H., and Hollinshead, W. H., Cholinesterases in sympathetic fibers and ganglia. *J. Neurophysiol.* **8**, 137 (1945).

389. Scanlon, J. W., Brown, W. V., Jr., Weiss, J. B., and Alper, M. H., Neurobehavioral responses of newborn infants after maternal epidural anesthesia. *Anesthesiology* **40**, 121 (1974).

390. Scanlon, J. W., Ostheimer, G. W., Lurie, A. O, Brown, W. V., Jr., Weiss, J. B., and Alper, M. H., Neurobehavioral responses and drug concentrations in newborns after maternal epidural anesthesia with bupivacaine. *Anesthesiology* **45**, 400 (1976).

390a. Schaffer, N. K., May, S. C., and Summerson, W. H., Serine phosphoric acid from diisopropylphosphoryl derivative of sel cholinesterase. *J. Biol. Chem.* **206**, 201 (1954).

391. Schaner, P. J., Brown, R. L., Kirksey, T. D., Gunther, R. C., Ritchey, C. R., and Gronert, G. A., Succinylcholine-induced hyperkalemia in burned patients. I. *Anesth. Analg. (Cleveland)* **48**, 764 (1969).

392. Schubert, J., Choline acetyltransferase; purification and effect of salts on the mechanism of the enzyme catalyzed reaction. *Biochim. Biophys. Acta* **122**, 470 (1966).

393. Schuberth, J., Sparf, B., and Sundwall, A., A technique for the study of acetylcholine turnover in mouse brain *in vivo*. *J. Neurochem.* **16**, 695 (1969).

394. Schuberth, J., Sparf, B., and Sundwall, A., On the turnover of acetylcholine in the brain. *In* "Drugs and Cholinergic Mechanisms in the CNS" (E. Heilbronn and A. Winter, eds.), p. 177. Försvarets Forskningsanstalt, Stockholm, 1970.

395. Shafai, T., and Cortner, J. A., Human erythrocyte acetylcholinesterase. II. Evidence for the modification of the enzyme by ion-exchange chromatography. *Biochim. Biophys. Acta* **250**, 117 (1971).

396. Shanor, S. P., Van Hees, G. R., Baart, N., Erdös, E. G., and Foldes, F. F., The influence of age and sex on human plasma·and red cell cholinesterase. *Am. J. Med. Sci.* **242**, 357 (1961).

397. Sharkawi, M., Effects of some centrally acting drugs on acetylcholine synthesis by rat cerebral cortex slices. *Br. J. Pharmacol.* **46**, 473 (1972).

398. Shen, S. C., Changes in enzymatic patterns during development. *In* "The Chemical Basis of Development" (W. D. McElroy and B. Glass, eds.), p. 416. Johns Hopkins Press, Baltimore, Maryland, 1958.

399. Shnider, S. M., Serum cholinesterase activity during pregnancy, labor and puerperium. *Anesthesiology* **26**, 335 (1965).

400. Shukuya, R., Kinetics of human blood cholinesterase. II. The temperature effect upon cholinesterase activity. *J. Biochem. (Tokyo)* **40**, 135 (1953).

401. Silver, A., Cholinesterases of the central nervous system with special reference to the cerebellum. *Int. Rev. Neurobiol.* **10**, 57 (1967).

402. Simpson, N. E. Polyacrylamide electrophoresis used for the detection of C5+ cholinesterase in Canadian caucasians, Indians and Eskimos. *Am. J. Hum. Genet.* **24**, 317 (1972).

403. Smith, J. C., and Deery, A. M., unpublished data (1966).

404. Smith, J. C., and Foldes, F. F.; An improved method for the recognition of atypical plasma cholinesterase. *Anesthesiology* **29**, 211 (1968).

405. Smith, J. C., and Foldes, F. F., The recognition of atypical plasma cholinesterase by relative substrate hydrolysis rate. *Biochim. Biophys. Acta* **289**, 352 (1972).

406. Smith, J. C., Cavallito, C. J., and Foldes, F. F., The inhibition of choline acetylase (ChA) by bisquaternary ammonium compounds. *Fed. Proc.* **25**, 320 (1966).

407. Smith, J. C., Cavallito, C. J., and Foldes, F. F., Choline acetyltransferase inhibitors: A group of styrylpyridine analogs. *Biochem. Pharmacol.* **16**, 2438 (1967).

408. Smith, J. C. Foldes, V. M., and Foldes, F. F., Distribution of cholinesterase in normal human muscle. *Can. J. Biochem. Physiol.* **41**, 1713 (1963).

409. Stedman, E., and Stedman, E., The purification of choline-esterase. *Biochem. J.* **29**, 2563 (1935).

410. Stefenelli, N., Die Korrelation zwischen dem Albumingehalt und der Cholinesteraseaktivität des Blutserums und ihre Beurteilung bei Albuminverlust. *Klin. Wochenschr.* **39**, 1019 (1961).

411. Stephenson, M., and Rowatt, E., The production of acetylcholine by a strain of Lactobacillus plantarum. *J. Gen. Microbiol.* **1**, 280 (1947).

412. Strelitz, F., Studies on cholinesterase. IV. Purification of pseudocholinesterase from horse serum. *Biochem. J.* **38**, 86 (1944).

413. Surgenor, D. M., and Ellis, D., Preparation and properties of serum and plasma proteins, plasma cholinesterase. *J. Am. Chem. Soc.* **76**, 6049 (1954).

414. Svensmark, O., Molecular properties of cholinesterases. *Acta Physiol. Scand.* **64**, Suppl. 245: (1965).

415. Svensmark, O., and Heilbronn, E., Electrophoretic mobility of native and neuraminidase-treated horse-serum cholinesterase. *Biochim. Biophys. Acta* **92**, 400 (1964).

416. Svensmark, O., and Kristensen, P., Isoelectric point of native and sialidase-treated human serum cholinesterase. *Biochim. Biophys. Acta* **67**, 441 (1963).

416a. Swift, M. R., and LaDu, B. N., A rapid screening test for atypical serumcholinesterase. *Lancet* **1**, 513 (1966).

417. Szentágothai, J., Einige Bemarkungen zur Struktur der peripheren Endausbreitung regetativer Nerven. *Acta Neuroveg.* **15**, 417 (1957).

418. Tether, J. E., Echothiopate in treatment of myasthenia gravis. (To be published.)

419. Thomas, J., and Mather, L. E., The maternal plasma levels and placental transfer of bupivacaine following epidural analgesia. *Br. J. Anaesth.* **41**, 1035 (1969).

420. Thompson, J. C., and Whittaker, M., Pseudocholinesterase activity in thyroid disease. *J. Clin. Pathol.* **18**, 811 (1965).

420a. Tobias, J. M., Lipton, M. A., and Lepinat, A., Effect of anaesthetic and convulsants on brain acetylcholine content. *Proc. Soc. Exp. Biol. Med.* **61**, 51 (1946).

421. Todrick, A., Fellowes, K. P., and Rutland, J. P., The effect of alcohols on cholinesterase. *Biochem. J.* **48**, 360 (1951).

422. Tsuji, F. I., Foldes, F. F., and Rhodes, D. H., Jr., The hydrolysis of succinyldicholine chloride in human plasma. *Arch. Int. Pharmacodyn. Ther.* **104**, 146 (1955).

423. Tuček, S., Subcellular localization of enzymes generating acetyl-CoA and their possible relation to the biosynthesis of acetylcholine. *In* "Drugs and Cholinergic Mechanisms in the CNS" (E. Heilbronn and A. Winter, eds.), p. 117. Försvarets Forskningsanstalt. Stockholm, 1970.

424. Usdin, E., Reactions of cholinesterases with substrates, inhibitors and reactivators. *In* "Anticholinesterase Agents" (A. G. Karczmar, ed.), Int. Encycl. Pharmacol. Ther., Sect. 13, Vol. I, p. 45. Pergamon, Oxford, New York, 1970.

425. Usdin, E., Mitz, M. A., and Killos, P. J., Studies on the mechanism of action of cholinesterase. *Abstr., Int. Congr. Biochem., 6th, 1964* Vol. 4, p. 184 (1964).

426. Van Der Meer, C., Effects of calcium chloride on choline esterase. *Nature (London)* **171**, 78 (1953).

427. Vicas, I. M., and Sharkawi, M., Synthesis of ^{14}C-acetylcholine by rat cerebral cortex. *Proc. Can. Fed. Biol. Soc.* **15**, 108 (1972).

428. Vizi, E. S., Kuze, S., and Foldes, F. F., The influence of temperature on neuromuscular transmission in the rat. (To be published.)

429. Wang, R. I. H., and Ross, C., Prolonged apnea following succinylcholine in cancer patients receiving AB-132. *Anesthesiology* **24**, 363 (1963).

430. Waser, P. G., Receptor localization by autoradiographic techniques. *Ann. N. Y. Acad. Sci.* **144**, 737 (1967).

431. Webb, E. C., The nomenclature of multiple enzyme forms. *Experientia* **20**, 592 (1964).

432. Weidemann, H., Die Serumcholinesteraseaktivität beim Leberparenchymschaden. *Med. Klin. (Munich)* **58**, 1795 (1963).

433. Wetstone, H. J., LaMotta, R. V., Middlebrook, L., and White, B. V., Studies of cholinesterase activity. IV. Liver function in pregnancy: Values of certain standard liver function tests in normal pregnancy. *Am. J. Obstet. Gynecol.* **76**, 480 (1958).

434. White, H. L., and Cavallito, C. J., Photoisomerization of styrylpyridine analogues in relation to choline acetyltransferase and cholinesterase inhibition *Biochim. Biophys. Acta* **206**, 242 (1970).

435. White, H. L., and Cavallito, C. J., Choline acetyltransferase, enzyme mechanism and mode of inhibition by a styrylpyridine analogue. *Biochim. Biophys. Acta* **206**, 343 (1970).

436. White, H. L., and Cavallito, C. J., Inhibition of bacterial and mammalian choline acetyltransferases by styrylpyridine analogues. *J. Neurochem.* **17**, 1579 (1970).

437. White, H. L., and Wu, J. C., Kinetics of choline acetyltransferases (EC 2.3.1.6.) from human and other mammalian central and peripheral nervous tissues. *J. Neurochem* **20**, 297 (1973).

438. Whittaker, M., The pseudocholinesterase variants: Esterase levels and increased resistance to fluoride. *Acta Genet. Stat. Med.* **14**, 281 (1964).

439. Whittaker, M., The pseudocholinesterase variants. A study of fourteen families selected via the fluoride resistant phenotype. *Acta Genet. Stat. Med.* **17**, 1 (1967).

440. Whittaker, M., The pseudocholinesterase variants. Differentiation by means of alkyl alcohols. *Acta Genet. Stat. Med.* **18**, 325 (1968).

441. Whittaker, M., Differential inhibition of human serum cholinesterase with n-butyl alcohol : recognition of new phenotypes. *Acta Genet. Stat. Med.* **18**, 335 (1968).

442. Whittaker, V. P., Specificity, mode of action and distribution of cholinesterases. *Physiol. Rev.* **31**, 312 (1951).
443. Whittaker, V. P., The binding of neurohormones by subcellular particles of brain tissue. *In* "Regional Neurochemistry" (S. Kety and J. Elkes, eds.), p. 259. Pergamon, Oxford, 1960.
444. Whittaker, V. P., Michaelson, I. A., and Kirkland, R. J. A., The separation of synaptic vesicles from nerve-ending particles (synaptosomes). *Biochem. J.* **90**, 293 (1964).
445. Whittaker, V. P., and Sheridan, M. N., The morphology and acetylcholine content of isolated cerebral cortical synaptic vesicles. *J. Neurochem.* **12**, 363 (1965).
446. Whittaker, V. P., and Wijesundera, S., The hydrolysis of succinylcholine by cholinesterase. *Biochem. J.* **52**, 475 (1952).
447. Wills, J. H., Toxicity of anticholinesterases and treatment of poisoning. *In* "Anti-cholinesterase Agents" (A. G. Karczmar, ed.), Int. Encycl. Pharmacol. Ther., Section 13, Vol. I, p. 355. Pergamon, Oxford, New York, 1970.
448. Wilson, C. W., Williams, J. P., and Miller, D. H., Hazard of cholinergic crisis during treatment of myasthenia gravis with octamethyl pyrophosphoramide. *Ann. Intern. Med.* **37**, 574 (1952).
449. Wilson, I. B., Acetylcholinesterase. XI. Reversibility of tetraethyl pyrophosphate inhibition. *J. Biol. Chem.* **190**, 111 (1951).
449a. Wilson, I. B., Bergmann, F., and Nachmansohn, D., Acetylcholinesterase X. Mechanism of the catalysis of acylation reactions. *J. Biol. Chem.* **186**, 781 (1950).
450. Wilson, I. B., Acetylcholinesterase. XIII. Reactivation of alkyl phosphate-inhibited enzyme. *J. Biol. Chem.* **199**, 113 (1952).
451. Wilson, I. B., The active surface of the serum esterase. *J. Biol. Chem.* **208**, 123 (1954).
452. Wilson, I. B., The mechanism of enzyme hydrolysis studied with acetyl-cholinesterase. *In* "The Mechanism of Enzyme Action" (W. D. McElroy and B. Glass, eds.), p. 642. Johns Hopkins Press, Baltimore, Maryland, 1954.
453. Wilson, I. B., Conformation changes in acetylcholinesterase. *Ann. N. Y. Acad. Sci.* **144**, 664 (1967).
454. Wilson, I. B., and Bergmann, F., Studies on cholinesterase. VII. The active surface of acetylcholine esterase derived from effects of pH on inhibitors. *J. Biol. Chem.* **185**, 479 (1950).
455. Wilson, I. B., and Bergmann, F., Acetylcholinesterase. VIII. Dissociation constants of the active groups. *J. Biol. Chem.* **186**, 683 (1950).
456. Wilson, I. B., and Cabib, E., Acetylcholinesterase: Enthalpies and entropies of activation. *J. Am. Chem. Soc.* **78**, 202 (1956).
456a. Wilson, I. B., and Ginsburg, S., Reactivation of acetylcholinesterase inhibited by alkyl phosphates. *Arch. Biochem. Biophys.* **54**, 569 (1955).
457. Wilson, I. B., and Quan, C., Acetylcholinesterase studies on molecular complementariness. *Arch. Biochem. Biophys.* **73**, 131 (1958).
458. Wilson, I. B., Harrison, M. A., and Ginsburg, S., Carbamyl derivatives of acetylcholinesterase. *J. Biol. Chem.* **236**, 1498 (1961).
458a. Winterberg, H., Über die Wirkung des Physostigmins auf das Warmblüterherz. *Z. Exp. Pathol. Ther.* **4**, 636 (1907).
459. Wolfers, P., Sensitivity to succinylcholine chloride. *Br. Med. J.* **2**, 778 (1952).
460. Wolleman, M., "Biochemistry of Brain Tumors" p. 134. Akadémiai Kiadó, Budapest, 1974.
460a. Wolleman, M., and Zoltan, L., Cholinesterase activity of cerebal tumors and tumorous cysts. *Arch. Neurol. (Chicago)* **6**, 161 (1962).
461. Yoshida, A., and Motulsky, A. G., A pseudocholinesterase variant (E Cynthiana) associated with elevated plasma enzyme activity. *Am. J. Hum. Genet.* **21**, 486 (1969).
462. Zajicek, J., Studies on the histogenesis of blood platelets. *Acta Haematol.* **12**, 238 (1954).
463. Zajicek, J., and Datta, N., Investigation on the acetylcholinesterase activity of erythro-

cytes, platelets and plasma in different animal species. *Acta Haematol.* **9**, 115 (1953).

464. Zarday, Z., Deery, A., Tellis, I., Soberman, R., and Foldes, F. F., Plasma and red cell cholinesterase activity in uremic patients before and after hemodialysis and after renal transplantation. *Abstr. Sci. Pap., Annu. Meet. Am. Soc. Anesthesiol.* p. 195 (1973).
465. Zarday, Z., Deery, A., Tellis, I., Soberman, R., and Foldes, F. F., Plasma and red cell cholinesterase activity in uremic patients.
466. Zarday, Z., Foldes, F. F., Deery, A. M., and Soberman, R., Butyryl-cholinesterase activity in plasma and urine of nephrotic patients. *Clin. Res.* **19**, 554 (1971).
467. Zech, R., and Engelhard, H., Acetylcholinesterase-Aktivität im Serum des Elektrischen Aals. *Hoppe-Seyler's Z. Physiol. Chem.* **348**, 735 (1967).
468. Zeller, E. A., and Bissegger, N. A., Über die Cholinesterase des Gehirns und der Erythrocyten. 3. Mitteilung über die Beeinflussung von Fermentreaktionen durch Chemotherapeutica und Pharmaka. *Helv. Chim. Acta* **26**, 1619 (1943).
469. Zsigmond, E. K., and Downs, J. R., Plasma cholinesterase activity in newborns and infants. *Can. Anaesth. Soc. J.* **18**, 278 (1971).
470. Zsigmond, E. K., Pyridostigmine: A safe and effective antagonist to d-tubocurarine in anesthetized man. *J. Clin. Pharmacol. Ther.* **13**, 155 (abstr.) (1972).
471. Zsigmond, E. K., Foldes, F. F., and Foldes, V. M., The in vitro inhibitory effect of LSD, its congeners and 5-hydroxytryptamine on human cholinesterase. *J. Neurochem.* **8**, 72 (1961).
472. Zsigmond, E. K., The effect of epinephrine and its congeners on human cholinesterases. *Arch. Int. Pharmacodyn. Ther.* **197**, 102 (1972).
473. Zsigmond, E. K., and Robins, G., The effect of a series of anti-cancer drugs on plasma cholinesterase activity. *Can. Anaesth. Soc. J.* **19**, 75 (1972).
474. Zupanĉic, A. O., Anionic centers of cholinesterases as possible cholinoreceptors. *Is. J. Med. Sci.* **1**, 1396 (1965).
475. Zupanĉic, A. O., Evidence for the identity of anionic centers of cholinesterases with cholinoreceptors. *Ann. N. Y. Acad. Sci.* **144**, 689 (1967).

Chapter 9

The Interaction of Anesthetic Agents with Hepatic Microsomal Enzymes

James M. Perel and Lester C. Mark

Introduction

Did it ever occur to you that a single dose of any drug would last a lifetime if the body did not have means of handling it? Fortunately, a variety of mechanisms is available to counter any invasion of the body by drugs or other foreign substances. These include distribution to and localization in body tissues, such as storage of unchanged drugs in adipose tissue and elsewhere (a temporizing measure, deferring but not avoiding the need for ultimate disposition), biotransformation mediated by enzymes located primarily but not exclusively in liver microsomes, and excretion where appropriate via the kidneys, biliary tract, lungs, skin, and various secretory glands and organs.

Renal excretion is an important mechanism, whose efficiency is limited by the physicochemical characteristics of the drug in question. The glomerular filtrate is essentially a protein-free replica of the blood plasma; hence, any drug arriving in the bloodstream at the glomeruli is quantitatively excreted into the renal tubules. Here, since the renal tubular cells act as a lipoid barrier, the fate of the drug is largely determined by the extent of its ionization at the existing urinary pH (highly ionized drugs do not cross the barrier) and the lipid solubility of its undissociated molecule. A highly lipid-soluble drug such as thiopental is completely reabsorbed during its transit through the kidney tubules, so that in man less than 0.5% of even a large dose appears unchanged in the urine (120). Conversely, a poorly lipid-soluble drug such as barbital is almost completely (95% or more) eliminated unchanged in the urine (120). Indeed, one convenient result of biotransformation of drugs and foreign chemicals by liver enzymes is the production of metabolites that are more polar (i.e., more strongly ionized) than the parent compounds and, therefore, more readily excreted by the kidney. Most drug therapy would be dangerous and impractical if urinary excretion were the major route of elimination, since the pharmacologic effects produced would be too long-lasting. This is due to the fact that most drugs are sufficiently lipid soluble in the pH range of the glomerular filtrate to be efficiently reabsorbed into the plasma from the renal tubules and thus persist in the body. If metabolic transformation did not occur, ethyl alcohol would have a half-life of 24 days (instead of a few hours), and quinine would remain in the body with a half-life of about 100 years (instead of a few days).

Some drugs are not metabolized, since they possess no sites susceptible to attack by drug-metabolizing enzymes. Examples include inorganic anesthetic gases, such as nitrous oxide and xenon, and some antibiotics, such as penicillin and tetracycline. Fortunately, both these groups are rapidly eliminated from the body by other routes, the former via the lung, the latter via the kidney.

Other compounds do offer potential sites for enzymatic alteration but are nevertheless excreted unchanged. Examples are quaternary nitrogen compounds (hexamethonium, decamethonium, gallamine), which are strong bases; strong acids (sulfonates, sulfate, and glucuronide conjugates); and high molecular weight polymers (e.g., alginates, polyethylene glycols). Such compounds are probably not metabolized because of their very low lipid solubility, which may prevent them from gaining access to intracellular sites of drug metabolism.

Finally, there are some drugs, e.g., barbital, acetazolamide (Diamox), and tolazoline (Priscoline), which are metabolized insignificantly, if at all, with no apparent reason.

It is important to realize that biotransformation of drugs is not necessarily equivalent to either detoxification or degradation and that several other consequences are possible. For example, drug metabolism may (1) transform an inert parent compound into a pharmacologically active metabolite, (2) convert an active compound into another substance of similar or greater activity, (3) alter the nature of the pharmacologic activity, (4) cause deactivation (and facilitate excretion) of the drug, (5) convert the drug into a toxic product, or (6) produce a larger, more complicated molecule (e.g., by the process of conjugation).

Metabolic studies have become an integral part of the preclinical and clinical evaluation of new anesthetic agents. In a comprehensive report on the application of metabolic data to the evaluation of new drugs, the Committee on Problems of Drug Safety of the Drug Research Board, National Academy of Sciences (43), promulgated the following requirements, which are equally applicable for studies with anesthetic drugs:

1. To establish the pharmacodynamics and pharmacokinetics of both the (anesthetic) agent and its metabolites, including rates and sites of absorption,

plasma protein binding, plasma and tissue levels, rates of plasma clearance and metabolism, and tissue accumulation.

2. To ascertain that the pertinent animal models used to study the metabolism of the anesthetics are qualitatively and quantitatively similar to man in this respect.

3. To determine which molecular species of the drug is responsible for the pharmacologic activity(s) and for any toxic reaction.

4. To elucidate in detail the variety of chemical changes (anesthetic) agents may undergo in the body, thus facilitating the design of safer and more effective molecules.

Pathways of Drug Metabolism

The pathways of anesthetic drug metabolism may be divided into two major categories (203): the first of these consists of Phase 1 enzyme reactions, including oxidation, reduction, and hydrolysis, which lead to the introduction of new functional groups into the anesthetic molecule or to the modification of an existing functional group; the second consists of Phase 2 synthetic processes catalyzed by various enzymes whereby a functional group of the molecule is masked by the addition of a new group, e.g., methyl, acetyl, sulfate, or glucuronic acid, or amino acids such as glycine or glutathione. Many anesthetic agents undergo both Phase 1 and Phase 2 reactions, e.g., glutethimide is completely inactivated by hydroxylation into two different metabolites which in turn are conjugated to a great extent (about 95%) into their respective glucuronides (99). With complex molecules such as the phenothiazines, many alternative pathways of metabolism are possible and indeed found.

Primary Pathways of Drug Metabolism

The following are examples of primary pathways of metabolism (Phase 1 reactions) (22, 24a, 28, 34, 42, 49, 55, 57, 58, 62, 65, 72, 78, 87, 104, 108, 118, 120, 140, 179, 189, 191) found with agents used in anesthesia. Unless otherwise stated, the responsible enzyme is located in the endoplasmic reticulum of the liver.

Phenobarbital (120)   p-Hydroxyphenobarbital

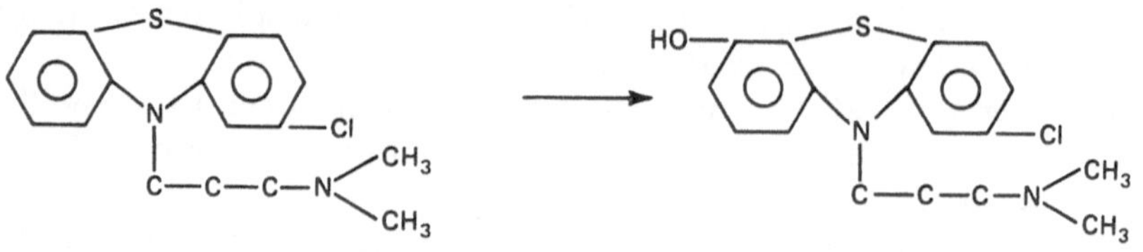

Chlorpromazine (60, 118, 140)

7-Hydroxychlorpromazine

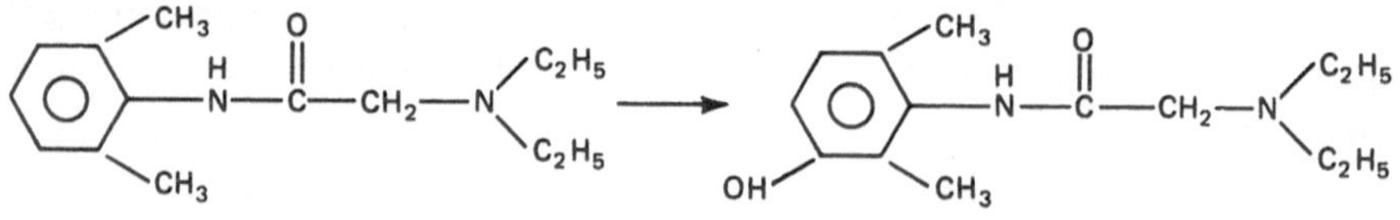

Lidocaine (55)

m-Hydroxylidocaine
(a minor metabolite)

Side Chain Oxidation

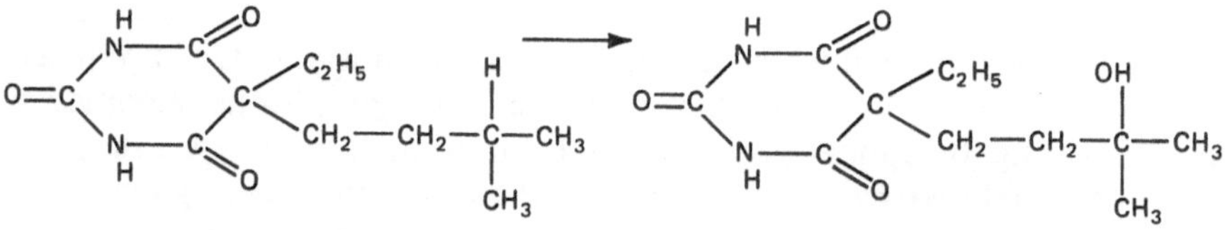

Amobarbital (120)

3'-Hydroxyamobarbital

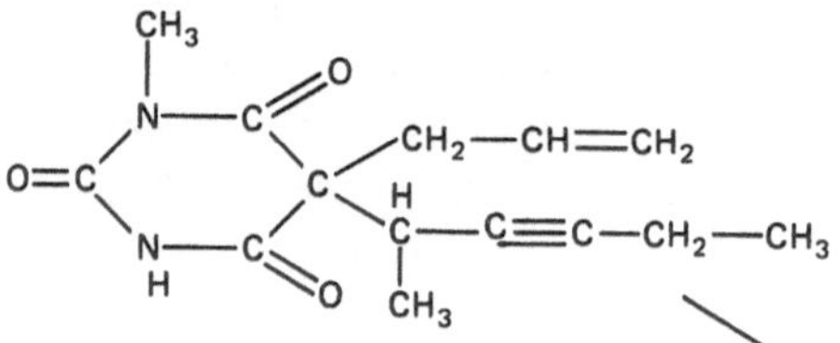

Methohexital (120)

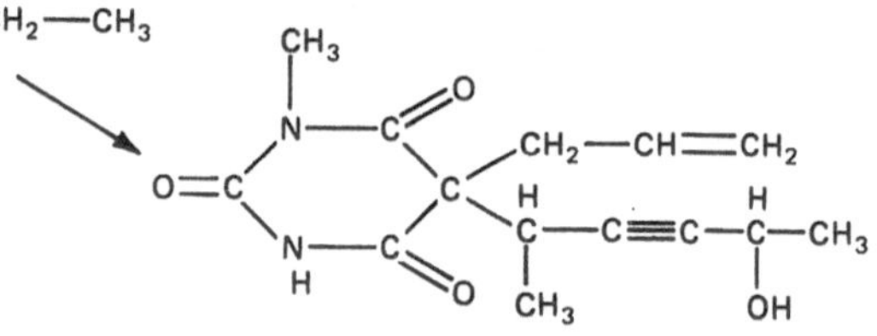

Methohexital Alcohol
(4'-Hydroxymethohexital)

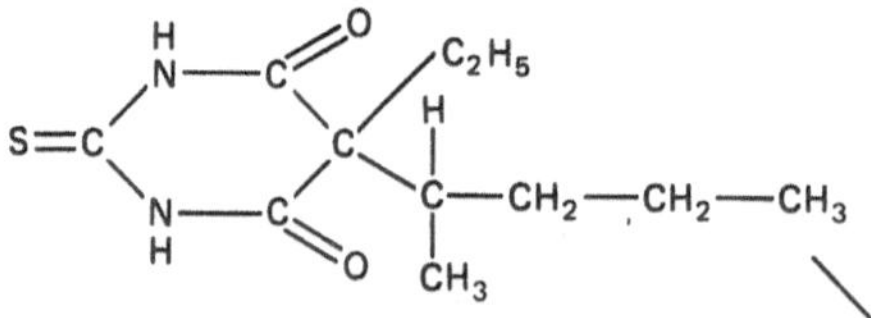

Thiopental (120)

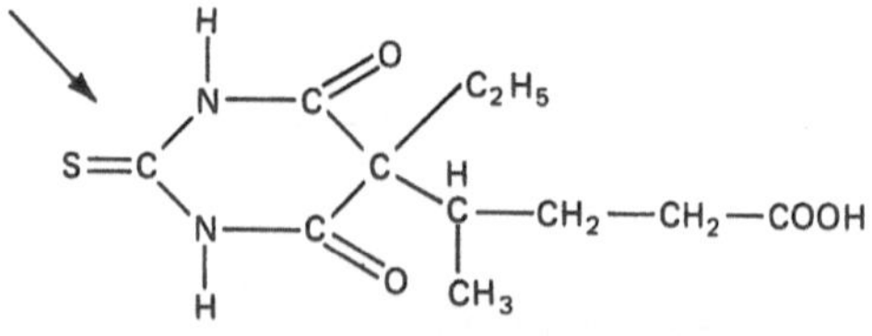

Thiopental carboxylic acid

172

Alicylic Hydroxylation

Hexobarbital (34, 87, 104, 120) → 3'-Hydroxyhexobarbital

Oxidative Dealkylation

Diazepam (189, 49) → N-Desmethyldiazepam

Lidocaine → Monoethylglycine xylidide (MEGX) + acetaldehyde

Mephobarbital (Mebaral) (120) → Phenobarbital

Oxidative Deamination

Amphetamine → Benzylmethylketone

173

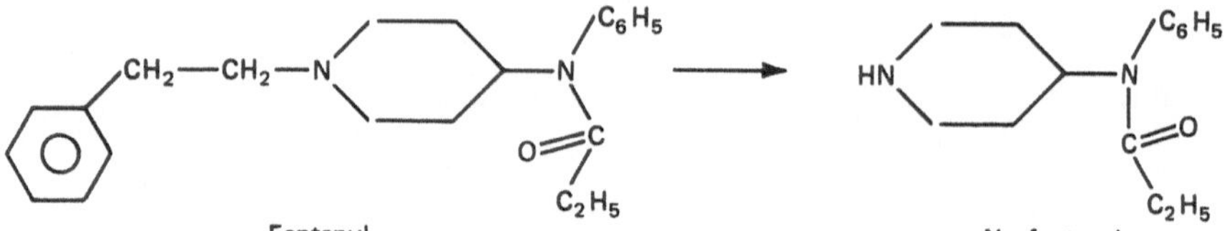

Fentanyl → Norfentanyl

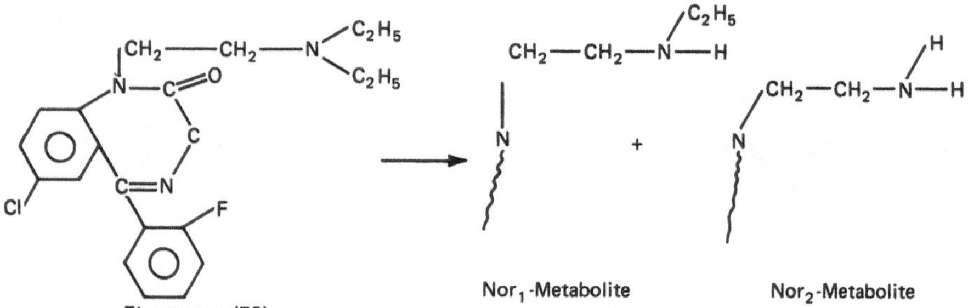

Flurazepam (79) → Nor$_1$-Metabolite + Nor$_2$-Metabolite

Desulfuration

Thiopental (120) → Pentobarbital

Oxidative Dehalogenation

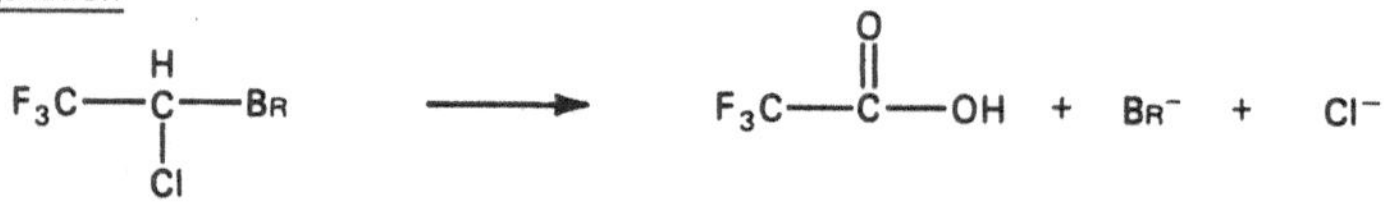

$$F_3C-\overset{\overset{\displaystyle H}{|}}{\underset{\underset{\displaystyle Cl}{|}}{C}}-Br \longrightarrow F_3C-\overset{\overset{\displaystyle O}{\|}}{C}-OH \ + \ Br^- \ + \ Cl^-$$

Halothane (28, 42, 179)   Trifluoroacetic Acid

$$H_3C-O-\overset{\overset{\displaystyle F}{|}}{\underset{\underset{\displaystyle F}{|}}{C}}-\overset{\overset{\displaystyle Cl}{|}}{\underset{\underset{\displaystyle Cl}{|}}{C}}-H \longrightarrow H_3C-O-\overset{\overset{\displaystyle F}{|}}{\underset{\underset{\displaystyle F}{|}}{C}}-COOH$$

Methoxyflurane (57, 108)

Methoxydifluoracetic Acid
+
oxalic acid
+
fluoride and chloride ions

$$H-C-Cl_3 \longrightarrow CO_2 \ + \ Cl^-$$

Chloroform

174

Miscellaneous Oxidations Drug	Product(s)	Enzyme Location
$F_3C-CH-O-CH=CH_2 \rightarrow$ Fluroxene[72]	$F_3C-COOH$ Trifluoroethanol	Liver endoplasmic reticulum
Ethyl ether $\longrightarrow$ Ethyl alcohol + acetaldehyde		Liver endoplasmic reticulum
Chloral hydrate $\longrightarrow$	Trichloracetic acid	Liver cytosol[a]
Ethanol $\xrightarrow{1}$ (1) microsomal alcohol oxidizing system: (2) NAD-alcohol dehydrogenase	Corresponding $\xrightarrow[\text{oxidase}]{\text{aldehyde}}$ Corresponding aldehyde acid	(1) Liver endoplasmic reticulum: liver cytosol (2) Liver cytosol

Hydrolysis Drug	Product(s)	Enzyme location
Cocaine (58) $\longrightarrow$	Benzoylecgonine	Blood plasma, liver
Procaine (24a) $\longrightarrow$	$\left\{ \begin{array}{c} p\text{-Aminobenzoic acid} \\ + \\ \text{diethylaminoethanol} \end{array} \right.$	Blood plasma, liver
		Applies to local anesthetics having ester linkages in the intermediate chain between the amino group and the aromatic ring
Succinylcholine $\longrightarrow$	Succinic acid + choline	Blood plasma, liver

[a]Liver microsomes are isolated by centrifugation of homogenized liver tissue at successively higher gravity (g) forces. Homogenized liver tissue is initially centrifuged for 10 min at 600 g; a pellet of dense material, primarily whole cells, cell debris, and cell nuclei, collects at the bottom of the tube. The supernatant is then centrifuged for 20 min at 9000 g, isolating less dense structures such as mitochondria and membrane fragments. Finally, the resulting supernatant is usually spun for 60 min at 100,000 g in an ultracentrifuge after which the microsomes are separated from the cytosol, or cell fluid.

Secondary Pathways of Drug Metabolism

Conjugations (Phase 2 Reactions). Conjugation of an anesthetic agent involves combination with some natural constituent of the body, such as the carbohydrate glucuronic acid, the amino acid glycine, or the tripeptide glutathione. In the presence of the appropriate enzyme these endogenous compounds react readily with drugs or metabolites possessing carboxy (COOH), sulfhydryl (SH), amino (NH_2) or hydroxyl (OH) groups. Some anesthetics which have these functional groups are handled initially by conjugation directly in the liver. For most drugs, however, conjugation is a second step that occurs after a Phase 1 reaction has taken place.

Glucuronide synthesis. Uridinediphosphoglucuronic acid (UDPGA) is produced in the soluble fraction of liver, renal cortex, intestinal mucosa and skin as follows:

Glucose 1-phosphate + uridine triphosphate $\longrightarrow$ UDPglucose + pyrophosphate

UDPglucose + 2NAD $\longrightarrow$ UDPGA + 2NADH

(NAD = nicotinamide adenine dinucleotide, a hydrogen acceptor)

An enzyme in the hepatic rough endoplasmic reticulum, glucuronyl transferase, transfers the glucuronic acid to an appropriate acceptor, such as an organic compound having a hydroxy, amino, sulfhydryl, or carboxylic group. These groups may be present in either the original drug or its metabolite. Examples include (a) trichloroethanol, which is a metabolite of chloral hydrate and also of trichloroethylene:

$$\text{UDPglucuronic acid} + Cl_3\overset{H}{\underset{H}{C}}-OH \longrightarrow Cl_3-\overset{H}{\underset{H}{C}}-O-\text{glucuronide}$$

$$\text{trichlorethanol} \qquad \text{urochloralic acid}$$

(b) the hydroxy metabolite of phenobarbital:

UDP-GA + *p*-hydroxyphenobarbital $\longrightarrow$ *p*-hydroxyphenobarbital *O*-glucuronide

(c) meprobamate, a parent drug

UDP-GA + Meprobamate (84, 116)
$\longrightarrow$ meprobamate mono-*N*-glucuronide

Sulfate synthesis. Conjugation with a sulfate group is a two step reaction. The first step, involving adenosine triphosphate, leads to the formation of 3'-phosphoadenosine 5'-phosphosulfate (PAPS). The sulfate is then transferred to a hydroxyl group in the anesthetic molecule (or its metabolite). For example, with the hydroxy metabolite of chlorpromazine the overall reaction is

7-Hydroxychlorpromazine + PAPS $\longrightarrow$ 7-hydroxychlorpromazine sulfate + H$_2$O

These reactions take place in the soluble fraction of the liver and other tissues.

Glycine conjugation. This is a general reaction of aromatic acids which requires

$$ATP + coenzyme\ A + aromatic—COOH \longrightarrow aromatic—\overset{\displaystyle O}{\overset{\displaystyle \|}{C}}—coenzyme\ A$$

$$\underset{glycine}{\overset{glycine\text{-}N\text{-acylase}}{\longrightarrow}} \quad aromatic—\overset{O}{\overset{\|}{C}}\ \overset{H}{\overset{|}{N}}—\overset{H}{\overset{|}{\underset{H}{C}}}—COOH$$

These reactions take place in the liver cytosol, which is apparently the only source of the acylase. Aromatic acids such as salicylic and benzoic acids are partially conjugated by this route. The neuroleptic agent haloperidol (61) is metabolized via Phase 1 reactions to *p*-fluorophenyl acetic acid which in turn is conjugated with glycine to form *p*-fluorophenyl acetylurenic acid

Acetylation. As with conjugations with the amino acid glycine, coenzyme A is required, but in this case to activate endogenous acetate rather than an exogenous acid. The acetyl group is then transferred to the nitrogen of an aromatic or a heterocyclic amine or of a hydrazide, for example:

Acetyl-CoA + mescaline $\rightleftharpoons$ *N*-acetyl mescaline

Acetyl-CoA + *p*-aminosalicylic $\rightleftharpoons$ *N*-acetyl *p*-aminosalicylic acid

This is an important mechanism of detoxication for drugs such as procaine amide, sulfonamides, and the antimalarial dapsone. A specific transacetylase (73) found in the Kupffer cells of the liver has been described for acetylation reactions with hydrazides, some of which are monoamine oxidase inhibitors (e.g., *p*-phenylethyl hydrazine), and another is a tuberculostatic agent, isonicotinic acid hydrazide (isoniazid, INH). These agents may be used as test substances to divide the population into rapidly and slowly acetylating individuals. High transacetylase activity is inherited as a single dominant gene. The rate of acetylation is an important factor in determining efficacy and toxicity of these drugs.

Methylation. N-Methylation of nicotinic acid to form *N*-methylnicotin-amide has been long recognized. In addition, O-methylation has also been demonstrated. For example, the following involves O-methylation:

Morphine ⟶ Codeine

Both types of methylation require *S*-adenosylmethionine as the donor of the methyl group. The most important examples of O-methylation are reactions involving the catecholamines:

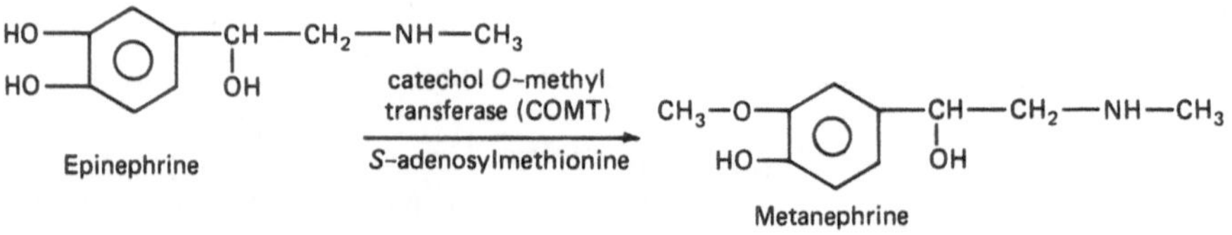

This is the primary mechanism of metabolism for injected epinephrine or norepinephrine. Oxidative deamination occurs secondarily.

Other important methylations are catalyzed by specific enzymes, e.g., N-methylation of histamine and O-methylation of hydroxyindole to melatonin. In addition to liver cytosol systems, nonspecific N-methylation of narcotic drugs may also occur in other tissues such as the lung.

The enzymes that effect these chemical changes are found in many tissues, particularly in the liver, gastrointestinal tract, lungs, and kidney. Within the liver, which is the most important organ for the metabolism of anesthetic agents, the enzymes are located mainly in the endoplasmic reticulum of the hepatic parenchymal cells. Homogenization and ultracentrifugation using liver tissue from different animal species and man have firmly established that the enzymes and enzyme systems participating in drug modification and detoxification are concentrated in the microsomes (both the "smooth" and the "rough" variety, but especially the former (151). The microsomes are small spherules, actually laboratory artifacts, produced during homogenization by disruption of the endoplasmic reticulum of the hepatocyte. Most of the enzymes catalyzing oxygenations are associated with the hepatic endoplasmic reticulum (i.e., the microsomal fraction) and are commonly known as the hepatic microsomal drug-metabolizing enzymes.

Properties of the Hepatic Microsomal Oxidase System

Brodie (24) and Axelrod (9) discovered the drug-metabolizing activity of liver microsomes. They showed that in the presence of reduced nicotinamide-adenine-dinucleotide phosphate (NADPH) and molecular oxygen, liver microsomes catalyze the hydroxylation of numerous drugs. The system was similar to that first described seven years earlier by Mueller and Miller (132), who had found that liver homogenates, supplemented with NADH or NADPH in the presence of oxygen, catalyzed the oxidative demethylation of the carcinogenic dye 4-dimethylaminoazobenzene to 4-aminoazobenzene and formaldehyde. Later it was found that microsomes from liver, adrenal cortex, testis, ovary, and placenta catalyze an NADPH and O_2-dependent hydroxylation of various steroids and that a steroid hydroxylating system is also present in adrenal-cortex mitochondria (56). The same system was found to be involved in the hydroxylation of aliphatic hydrocarbons and the ω-1 oxidation of fatty acids. The hydroxylating enzyme has been termed a "mixed function oxidase" or "monooxygenase," since in the course of the reaction one of the oxygen atoms of O_2 is incorporated into the compound undergoing hydroxylation and the other into H_2O. The process may be described by the general equation:

$$RH + NADPH + H^+ + O_2 \longrightarrow ROH + NADP^+ + H_2O$$

where RH is the substrate and ROH the product of the hydroxylation reaction.

Phase 1 oxidation or hydroxylation reactions are catalyzed by the microsomal drug-metabolizing enzymes which, as indicated above, require both molecular oxygen and reduced NADPH. The reactions involve the participation of an electron transport chain which terminates in an oxygen-transferring enzyme, cytochrome P-450. The latter is a hemoprotein complex characterized by its formation of a ligand compound with carbon monoxide. This reaction inactivates the enzyme and gives it a characteristic absorption maximum at 450 nm, hence its name (71).

In 1963, Estabrook, Cooper, and Rosenthal (56) detected cytochrome P-450 in the microsomes of the adrenal cortex, which hydroxylate progesterone at position C-21. Subsequently, other investigators found the same cytochrome involved in drug hydroxylations in other organs and tissues (153). The highest content was found in the liver, with much smaller amounts occurring in kidney, lung, intestinal mucosa, and skin (186). The dominance of the hepatic fraction of cytochrome P-450 activity is shown by the fact that on wet weight basis the kidney contains only $1/7$ of the amount found in the liver. Since the liver is 6 times greater in weight than the kidneys, the latter contribute only 2–3% of the total hydroxylation of a drug in the body. Other tissues contribute even less. Cytochromes P-450 and b_5 were first demonstrated in human hepatic microsomes in 1969–1970 (1, 7) and found to have spectral characteristics similar to those in animals.

According to the current understanding of the mechanisms of these enzymes, equivalent amounts of drug, oxygen, and NADPH are utilized during the oxidative reaction. The drug substrates combine with the oxidized form of cytochrome P-450 to form complexes that are reduced by transfer of an electron from NADPH-cytochrome c reductase. Until recently, it was generally accepted that the reduced cytochrome P-450 substrate complexes next react with oxygen to form oxygenated complexes which in turn accept a second electron to form "activated oxygen" complexes (68, 71). However, recent data indicate that in mammalian systems the reduced cytochrome P-450 substrate complexes can accept the second electron *before* reacting with

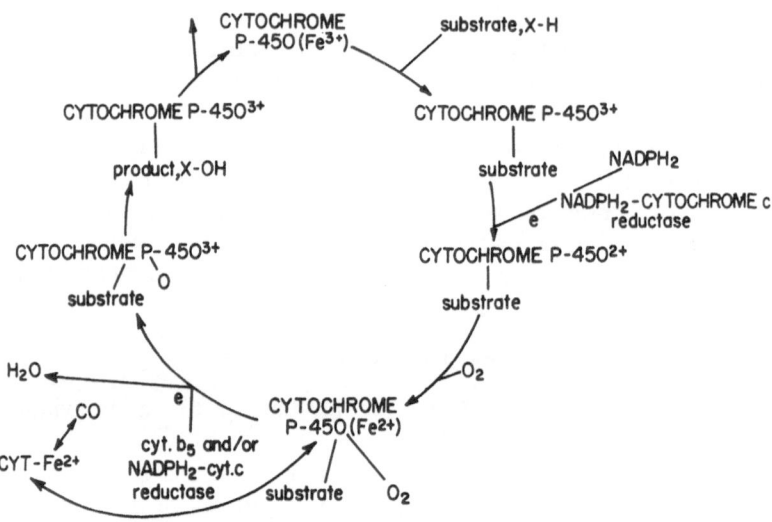

Figure 9-1.   Scheme of electron transport from NADP to cytochrome *P*-450 concurrently with the hydroxylation of drug substrate X.

oxygen (11). The source of the second electron is either from NADPH by way of NADPH-cytochrome *c* reductase or from NADH by way of NADH-cytochrome b_5, both of which are present in the microsomes. Once the "activated oxygen" cytochrome *P*-450–substrate complexes are formed, they rearrange themselves to form oxidized cytochrome *P*-450 and oxidized products. A scheme of the multiple electron transport involved in the hydroxylation of drugs by cytochrome *P*-450 is shown in Figure 9-1.

Changes in hydroxylase activity usually correlate with cytochrome *P*-450 content (135), but exceptions do occur. For example, anomalies concerning the rate (88) and species variation (50) of ethylmorphine demethylation suggest that in this instance cytochrome *P*-450 reductase is the rate-limiting enzyme. Since cytochrome *P*-450 substrate complexes produce two different types of spectral change, with differences in the kinetics of *P*-450 reduction in the two groups (67) (see below), it has become apparent that the rate of oxidation may be altered by affecting any one of the components, including the amount of cytochrome *P*-450, qualitative variations in cytochrome *P*-450 (e.g., *P*-448) (85, 119), or the rate of cytochrome *P*-450 reduction in both the presence and absence of substrate. It is not surprising, therefore, that changes in the activity of the system are not always related to any given component, such as the cytochrome *P*-450 content.

Human liver contains less cytochrome *P*-450 than that of other species. From available data it can be estimated that about 10 to 20 nmoles of cytochrome *P*-450 is present in 1 g of human liver. This is 1/3 to 1/2 of the amount found in 1 g of rat liver. However, taking into account the smaller liver/body weight ratio in man (2% as compared to 4% in rats), rats have 4 to

6 times more cytochrome P-450 than man. This corresponds to the observations of many researchers that man metabolizes drugs *in vivo* at rates 2 to 10 times slower than does the rat (145).

In summary, the microsomal hydroxylase located in the endoplasmic reticulum of liver cells belongs to the group of mixed-function oxidases which utilize two electrons, one for the reduction of one atom of oxygen to water, while the second oxygen atom is incorporated into the substrate. The stoichiometric equation for the overall process is

$$R + 2e^- + 2H^+ + O_2 \longrightarrow R:O + H_2O$$

The reaction catalyzed by the hydroxylating system involves a binding of the substrate with the oxidized form of cytochrome P-450, thereby causing small but significant changes in its absorbance spectrum (142, 164). The lack of substrate specificity is a highly unique feature, contrary to the behavior of almost all other enzyme systems in the body. Some substances, such as hexobarbital, aminopyrine, and ethylmorphine, when added to microsomal suspensions cause a decrease (trough) in the spectrum at about 417 nm and an increase (peak) at 391 nm, as determined by ultraviolet difference spectrophotometry. Such substances are called Type I compounds. Other substances, such as nicotinamide and aniline, produce a trough in the spectrum at 418 nm and a peak at 423 nm. These are called Type II compounds (154, 165). Of the two types of spectral changes, Type I is generally believed to be due to a binding of the substrate to the apoprotein of the cytochrome, whereas Type II may, in part at least, be due to an interaction between the substrate and the heme prosthetic group. Only the Type I spectral change appears to be directly related to the hydroxylating activity. Strong evidence for this relationship comes from the finding that in the hydroxylating systems the binding constants (K_s) of various substrates, defined as the concentration giving half-maximal Type I spectral change, closely parallel their Michaelis constants (K_m) (66, 67, 69, 92, 154).

Because of the complexities of these spectral changes and since other endogenous compounds such as steroids and fatty acids can also alter the absorbance spectrum of P-450, the interpretation of the apparent binding constants and maximal values for the spectral changes is frequently difficult. Nevertheless, it is usually found that Type I compounds are metabolized more rapidly than Type II because the oxidized cytochrome P-450 complexes of Type I compounds are reduced by NADPH more rapidly than are those of Type II compounds (67).

Factors Affecting Metabolism of Anesthetic and Adjuvant Drugs

Hepatic Enzyme Induction

A great variety of lipid-soluble drugs can accelerate their own metabolism and also the metabolism of other compounds, whether pharmacologically or chemically related or unrelated, by stimulating the drug-hydroxylating enzyme systems in the endoplasmic reticulum of the liver. This phenomenon, termed enzyme induction, was first described by Brown and associates (27) who found that 3,4-benzpyrene increased the activity of the enzyme benz-pyrene hydroxylase. Remmer (150) noted independently that phenobarbital caused an increase in phenobarbital hydroxylase activity. These increases were subsequently demonstrated to be due to corresponding increases in the actual amount of enzymes present, which, in turn, have been shown by means of isotopic incorporation studies to represent *de novo* synthesis. Induction of the microsomal enzymes involves increases in the biosynthesis of both protein and heme. The first enzyme to increase is δ-aminolevulinic synthetase (17) (also found in the mitochondria), which catalyzes the rate-determining reaction for heme biosynthesis. This is followed sequentially by increases in hepatic microsomal heme and cytochrome *P*-450, and in the specific NADPH-dependent cytochrome *c* reductase (12). The concentration of other enzymes not involved in drug hydroxylation, such as cytochrome b_5, increases later to a much smaller extent. After a single injection of phenobarbital, the induction of cytochrome *P*-450 is maximal in 24 hr (but repeated dosing is more effective, as explained below). The newly formed enzyme has the same half-life as normal cytochrome *P*-450, about 1 day (152).

Several barbiturates have been shown to increase the level of δ-aminolevulinic acid synthetase activity, probably by removing heme for the induced synthesis of cytochrome *P*-450 (130). This leads to a decreased net concentration of heme, a repressor of the synthetase, and thereby causes a reduction of the normal feedback control. Curiously, allobarbital and other barbiturates containing allyl groups (all these compounds are potent porphyrogenic agents) have been shown to cause destruction of cytochrome *P*-450, probably due to the action of a metabolite in which the allyl side chain has undergone biotransformation (111).

Repetitive administration of an inducing agent causes a progressive increase of cytochrome *P*-450 concentration reaching a maximum in 3–5 days. The continued presence of the inducing agent in the liver cells enhances the rate of synthesis of both the heme and the protein moiety; hence, inducing agents which undergo slower rates of metabolism should presumably be more potent stimulators. The actual mechanism of enzyme induction is not fully understood, although the phenomenon is apparently DNA-dependent since both cycloheximide and puromycin, which act at the ribosomal level, and also actinomycin D (6), which prevents transcription, block both induction and regression after induction. Paralleling the increased microsomal enzyme activity, a significant increase of the smooth endoplasmic reticulum of the liver cells becomes visible with electron microscopy (152). As the growth of the endoplasmic reticulum becomes obvious, the liver itself enlarges. The increase in metabolizing activity appears to be due to the increased quantity of enzymes rather than to any basic change in the oxidative systems. The Michaelis and inhibitor constants for the induced and noninduced systems are similar, confirming a quantitative rather than a qualitative change.

Nevertheless, studies have indicated that the stereospecificity of the induced oxidative enzymes is less than that of the noninduced enzymes when reacting with optically active substrates or forming optically active centers (126). The hydroxylation of ethylbenzene, a Type I substrate, by liver microsomes is a stereochemically controlled reaction whose product, R-($+$)-methylphenylcarbinol, has an optical purity ranging from 81 to 86%. Pretreatment with phenobarbital, m-bromophenobarbital, or chlorcyclizine reduces the stereospecificity of the reaction by yielding R-($+$)-methylphenylcarbinol with an optical purity of only 61 to 67%. In other words, stimulated microsomal enzymes are less stereospecific than noninduced enzymes (126). Consequently, pharmacologic and clinical studies of the effects of racemic anesthetic agents where one of the optical isomers (e.g., S-($-$)-pentobarbital, S-($-$)-secobarbital) is more active (121, 122) could show even greater individual variability than expected in a heterogeneous population because of differences in hepatic microsomal enzyme activities.

More than 200 substances, including barbiturates, polycyclic hydrocarbons, steroids, polychlorinated insecticides, and even the terpenes found in wood chips used in animal bedding have been found to enhance the metabolism of drugs by the cytochrome P-450 systems. A partial list of such compounds with their relative inducing effects is shown in Table 9-1.

These substances increase the activity of the enzyme systems in different ways. Pretreatment with phenobarbital causes increases in the amounts of both cytochrome P-450 and NADPH-cytochrome c reductase, while pretreatment with spironolactone causes little change in cytochrome P-450 but does increase the amount of NADPH-cytochrome c reductase (70). Pretreatment of animals with aromatic polycyclic hydrocarbons such as 3-methylcholanthrene and 3,4-benzpyrene stimulates the formation of a new hemoprotein (or a new variant) called P-448, but has little or no effect on the activity of NADPH-cytochrome c reductase. Cytochrome P-448 differs from cytochrome P-450 not only in the absorbance spectrum of its complex with carbon monoxide (maximum $=$ 448 nm) but also in its ethyl isocyanide difference spectra (maximum $=$ 455 nm) and drug-complexing properties (5, 119, 171). Not surprisingly, it has a distinct substrate specificity. Thus, pretreatment with 3-methylcholanthrene results in marked increases in the metabolism of some substances, such as the hydroxylation of 3,4-benzpyrene, but has little or no effect on others, such as the hydroxylation of barbiturates and the N-dealkylation of ethylmorphine and diazepam (174). Another difference is that enzyme activity is induced much more rapidly (less than 24 hr) with these aromatic hydrocarbons than with phenobarbital and other cytochrome P-450 inducers, which ordinarily require several days of pretreatment to be effective.

Although much is known about enzyme induction in laboratory animals, current knowledge concerning the induction process in normal man or in the presence of hepatic disease is meager. Species differences in drug metabolism prevent the indiscriminate transfer to man of data derived from experimental

Table 9-1.   Partial list of drugs which cause enzyme induction in man

Pharmacologic classification	Enzyme stimulator	Effect[a]
Anesthetic Gases	Halothane	+ +
	Methoxyflurane	+ +
Hypnotics and Sedatives	Barbiturates	+ + + +
	Chloral Hydrate	+ +
	Glutethimide	+ + +
	Triclofos (trichloroethyl sodium phosphate)	+ + +
	Ethanol	+ +
	Methaqualone	+ +
	Methyprylon	+ +
Anticonvulsants	Diphenylhydantoin	+ + +
	Primidone	+ + +
	Methylphenylethylhydantoin	+ + +
	Carbamazepine	+ + +
Anti-inflammatory agents	Phenylbutazone	+ + +
	Aminopyrine	+ +
Tranquilizers	Meprobamate	+ + +
	Chlordiazepoxide	+
Hypoglycemic agents	Tolbutamide	+ + +
Antihistaminics	Chlorcyclizine	+ +

[a]Compounds listed (+ + +) or higher cause induction if prescribed for several days. Drugs marked (+ +) cause induction after high doses or long-term administration. (+) denotes evidence of induction suggestive but not conclusive [based on Remmer (152)].

animals. The problem is compounded by the occurrence of genetic variations in drug metabolism and by the wide variation in dosages employed; doses in laboratory studies are commonly much higher than those used therapeutically in man.

The following considerations are relevant to the induction of microsomal oxidases in man. First, most lipid soluble compounds metabolized by these microsomal enzymes have some inducing properties. An interesting example is ethanol, which, metabolized by both microsomal and mitochondrial enzymes, is a potent enzyme-inducing agent in man (113, 114, 157, 158). Enzyme induction is the most likely explanation for the reduction in the plasma half-life of coumadin, diphenylhydantoin, and tolbutamide seen in chronic alcoholics (97). The mechanism for the more rapid elimination of diazepam from plasma in chronic alcoholic patients is complex, with an increased volume of distribution as well as increased hydroxylation of diazepam to oxazepam (167). In addition, some drugs, such as barbital, which are not metabolized in measurable quantities during the period of testing, can also produce stimulation of enzyme activity. Second, the inducing action is dose-dependent (21). Some drugs are active inducers only in high (nearly toxic) doses; others, such as barbiturates, can induce the hydroxylating enzyme while being used in ordinary therapeutic doses. Third, an appreciable concentration of the inducing agent must be maintained in the liver cells for a certain minimum period of time. For this reason, a short-acting drug such as hexobarbital causes an increase in enzymes only after repeated administration. Indeed, as a general rule, the more rapidly a barbiturate is metabolized in man, the higher the daily dose required to induce drug-metabolizing enzymes. Conversely, microsomal enzymes are induced preferentially by barbiturates or other drugs with a relatively long half-life and thus accumulating to some extent in the organism. These observations provide a clue to understanding of many inconsistencies in reports on enzyme induction. In the case of phenobarbital, a well-established inducing agent, Remmer (151) calculated that significant induction should be expected only in patients in whose plasma this slowly metabolized drug reaches a steady-state level of 10 to 30 $\mu g/ml$. Examples of phenobarbital-mediated induction are the observed reduction in plasma levels of griseofulvin (33) and diphenylhydantoin (103, 131). The latter has important consequences in the management of epileptics. An indirect indicator of enzyme induction in man is the extent of urinary excretion of D-glucaric acid, a metabolite of glucuronic acid. Thus, Hunter and associates (91), demonstrating an elevated level of D-glucaric acid in the urine of epileptic patients after prolonged anticonvulsant therapy, concluded that this reflected enzyme induction. They also found a significant correlation between the rise in urinary D-glucaric acid excretion and the fall of serum bilirubin levels in both normal volunteers and patients with Gilbert's syndrome (a relative deficiency of bilirubin UDPglucuronyl transferase) treated with phettharbital, a nonhypnotic barbiturate (91).

Other important examples concern the interactions between sedatives and oral coumarin-type anticoagulants. Phenobarbital has been shown to increase the rate of metabolism of coumarin-type anticoagulants. Phenobarbital has been shown to increase the rate of metabolism of coumadin and bishydroxycoumarin (19, 47, 117), thereby lowering the plasma concentrations attained with a given dosage. Consequently, patients in the hospital who nightly receive hypnotic drugs, most of which are potent enzyme inducers, will require larger doses of anticoagulants to reach and maintain therapeutic levels. Moreover, if on discharge they stop taking the hypnotics but continue with the same doses of the anticoagulants, the rate of metabolism of the latter will fall and their plasma levels will rise. Hemorrhage from hypoprothrombinemia may follow. A fatal outcome in one such case involving the use of chloral hydrate (the inducing agent) has been reported (48), although it should be mentioned that an alternate explanation has been offered, postulating the displacement of the anticoagulant from plasma proteins by trichloroacetic acid, a metabolite of chloral hydrate (166).

The effects of some commonly used barbiturate hypnotics were evaluated in patients receiving long-term therapy with coumadin. Amobarbital administered in doses of 200 mg daily caused significant decrease in plasma coumadin levels with simultaneous change in anticoagulant control (156). Measurement of another index of enzyme induction in man, the urinary excretion of 6β-hydroxycortisol, a polar metabolite of cortisol formed in the endoplasmic reticulum of the liver, likewise showed marked increases. In the rat, too, pretreatment with amobarbital for 4 days caused a significant decrease in sleeping and paralysis (absent resting reflex) times and half-life of ^{14}C-pentobarbital and a concomitant increase in the maximal velocity (V_{max}) of N-demethylation of ethylmorphine (44). Similar findings have been reported with secobarbital in patients receiving anticoagulant therapy (20). Chlordiazepoxide (79) 5 mg three times daily also caused a slight (equivocal) fall in plasma warfarin levels, with a concomitant increase in urinary excretion of 6β-cortisol (20). These changes were less significant than those observed with the barbiturates. Two other chemically related sedatives, nitrazepam and flurazepam, changed neither plasma coumadin levels nor anticoagulant activity in man or rat (20, 155). Other workers studied the influence of phenobarbital, glutethimide, chlordiazepoxide, and chloral hydrate on the plasma disappearance rate of the anticoagulant ethyl bicoumacetate in 16 patients. They found marked reduction in the half-life of the anticoagulant with both phenobarbital and glutethimide but no significant changes with either of the two other sedatives (188).

The deliberate utilization of enzyme induction to alter the concentration of an endogenous substrate was first reported by Werk and associates (201). They obtained clinical and biochemical improvement in nonneoplastic Cushing's syndrome by using diphenylhydantoin to enhance the 6β-hydroxylation of cortisol (45). In another area, both phenobarbital and glutethimide have been administered to pregnant women prior to parturition (96, 124) or to neonates (185) to diminish neonatal jaundice, presumably by enhancing the conjugation and excretion of bilirubin. In this instance, enzyme induction also increases the formation of heme and newly synthesized bilirubin, with concomitant reduction in morbidity and mortality. The salutary effects are due, not only to the induction of bilirubin glucuronyl transferase, but also to an increase in the hepatic intracellular bilirubin "Y" protein (59).

As a rule, the induction of drug-metabolizing enzymes should decrease the pharmacologic effects of a drug by hastening its transformation to an inactive product. However, as previously indicated, some drugs are metabolized into pharmacologically active compounds. The pharmacologic and toxic effects of such compounds may be increased by phenobarbital pretreatment. An example occurs in genetically predisposed persons and drug abusers who convert a small fraction of a dose of phenacetin into the methemoglobin-forming compounds 2-hydroxyphenetidin and its N-hydroxy metabolites (67, 169). Phenobarbital stimulates these reactions, thus increasing the urinary excretion of the abnormal metabolic product, a finding that is related to an increase in methemoglobinemia and Heinz body formation (168). Another example of induced toxicity observed in animals is the increased rate of transformation of bromobenzene to a hepatoxic metabolite produced by phenobarbital pretreatment (23).

Enzyme induction leads to an interesting change in pharmacologic activity of diazepam which normally has a biologic half-life of 54 hours in man. Since the apparent half-life of *N*-desmethyldiazepam, the principal metabolite of diazepam, is longer, 92 hr (86), chronic administration of the parent drug leads to an accumulation of this metabolite (189). A close correspondence has been demonstrated between the serum levels of diazepam and its clinical effects, with 400 ng/ml representing the minimal effective plasma concentration in the steady state (49). On the other hand, a plasma concentration of 300 ng/ml of the desmethyl metabolite causes unwanted side effects. Pretreatment with barbiturates presumably leads to increased N-demethylation and lower concentrations of diazepam with diminution in its therapeutic effects accompanied by an increase in the untoward effects of the desmethyl metabolite.

The metabolic transformation of local anesthetics may be similarly increased by enzyme induction. Heinonen (83) reported a greater than normal rate of disappearance of lidocaine from the plasma of epileptic patients receiving long-term therapy with anticonvulsants. Subsequent studies in dogs (55) showed that phenobarbital stimulates the metabolism of lidocaine, causing faster hepatic removal of the parent compound and increased formation and excretion of its metabolite, glycinexylidide. Furthermore, concentrations of the intermediate metabolite, monoethylglycinexylidide, are markedly decreased after phenobarbital pretreatment. Since this metabolite has marked antiarrhythmic activity in animals, it may well contribute to the overall action of lidocaine.

The importance of the alterations in lidocaine metabolism due to enzyme induction has been recently accentuated by an elegant clinical study designed to evaluate the pharmacologic activity and kinetics of the metabolism of glycinexylidide in patients treated with lidocaine infusion for 24 hr (180). For purposes of comparison, pharmacokinetic analyses were also made in subjects receiving the metabolite alone. The half-life of glycinexylidide was found to be much longer (10 hr) than the reported $1\frac{1}{2}$ hr value for lidocaine. The metabolite has one-fourth the antiarrhythmic activity of the parent drug and adverse effects on mental performance were observed after its administration. This metabolite would seem to play a substantial role in the production of adverse clinical responses to lidocaine. Consequently, with any alteration in its plasma concentration due to the influence of drug interactions, renal or liver disease may have significant side-effect liability, especially in patients being treated with lidocaine infusions for more than 24 hr.

An interesting human study showed that halothane can induce its own biotransformation to sodium trifluoroacetate. When single test doses of halothane were administered to 5 practicing anesthesiologists and to 5 control subjects, their rates of urinary excretion of trifluoroacetate differed markedly. The anesthesiologists as a group excreted more of the metabolite and showed greater intragroup variation than the controls. Repeating the study with the same anesthesiologists after 1 year of additional exposure to the halothane present in low concentrations in the operating room air showed further increases of the trifluoroacetate metabolite, confirming that halothane is a self-inducing agent (36, 39). A recent study in man showed that halothane administration leads to the production of reactive intermediates, as evidenced by the urinary products. Three major metabolites were identified, trifluoro-acetic acid and both the ethanolamine and cysteine conjugates of halothane (42). The presence of these conjugates implies that intermediates are co-valently bound to proteins and phospholipids in the liver during anesthesia. Since halothane stimulates its own metabolism which, in turn, may be induced by barbiturates in man,* accumulation of the reactive intermediates may provide an explanation for the observed hepatoxicity.

Halogenated anesthetics are mostly excreted unchanged via the lungs. The extent of metabolism by hepatic microsomal enzymes in man varies, but is at most 25% (38, 65). The pathways involved are dehalogenation, ether cleavage, and oxidation. Some of this biotransformation may result in toxic inter-mediate or end products. Thus, renal damage after methoxyflurane anesthesia is a reproducible, dose-dependent phenomenon related directly to the levels of inorganic fluoride in the patient's serum (181). The critical serum threshold for fluoride is between 80 and 100 μmoles/liter; when fluoride levels rise higher than 150–200 μmoles/liter, clinical evidence of renal damage is seen (125). Under ordinary circumstances, the small amounts of methoxyflurane used in anesthetic procedures will not produce toxic concentrations of fluo-ride. Larger total amounts of methoxyflurane administered during long oper-ations, or smaller amounts in patients whose enzyme activity has been induced, may result in the production of toxic levels of fluoride (57). Similarly trifluoroethanol, the principal metabolite of fluroxene, has been implicated in toxicity with this agent in mice. Since the metabolism of both methoxyflurane and fluroxene is induced by pretreatment with agents such as phenobarbital, use of these anesthetics for patients receiving chronic drug therapy is poten-tially hazardous, even if they are administered in low concentrations (15, 37, 108, 133).

It is also interesting that chronic exposure to subanesthetic concentrations of halothane is capable of inducing the drug-metabolizing hepatic microsomal enzyme systems. This resulted in rats in enhanced *in vitro* metabolism of aminopyrine (*N*-demethylation) and reduced hexobarbital sleeping time (14).

---

*Reports on the effects of phenobarbital pretreatment on halothane hepatoxicity in rats are conflicting. Stenger and Johnson (177) observed mutiple discrete foci of parenchymal necrosis whereas Brown *et al.* (29) and Davis *et al.* (51) did not observe alteration of the morphologic changes produced by halothane.

Indeed, inhalation anesthetics (except for nitrous oxide and cyclopropane) have been shown capable of stimulating drug metabolism (115). In general, such stimulation resembles the induction seen with phenobarbital, as described above. Methoxyflurane is an exception. It only enhances biotransformation of certain drugs by a nonspecific stimulation which may be related to decreased catabolism of NADPH-cytochrome c reductase (28).

Inhibition of Drug-Metabolizing Enzymes

Any anesthetic agent subject to oxidation by the hepatic microsomal system can in theory compete with another drug for the enzyme. Depending on the tissue affinity and on the concentration of each drug in the liver, inhibition of enzyme activity can also occur. Even ethyl alcohol, with a low affinity for cytochrome P-450 [modified Type II binding spectrum (159)], can slow the metabolism of barbiturates by competitive inhibition because of the extremely high concentration of alcohol which may be reached in the liver (93, 123, 128). The alcohol concentration of the liver may exceed that of barbiturates about 100-fold if the alcohol concentration in the blood reaches a level of 100 mg%. Indeed, any drug that is only effective in high doses is a potential inhibitor of drug-metabolizing enzymes and may interfere with the metabolism of a second drug administered simultaneously. This can lead to an increase in the intensity and duration of the effect of the second drug. The drug-metabolizing activity of hepatic microsomal enzymes may be inhibited by a wide variety of compounds which may or may not themselves be substrates for these enzymes. Valuable information may be obtained by comparing the Michaelis constant for metabolism (K_m) of a drug with its competitive inhibition constant (K_i) for the metabolism of another drug. Finding that the two are identical or very similar indicates that the two substances are metabolized by the same enzyme (8).

Although many drugs may competitively inhibit each other's metabolism *in vitro*, they rarely do so in *in vivo*. This is explained by the fact that, in the therapeutic dose range of most drugs, the concentration of free drug in the body is considerably below that required for half-saturation of the enzyme (K_m). Mutually competitive inhibition between any two drugs becomes negligible when the concentrations of the two are below their K_m values for the mixed-function oxidases. Indeed, if the *in vivo* concentrations of both drugs are as high as their K_m values, their rates of metabolic transformation will only be decreased by one-third. Therefore, competitive inhibitors must have inhibitory constants of the same magnitude as the levels of the inhibitors expected to be present in the body. Most substances known to block drug metabolism *in vivo* require relatively low concentrations to effect 50% decreases in the rates of specific metabolic conversions (K_i values less than 10^{-5} M), or are strongly bound to nonenzymatic sites of the microsomal system (noncompetitive inhibition). Here inhibition can be demonstrated by pretreating the living animal with the suspected inhibitor and then assaying the activity of the drug-metabolizing enzyme in the microsomal systems.

Saturation of drug-metabolizing enzyme systems by inhalation anesthetics, with consequent alteration of the rate of biotransformation of other drugs, has been investigated in animals. One set of *in vitro* studies with rat liver slices preincubated in ethyl ether showed impairment in the dehalogenation of methoxyflurane and halothane. Other studies with rat liver slices showed that, with exposure to fairly high concentrations, inhalation anesthetics inhibit their own metabolism (190). *In vivo* studies by Sawyer and associates (163), utilizing miniature swine with sampling catheters chronically implanted in hepatic vessels, demonstrated saturation of liver drug-metabolizing enzymes with halothane. They found that the higher the concentration of halothane in the liver the lower the metabolism of other compounds. The metabolism of pentobarbital in rats was also shown to be significantly diminished during diethyl ether anesthesia (10). In analogous studies of drug metabolism by rat liver microsomes of a variety of drugs, including hexobarbital, amobarbital, and pentobarbital, inhibition by anesthetic concentrations of halothane and methoxyflurane was demonstrated to occur in a dose-dependent manner (30), implying saturation of the enzyme system, probably by binding to cytochrome P-450.

The mechanism of inhibition of cytochrome P-450 enzymes has not been fully elucidated. It has been tentatively postulated that Type I (see page 181) substances competitively inhibit each other's metabolism and that Type II compounds do likewise, but that Type II substances also inhibit the metabolism of Type I substrates by a noncompetitive mechanism. Experimental data have only partially confirmed these hypotheses. Thus N-demethylation of ethylmorphine, a Type I substrate, is competitively inhibited by a variety of other Type I substances, including hexobarbital, chlorpromazine, and the classic microsomal inhibitor, SKF525-A (160). We have found that methylphenidate, a Type I substance, competitively inhibits the metabolism of imipramine, phenobarbital, diphenylhydantoin, and phenylbutazone, which are also Type I substrates (52). However, no consistent mechanism for the inhibition of the metabolism of Type I compounds by Type II substances has been shown. For example, while nicotinamide, which is a Type II compound, in low concentrations inhibits N-demethylation of ethylmorphine noncompetitively, at high concentrations the inhibition is mixed, i.e., both competitive and noncompetitive. Other Type II inhibitors, such as 2,4-dichloro-6-phenylphenoxyethylamine (DPEA) and SKF26754-A, produce similar effects in the ethylmorphine system (162). These findings suggest that Type II substances inhibit the oxidation of Type I substances not only by slowing the reduction of cytochrome P-450 but also by altering the affinity of cytochrome for a Type I substrate. Since both effects occur at different concentrations of nicotinamide and other Type II inhibitors, they are probably caused by independent mechanisms.

Another interesting aspect of inhibition of drug metabolism is that the inhibitor itself may be metabolized. The relative antagonistic effects of the inhibitor and its metabolite on the metabolism of the anesthetic agent may influence the duration of the inhibitory effect. Thus, certain inhibitors appear to prolong hexobarbital sleeping time less in male rats than in female rats, probably because they are metabolized to inactive products more rapidly in males than in females.

Clinical Implications of Inhibition of Hepatic Microsomal Enzymes

Determination of the extent of inhibition of drug metabolism produced by a particular agent in man is complicated by factors such as genetic differences and exposure to other drugs. In subjects previously exposed to a stimulator of drug metabolism, the effects of the subsequent administration of an inhibitor may not be rapidly discernible. Another complicating factor is the biphasic effect of certain drugs on the metabolism of other drugs. For example, phenylbutazone at first inhibits then stimulates the metabolism of aminopyrine (41). A biphasic interaction has also been reported between phenobarbital and diphenylhydantoin (44). Nevertheless, inhibition of drug metabolism in man and release from stimulation can be clinically more dramatic than the more familiar process of induction. For example, enzyme inhibition can lead to excessive blood and tissue concentrations of barbiturates, anticonvulsants, or hypoglycemic agents and cause severe intoxication. Unquestionably, the whole subject of inhibition of drug metabolism in man as a major mechanism of toxicity merits intensive study. At the present time no estimates are available on the prevalence of this form of drug toxicity.

The findings of Kutt *et al.* (106) and Buchthal *et al.* (31) concerning the relationships among diphenylhydantoin plasma levels, antiepileptic efficacy, and toxicity have facilitated the evaluation of the effects of other drugs on the metabolism and clinical effects of diphenylhydantoin. They established that control of seizures is best achieved at a therapeutic concentration "window" of about 8 to 20 μg/ml of diphenylhydantoin in plasma. Lower levels lead to insufficient control, and higher ones commonly cause toxic manifestations such as nystagmus, ataxia, and lethargy. Interference with diphenylhydantoin metabolism in man by a number of other drugs given concurrently has been reported. Examples include halothane, methylphenidate, disulfiram, isoniazid, phenothiazines, and propoxyphene (105). Usually, the interference has been discovered through observation of signs of diphenylhydantoin toxicity following the addition of the other drugs in patients who had previously been free of adverse effects. Elevated diphenylhydantoin blood levels were detected in every case of inhibition.

The mechanism of inhibition by the antitubercular agent isoniazid is of special interest because it demonstrates how an inherited anomaly of drug metabolizing enzymes can increase drug toxicity. About 10% of a group of patients receiving both diphenylhydantoin and isoniazid developed toxic symptoms. In all of the affected patients, the inactivation of isoniazid through acetylation, a reaction under genetic control, was found to be unusually slow. The resulting higher and persistent isoniazid levels markedly inhibited diphenylhydantoin metabolism (105).

A careful study by Vesell and Passananti (196) has shed additional light on the phenomenon of inhibition of drug metabolism. They determined the half-life of the microsomal marker antipyrine in a longitudinal design before and after treatment with test agents for periods ranging from 4 to 14 days. Disulfiram, prazepam (a long-acting analogue of diazepam), nortriptyline, Δ^9-tetrahydrocannabinol, L-dopa, L-α-methyldopa hydrazine (a decarboxylase inhibitor), and allopurinol all prolonged the plasma half-life of antipyrine without affecting its volume of distribution. Although the confirming experiments have not been done, it is likely that anesthetic agents metabolized by these microsomal enzyme systems would also be inhibited in similar fashion, introducing a potential hazard in patient care. Recent human studies of similar longitudinal design have shown that haloperidol, chlorpromazine, and perphenazine (but not the benzodiazepines) inhibit the hydroxylation of nortriptyline (74–76). Other reported examples of inhibition of drug metabolism are the interactions of the pairs methylphenidate–phenylbutazone (52), diazepam–diphenylhydantoin (18, 187) and probably methylphenidate–barbiturates (139).

The paucity of clinical information on inhibition of metabolism of anesthetic and adjuvant drugs gives us license to suggest some studies and perhaps predict some data. It would be of great importance to determine the nature of the interactions between narcotic analgesics and anesthetics. Most available animal data suggest that morphine, meperidine, and pentazocine function as inhibitors of microsomal systems *in vitro*. If this also occurs under clinical conditions, narcotic analgesics might potentiate the effects of anesthetic agents such as barbiturates, halothane, and methoxyflurane. On the other hand, since the latter may act as inducing agents for the metabolism of analgesic drugs, e.g., the stimulation of methadone metabolism by phenobarbital, they might in turn decrease the potency of the analgesics. Such reciprocal interactions could have clinical significance. It is thus important for the clinician to be aware of the possible effects of both enzyme induction and inhibition on the intensity and duration of the actions of anesthetic and adjuvant drugs.

Influence of Biologic Factors on Hepatic Microsomal Metabolism

Alterations and variations in anesthetic drug metabolism are of considerable importance from the perspective of therapeutics. Very often, as noted above, data on the metabolism of a drug obtained from one or more animal species may not be directly applicable to man. For example, the half-life of hexobarbital is 19 min in the mouse, but 360 in man; meperidine is metabolized at the rate of 20%/hr in man, but in the dog 90% disappears after 1 hour. In other cases, species differences in metabolic pathways make extrapolation to man totally invalid. Of even greater importance are variations in the rate of metabolism of an agent among different individuals. Nearly 30 years ago, Brodie and associates (26, 32) described large variations among otherwise drug-free human subjects in the clearance of the same dose of a drug from blood, e.g., three- and tenfold variations in plasma half-lives of phenylbutazone and ethyl biscoumacetate, respectively. Since then, numerous studies have documented wide variability in plasma concentrations of drugs after the administration of the same dose, especially with psychoactive agents. For example, after oral ingestion of 25 mg desmethylimipramine t.i.d., steady-state plasma levels ranged from 8 to 295 ng/ml (172).

Thus far in this review we have endeavored to show how alterations of hepatic microsomal enzyme activity may cause significant changes in drug metabolism and in clinical response to therapeutic agents. However, it is important to note that much of this variability may result from the body's handling of the drug, rather than merely the drug's action on the body. Let us now examine the relative contribution of some physiologic factors to the large individual differences in *uninduced* rates of drug elimination.

Pharmacogenetics. The environmental and hereditary components of individual variations in rates of drug metabolism have been investigated in otherwise nonmedicated healthy human twins (193). This method compares age- and sex-matched individuals and depends on the fact that, in identical twins, all of the genes are identical, and, in fraternal twins, only half. If the metabolism of a given drug in identical twins exposed to different environmental conditions is nearly identical, but large differences exist in the metabolic transformation of the same drug between fraternal twins, then it may be assumed that the metabolism of the drug is controlled by genetic factors. This variation can be mathematically expressed by the hereditary factor H whose value can vary from 0.0 to 1.0. An H factor of 1.0 means virtually complete genetic control whereas 0.0 indicates negligible genetic and complete environmental control of the metabolism of the drug in question (192). An approximate estimate of H can be made by determining the mean variance within the sets of identical (I) and fraternal (F) twins according to the following formula:

$$H = \frac{\text{(variance within pairs of fraternal twins)} - \text{(variance within pairs of identical twins)}}{\text{(variance within pairs of fraternal twins)}}$$

Variance within pairs is calculated as follows:

$$\frac{\Sigma(\text{difference between twins})^2}{2n}$$

Theoretically, solely on the basis of genetic control, fraternal twins, having in common approximately half of their total number of genes, should have an intraclass correlation coefficient of 0.5, whereas the value for identical twins should be 1.0.

The main assumption in all twin studies is that the environment remains the same for all twins. Single-dose studies clearly indicate that for the microsomal metabolism markers, phenylbutazone, antipyrine, bishydroxycoumarin, ethanol, intratwin differences in half-lives are appreciably greater between fraternal than identical twins. Thus, the intraclass correlation coefficients range from 0.33 to 0.66 with fraternal twins, but from 0.82 to 0.85 with identical twins. Since the H values range from 0.97 to 0.99, virtually complete hereditary control is indicated; therefore, it is apparent that the major mechanisms responsible for the transformation of these drugs are genetically rather than environmentally determined (194). This in turn implies that, in the absence of secondary modifying factors, drug metabolizing capacity is constant in each person. Since the number of available metabolic pathways is limited it seems reasonable to expect that persons metabolizing one drug at a fast rate would also be fast metabolizers of other drugs using the same pathway. If such a relationship in the metabolism of different drugs in the same person is indeed valid, it would be possible to predict from the effects of one to those of a second drug in the same patient. Recent findings seem to support this hypothesis. Thus, the metabolic rates of the hypnotics glutethimide and amobarbital in ten individuals (not twins) yielded a correlation coefficient of $r = 0.69$ and a P value of < 0.05, even though the range of serum half-lives was 8.1 to 17.9 hr with glutethimide and 13.7 to 42.2 hr with amobarbital (95). Similarly, good correlation was found between steady-state plasma concentrations of imipramine and the half-lives of phenylbutazone in 29 hospitalized patients (141). Good correlation has also been obtained among the rates of metabolism of nortriptyline, desmethylimipramine, and oxyphenylbutazone, although the main pathway of biotransformation of the first two is hydroxylation and that of the third is glucuronidation (81).

It should be realized, however, that the relative importance of genetic and experimental factors in the large individual differences observed in rates of drug clearance from plasma varies with the drug under study. At present, it is still necessary to determine independently for each therapeutic agent the relative contributions of the above cited factors. Later on, when enough data have been accumulated for several commonly used anesthetics, the correlation approach may well become applicable.

The biotransformation of halothane in volunteer twins is illustrative of these considerations. After a single intravenous injection of 3.4 mg of radioactive halothane, an almost fourfold variation in its metabolism between twins was encountered (38). These large individual differences were found to be controlled predominantly by genetic factors. The authors' lower normalized H value of 0.88, however, indicated that environmental factors are relatively more important determinants of the rate of metabolic transformation of halothane than that of some previously discussed compounds (38). The intraclass correlation coefficient was 0.52 for identical and 0.36 for fraternal twins.

In similar studies with nonmedicated, nonhospitalized twins in whom steady-state plasma levels of nortriptyline were produced, the metabolic transformation rate of this compound varied over a tenfold range for the whole group. In identical twins within the group the H value was 0.98, indicating an almost complete genetic control of nortriptyline metabolism (2).

In summary, the limited number of studies to date indicate that large differences in rates of drug metabolism among healthy, nonmedicated volunteers are primarily controlled by genetic factors. The mode of inheritance seems likely to be polygenic, i.e., controlled by unknown numbers of allelic genes at unknown numbers of loci.

The genetic control of drug metabolism can be observed even after stimulation of the microsomal systems. Thus, the plasma half-life of antipyrine, used as a microsomal marker in twins pretreated with phenobarbital for 2 weeks, showed a marked but variable decrease in all cases. Here again, intratwin differences were small between identical twins, but appreciable between fraternal twins (195). These results suggest that individual differences in response to induction may also be predominantly under genetic control. In the case of nortriptyline, induction by administration of barbiturates for several months caused the expected decrease in steady-state levels. Although after induction intrapair differences were accentuated in all the twins, they were more pronounced in fraternal twins. It was concluded from these studies with nortriptyline that although variability in steady-state plasma levels in nonmedicated persons is genetically controlled, the decrease of these levels caused by exposure to hypnotics was due to both genetic and environmental factors (2).

Other hereditary variations in hepatic microsomal metabolism concern the O-dealkylation of phenacetin and the hydroxylation of dipheylhydantoin. As previously mentioned the main pathway of phenacetin metabolism in man is O-deethylation to form N-acetyl-p-aminophenol, followed by conjugation with glucuronic acid or sulfate. Some individuals are genetically deficient in dealkylating microsomal enzymes, which forces the metabolism of phenacetin into alternative routes. In such individuals a substantial portion of the drug is deacetylated and hydroxylated to form 2-hydroxyphenetidine, a substance causing hemolysis of red blood cells both *in vitro* and methemoglobinemia *in vivo*. As much as 50% of the hemoglobin may be converted to methemoglobin; hemolysis may be sufficiently extensive to require blood transfusions (169). Methemoglobinemia also occurs in some persons after administration of antimalarials, sulfonates, or nitrites. Here, in addition to differences in hepatic microsomal enzymes, the activity of erythrocyte glucose-6-phosphate dehydrogenase is also reduced.

The hydroxylation of diphenylhydantoin is also genetically controlled, the distribution in the general population being polymodal. In each of several families, some nonmedicated members showed limited metabolizing capacity, being unable to metabolize more than 1 to 3 mg/kg of diphenylhydantoin per day (107). Conversely, other members of these families and also some patients demonstrated unusually rapid metabolism, which was probably also genetically determined.

Although most studies on the distribution of responses to the same dose of a drug administered to many individuals yield unimodal Gaussian curves, the contemporary interpretation is that the responses are regulated by genes at multiple genetic loci. Variations in polygenically controlled traits, such as height, weight, shape of body structures, and levels of some proteins and of most drugs, are generally quantitative. They are considered to be under polygenic control because each of the several genetic loci contributes a given quantity to the total trait or character. The measurable amount of the trait or character is the arithmetic sum of all the variable contributions from each gene at each of the various loci. The mode of inheritance of polygenically controlled traits differs from the classic borderline pattern in that offspring exhibit a value for the trait midway between the values of their parents. It now appears that rates of drug metabolism are also polygenically determined. The proteins controlling these different rates are situated in the mixed-function oxidase system of hepatic microsomes. Thus, modern genetic theory interprets the unimodal distribution curves of response to drugs and of their half-lives, not as manifestations of genetic uniformity, but rather as signs of genetic diversity for polygenically controlled traits.

Sex Differences. Numerous investigators have demonstrated that the actions of a variety of drugs are more pronounced and longer lasting in female than in male rats. For example, Quinn, Axelrod, and Brodie (144, 145) showed that female rats are more susceptible than males to the action of hexobarbital and other drugs. Female rats given hexobarbital slept about 4 times as long as the males and a comparison of the plasma levels showed that this barbiturate is metabolized more rapidly in males than in females. Correspondingly, the hexobarbital metabolizing enzyme system of males is considerably more active *in vitro* than that of females. Again, in female rats masculinized by testosterone therapy, hexobarbital sleep time became shorter and microsomal enzyme activity greater, resembling the findings in males. In contrast, in male rats feminized by estrogen therapy, the reverse was true (98). The disconcerting feature about sex differences in drug metabolism is that they have been unequivocally found *only* in rats (152). Other animals such as mice, rabbits, and guinea pigs do not show similar sex specificity.

Only a few investigations have been carried out on the possible effects of sex on drug metabolism in man. To date no significant sex differences have been found in humans, though a recent study with antipyrine in healthy, age-matched volunteers (34 males, mean age 27.4 years; 27 females, mean age 25.3 years) from similar working environments showed the mean half-life of antipyrine to be 30% longer in males than in females (134). The difference was statistically significant, suggesting a slightly lower rate of metabolism of this compound in males (134) (see below). On the other hand, Whittaker and Price Evans (202) found no sex difference in the rate of phenylbutazone metabolism in a large group of normal subjects. Since drug metabolism is known to be under genetic control, and is also influenced by drugs and environmental factors, it is possible that significant sex differences can only be observed in studies of homogeneous population samples.

Since much of the pharmacologic screening of anesthetic agents is carried out in rats, the results and theories on the influence of sex on the rates of drug metabolism in this species will be described briefly.

To begin with, the sex-related differences demonstrated in hepatic metabolism of barbiturates have also been observed with other drugs. Thus, liver microsomes of male rats N-demethylate some narcotics such as morphine, methadone, and meperidine more rapidly than do those of female rats (98). The activity of hepatic microsomal enzymes is low in newborn and immature rats; clear sex differences are not observed until 30 days of age. By 50 days of age, enzyme activity has become significantly higher in male than in female rats. The appearance of these differences coincides with the beginning of sex maturation and it persists in normal rats at least up to 600 days of age. Seemingly, male rats produce an endogenous inducing agent, probably a steroid formed during the metabolism of adrenal or sex hormones, by a reaction unique to the rat. Some of the metabolic pathways are androgen-dependent, for example, oxidative capacity is reduced by castration and restored by androgen treatment. In turn, the effect of androgens is antagonized by estrogens. The hydroxylation of barbiturates and N-demethylation of aminopyrine are androgen-dependent, whereas aniline hydroxylation is not (77).

The hypothetic endogenous inducer in male rats has very different properties from those of the well-known inducing agents previously described. It is able to increase the oxidation of Type I but not that of Type II substances (165). The small effects on the oxidation of Type II compounds correspond to a moderate increase in cytochrome P-450. On the other hand, the sex-dependent difference in the oxidation rate of Type I compounds due to this postulated inducer cannot be accounted for by increases in either the absolute amount of cytochrome P-450 or the relative activity of the cytochrome P-450 reductase already present (50). It is probably dependent on the much stronger binding of Type I compounds to cytochrome P-450 in microsomes from male rats. This is demonstrated by the finding that hexobarbital causes about 3 times greater spectral change when added to microsomes of male, rather than female rats (152). Recent work by Holtzman and Rumack (89) indicates that there may be as many as four activation sites in male and three in female microsomes capable of binding ethylmorphine, a Type I substrate, in some fashion during drug metabolism.

On the basis of this information some tentative guidelines may be postulated to extend to humans the significance of sex differences observed in animals. Clear sex differences in potency and/or duration of action of an anesthetic observed in rats but not in mice (or any other species) could be attributed to differences in its metabolism peculiar to the rat, and hence unlikely to occur in man. On the other hand, if these attributes of a drug showed clear sex differences in both rats and mice, then the differences might be related not only to its metabolism but also to other factors, such as absorption, volume of distribution, tissue and/or receptor sensitivity, and even anatomic differences. With such a drug, sex differences could also be expected to occur in man. It is of interest that a recent study with the antibiotic cephradine in humans clearly demonstrated slower absorption and lesser bioavailability in females than in males after intramuscular injection, probably due to sex differences in vascularity and whole body fat distribution.

Age. It has long been apparent that the human fetus and the newborn infant are much more sensitive than adults to many drugs. Most anesthetic and adjuvant agents can easily cross the placental barrier. For instance, barbiturates or morphine can accumulate in the infant's tissues and cause respiratory depression. The explanation for the sensitivity of infants to depressant and other drugs has emerged in recent years from a number of studies of the maturation of the capacity to oxidize and conjugate drugs. The reports emphasize that the mammalian fetus and newborn possess a limited capability to metabolize drugs, which increases after birth at a rate that varies with the species, the type of reaction, and the drug. The initially low rates of drug metabolism result in increased intensity and duration of drug action. For example, newborn mice treated with 10 mg/kg of hexobarbital slept more than 6 hours, where adult mice given 100 mg/kg slept less than 1 hr.

Because the activity of the cytochrome P-450 enzyme systems in immature laboratory animals is low, it was assumed for many years that human fetuses would also metabolize drugs very slowly. However, liver microsomes from human fetuses perform reasonably well in metabolizing such substances as testosterone, laurate, ethylmorphine, aminopyrine, aniline, hexobarbital, and desmethylimipramine (71). In man the rates of these reactions in relation to microsomal protein content are about 30–50% lower in fetal than in adult liver microsomes (138). It is generally accepted, however, that the liver represents about 4–5% of body weight in the fetus but only about 2% in the adult. Thus, in relation to body weight, the activity of these enzymes in the fetus may actually approach that of adults.

Concomitant with lower drug-metabolizing capacity, the quantity of cytochrome P-450 and the activity of NADPH-cytochrome c reductase are also lower in human fetal than in adult liver. Fetal liver microsomes contain 27% and 49% and liver homogenates 29% and 30% of cytochrome P-450 and NADPH-cytochrome c reductase, respectively, of the amounts found in adult livers. (It must be noted that these are average values, with considerable variability in all parameters.) In contrast, however, the activity of 3,4-benzpyrene hydroxylase in human fetuses is only 2.5% of that of adults. A possible explanation may be the lack of the polycyclic aromatic hydrocarbon-specific form of microsomal hemoprotein cytochrome P-448 in the fetus (138).

Ultrastructural studies of human fetal liver have shown that a differentiated endoplasmic reticulum is already present during the first half of the gestational period (204). In the seventh to ninth weeks, the endoplasmic reticulum has the appearance of a tubular system with ribosome-studded membranes. Around the third month, however, there seems to be an increase in agranular endoplasmic reticulum membranes and deposition of glycogen and iron in the hepatocytes. These changes may imply a function (drug metabolic) differentiation at this stage of development.

Although the ability of human liver to metabolize foreign substances *in vitro* is lower in the fetus than in the adult, additional factors must be considered regarding drug disposition and elimination in the fetus. In general, the transfer of metabolites of a drug across the placenta is slower than that of the parent compound. For this reason, drug metabolism in the fetus may lead to an accumulation of the metabolites on the fetal side, since the only significant route of their elimination from the fetus is via the placenta.

If the metabolites are inactive, the accumulation is unimportant; but if they are either active or toxic (or both), the consequences could be serious. Another factor is that fetal liver preparations, unlike those of adults, do not catalyze the glucuronidation of hydroxylated substrates to any significant extent. In fact, Dutton (55a) found that the glucuronidation of O-aminophenol occurs at a more rapid rate in the human fetal kidney than in the fetal liver. These results suggest that in the human fetal liver the cytochrome P-450 system develops before the UDPglucuronyl transferase system so that, in the fetus, Phase 1 reactions are more important than Phase 2 reactions.

Unfortunately, systematic studies of the age dependence of hepatic drug-metabolizing enzyme activity and its pharmacodynamic applications in man are lacking. There are scattered reports that children, 1 year and older, metabolize drugs faster than adults. Diazoxide (Hyperstat) has recently been introduced into clinical practice as an intravenous antihypertensive agent, used primarily to terminate acute bouts of hypertension in adults. On an investigational basis it has also been administered orally to children for the treatment of hypoglycemia. The half-life of diazoxide in children was found to range from 9.5 to 24 hr, compared to 20 to 72 hr in adults (Table 9-2) (53), while conversely, the concentration of unbound diazoxide in plasma was

Table 9-2. Effect of interspecies differences and aging on the metabolism of diazoxide[a]

Species	Half-life[b] (hr)	Binding[c] to		Extent of metabolism (%)
		Plasma (%)	Albumin (%)	
Adult man	23–36	90–93	90–93	54–60
	21–35			
	22–31		90	
	48–72			
	20–53			
		90–90		
Children	9.5–24	85–90		
Monkey	17–23			10–70
Dog	2–3.5	55	68	30–35
Rabbit	1.1–2.4	90–99		4
Rat	1–2	75		2
Guinea pig				2

[a]From Dayton et al. (53).

[b]Half-life was determined during anesthesia in rats, rabbits, and dogs.

[c]Binding to human serum varies from 70 to 90%, depending on concentration of diazoxide.

higher in children than in adults. The latter finding presumably leads to more rapid onset of action in children. Since one of the metabolites is anti-hypoglycemic but not antihypertensive, an explanation may now be provided for the differences observed in the pharmacologic effects of the compound between children and adults. The antihypoglycemic effect of diazoxide in children is increased, not only because of its higher free plasma level, but also

Table 9-3. Pharmacokinetic analysis of neuromuscular blocking effect of succinylcholine in nine infants[a]

Patient	Age (days)	Duration (min)	Rate of decline (% min^{-1})
11	28	5	40
10	24	6	40
14	37	6	40
15	38	6	40
7	10	7	20
4	3	9	27
2	1	12	27
5	10	14	16
8	20	26	9

[a]From Levy (112).

because of its more rapid biotransformation to the antihypoglycemic metabolite.

An excellent correlation between plasma levels and duration of action of succinylcholine in infants was demonstrated by Levy (112) (Table 9-3). The unusually protracted effect of succinylcholine in infants 5 and 8 (Table 9-3) is due to its slow rate of decline in these two.

A recent study by Alvares and associates (4) has yielded much information on the differences in drug-metabolizing capacity between adults and children (1 to 8 years of age). Two test drugs, antipyrine and phenylbutazone, were used to type the rates of metabolism of the two groups. The model drugs are metabolized primarily by the hepatic cytochrome *P*-450-dependent mixed function oxidase system. As shown in Figure 9-2, children in the age range of 1 to 8 years metabolized the drugs almost twice as fast as adults. The mean half-life of antipyrine in children and adults, respectively, was 6.6 and 13.6 hr, and that of phenylbutazone, 1.7 and 3.2 days.

A cytochrome *P*-450 system of greater activity in children than adults has considerable significance in the calculation of proper dosages of drugs. Since renal excretion of drugs in children is qualitatively and quantitatively similar to that in adults, differences in renal elimination cannot account for the faster decline of plasma levels of drugs in children. The explanation may lie in the greater liver/body weight ratio in children than in adults. The relative weight of the liver of a 2-year-old child is approximately 40 to 50% greater than that of the adult, and of a 6-year-old child, 30% greater (3). In contrast, the

Figure 9-2. Differences in antipyrine and phenylbutazone in children and adults. Horizontal bar represents mean value for 10 subjects (4).

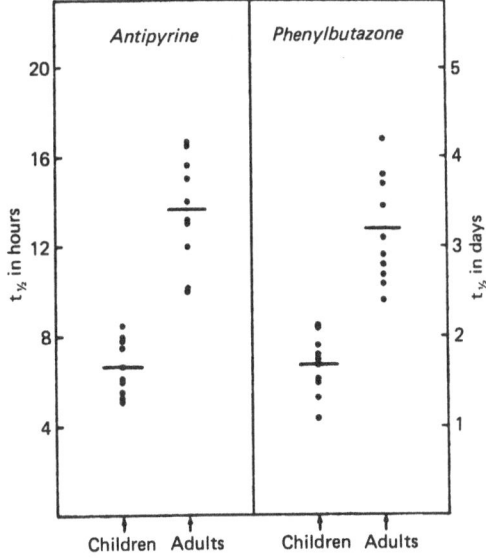

adipose compartment in children is relatively smaller than in adults. This reduces the volumes of distribution of lipophilic drugs, resulting in higher plasma and tissue concentrations, especially in organs such as the brain, heart, and liver, which have excellent blood supply. Clearly, then, the administration of the same mg/kg doses should result in higher levels, speedier onset of action, faster metabolism, and shorter duration of therapeutic activity in children than in adults. In the light of these assumptions, let us now examine the fate of several typical drugs in some detail.

Both the oxidation and the conjugation of acetanilide were shown to be slower in early life than later (197). When N-acetyl-p-aminophenol (acetaminophen), a metabolite of acetanilide, was measured in the plasma of newborn infants and older children after oral administration of the parent drug, peak plasma concentrations of both the metabolite and its glucuronide were encountered later in newborn infants than in older children (197). A more recent study (129) found no age-related differences in the overall rate of elimination by neonates, children, or adults, but major age-related differences in glucuronic acid and sulfate conjugation. Acetaminophen elimination by adults is primarily via conjugation with glucuronic acid. This "adult" pattern was evident at 12 years of age but, in children up to the age of 9 years and in newborn infants, metabolism was mainly by sulfate conjugation.

Sulfisoxazole, phenobarbital, clindamycin, sulfamethoxine, nalidixic acid, aminopyrine, imipramine, salicylate, mepivacaine, amobarbital, and diazepam all have age-dependent rates of elimination, i.e., all show a progressive increase with age in their rates of disappearance from plasma. The age at which the plasma half-life approaches that of adult subjects varies with different drugs. On the other hand, the rate of elimination of some drugs, e.g., diphenylhydantoin (146), shows no apparent age-dependency.

Garattini (63) measured blood levels of diazepam and its metabolites 7 to 81 days after birth in premature infants, within 48 hr after delivery in full-term neonates, in children 3 to 6 years of age, and in adults. Full-term newborns metabolize diazepam at a very low rate; only after 8 to 10 days of age do they excrete hydroxylated metabolites in the urine. The apparent plasma half-life of diazepam was longer in full-term neonates than in older children and adults, but shorter than in premature infants.

A comparison of the metabolism of mepivacaine in human neonates (first 30 hr of life) and in adults showed selective deficiencies of metabolism in the former (127): aromatic hydroxylation did not occur and they excreted extremely small quantities of the phenolic metabolites. In contrast, adults did excrete considerable quantities of these metabolites. (On the other hand neonates produced appreciable amounts of 2′,6′-pipecoloxylidide, a metabolite resulting from N-demethylation.)

Studies of drug metabolism in geriatric subjects are scanty. In the older patient, the metabolism and excretion of drugs in general may be slowed, requiring reduction in dosage to avoid toxicity. The half-lives of antipyrine and phenylbutazone were compared in elderly patients of both sexes (134, 178), with a mean age of 77.6 years, and in healthy volunteers with a mean age of 26.0 years. The geriatric patients were all ambulatory but had variable degrees of cerebrovascular and other pathologic disorders. Only those patients whose serum alkaline phosphatase, plasma proteins, bilirubin, hemoglobin, and blood urea were all within normal limits were included in the study. The mean plasma half-life of antipyrine was 17.4 ± 6.8 hr in the geriatric subjects and 12.0 ± 3.5 hr in the young controls. The half-life of phenylbutazone was 29% longer in the geriatric patients. The difference in antipyrine metabolism between the two groups was statistically significant ($P < 0.01$, Wilcoxon Rank sum test), but for phenylbutazone it was only suggestive. The volume of distribution of antipyrine was similar in the two groups but with phenylbutazone the mean value was 21% greater in the elderly subjects than in the control group. [The differences were statistically significant (t test) at the $P < 0.01$ level.] The clinical significance of these findings is not clear, since drug clearance was not determined in either age group. It is possible that partial to complete compensation for the increase in drug half-life seen in the elderly subjects might result from an increased volume of distribution (184).

Two groups of clinically healthy adult male subjects participated in a study to evaluate the effect of increasing age on the hydroxylation of amobarbital (94). One group of eight subjects ranged in age from 20 to 40 years, while the other group of eight was over the age of 65 years; in all cases the creatinine

Table 9-4.   Total urinary excretion of 3'-hydroxyamobarbital in elderly and young groups of subjects[a, b]

Percentage of oral dose	
Elderly group	Young group
0.6	9.9
2.2	10.3
3.2	12.4
3.9	13.3
4.0	13.3
5.2	17.2
6.7	18.6
8.4	19.2
Mean = 4.3%	Mean = 14.2%

[a]From Irvine et al. (94).
[b]Twenty-four hr urinary excretion after oral administration 200 mg sodium amobarbital.

clearance was greater than 70 ml/min. Each subject received a single oral dose (200 mg) of sodium amobarbital. Plasma concentrations of amobarbital and urinary excretion of 3'-hydroxyamobarbital were determined. The geriatric subjects showed significantly higher plasma levels of the parent compound at 24 hr and much lower urinary output of the metabolite, indicating a reduced rate of hydroxylation of amobarbital in older subjects (Table 9-4).

Plasma levels of the β-adrenoceptor blocking agents, propranolol and practolol, were also compared in groups of young and elderly subjects (40). Propranolol is eliminated mainly via hepatic metabolism whereas the route of elimination of practolol is renal. The mean plasma drug concentrations were substantially greater after propranolol administration in the elderly subjects than in the young, but with practolol did not differ significantly between the two groups for the first 2 hours after administration.

Differences in absorption or plasma protein binding between young and elderly groups may influence plasma concentrations of drugs. Direct comparisons of half-life or steady-state plasma concentrations between such individuals may be misleading; estimations of clearance* (which takes account of the volume of distribution) then provide a better method of comparison. Thus, for diphenylhydantoin (82), binding decreased with increasing age from a mean of 9.9% unbound at 17 years to a mean of 12.7% unbound at 53 years. If due allowance is made for this increase in "free" drug, which is probably due to lower plasma albumin concentration in the older group, then no apparent change in hepatic diphenylhydantoin metabolism with age is discernible.

In normal individuals, the plasma half-life of diazepam exhibited a striking age dependence (101): at 20 years the half-life was about 20 hr but it increased with age to about 90 hr at 80 years. Despite this profound decrease of plasma half-life of diazepam with age, the plasma clearance showed no significant age dependence because of a simultaneous increase in the initial volume of distribution, which also increased linearly with age. The constancy of clearance indicates no greater accumulation of the drug in the old than in the young, making unnecessary any modification of chronic dosage based on pharmacokinetic considerations.

Pathologic and Environmental Factors. The most important pathologic factors influencing the metabolism and elimination of drugs are liver and kidney diseases. The activity of the microsomal oxidases may be decreased in acute hepatitis if the accompanying cell damage is comparable to that seen

---

*Relative clearance refers to the rate at which an apparent volume of distribution (liters) is cleared of the drug, per unit time (hr), and per kg of body weight. It is calculated as follows: Relative clearance = elimination rate constant × volume of distribution; the elimination rate constant, in turn, is equal to 0.693/half-life of the terminal slope of a plasma level disappearance curve. The magnitude of the volume of distribution depends on factors such as "free" or unbound drug concentration and relative lipid solubility.

after administration of small doses of carbon tetrachloride (151). This compound is metabolized by the microsomal enzyme systems into a free radical fragment which reacts predominantly with the phospholipid moieties of cytochrome P-450 and glucose-6-phosphatase, but leaves the protein portion of the enzyme intact (161, 176). Lipid peroxidation is another reaction leading to the formation of free radicals which react specifically with phospholipids (175). A recent study in which ^{14}C-halothane was administered to eight individuals proved conclusively that this compound forms covalent bonds with normal lipid constituents of hepatic cell membranes (42). Among the nonvolatile urinary metabolites found were trifluoroacetic acid, the N-acetyl cysteine conjugate of the dehydrofluorination product of halothane, 1-bromo-1-chloro-2, 2-difluoroethylene, and N-trifluoroacetyl-2-aminoethanol. The two latter metabolites are breakdown products, respectively, of the initial conjugation product of halothane with glutathione and that of trifluoroacetic acid with phosphatidylethanolamine, a normal lipid constituent of cell membranes. All these reactions involve the hydroxylating system. The higher the induction of this system by agents such as phenobarbital, the greater the possibility of production of deleterious free radicals, with subsequent liver damage. A similar mechanism may account for liver injury produced by a variety of drugs such as phenothiazines, tetracyclines, and sulfonamides.

Since nearly all sedatives, hypnotics, tranquilizers, and other psychoactive drugs are converted to inactive metabolites in the liver, one might expect their action to be intensified and prolonged in patients with liver disease. This is not necessarily true. Animal studies indicate that impairment of drug-metabolizing activity is not reliably reproducible until at least 90% of the liver substance has been removed by hepatectomy (170). The clinical impression that patients with liver disease tolerate drugs less well than normal persons does not necessarily prove impairment of drug metabolism. Such patients tend to be debilitated, hence unduly sensitive to the effects of depressant drugs because of their poor general condition rather than the local liver disease itself. The few reports published do not indicate that drug metabolism is generally impaired in the presence of hepatocellular pathology. Indeed, drug oxidizing capacity is only rarely decreased in patients with liver cirrhosis. On the other hand, several patients with mild forms of hepatitis metabolized drugs more slowly than would be expected.

These findings can be explained in part by previous drug administration. Thus, the half-life of phenylbutazone was found to be normal in cirrhotics who had previously received barbiturates leading to induction, but prolonged in those who had not (110). The nearly normal rates of drug metabolism observed in the medicated cirrhotic patients were due to the induction of microsomal oxidase activity in the parts of the liver still functioning. This may also occur in some types of hepatitis. It is known that the smooth endoplasmic reticulum proliferates during inflammatory processes of the liver. As long as the membranes in one part of the liver remain hyperactive, their enhanced function may compensate for the decreased metabolism

occurring in other parts of the liver where the progression of the inflammatory process has converted the hypertrophic endoplasmic reticulum into a
hypoactive form.

An alternate interpretation may be based on the concept of classifying
drugs according to their hepatic clearances (133a). This physiologic concept
of drug elimination (202b) has been defined as the volume of blood from
which drug is completely removed per unit time, and is equal to the product of
hepatic blood flow and the hepatic extraction. "High clearance drugs" are
those undergoing greater than 70% extraction by the liver, so that they exhibit
intrinsic hepatic clearances greater than liver blood flow. Since the term
"intrinsic clearance" characterizes the ability of the liver to remove drug
irreversibily, for high clearance drugs blood flow is a prime determinant of
disposition, while changes in enzymatic activity will be less important. In this
category are the drugs such as propranolol, lidocaine, meperidine, propoxyphene, and tricyclic antidepressants. For drugs whose intrinsic hepatic
clearance is low, the clearance is quite independent of blood flow, being
instead determined by microsomal enzymatic activity. Examples of low
clearance drugs include diazepam, diphenylhydantoin, coumadin, and phenylbutazone. Thus, the effects of liver disease on drug elimination are selective,
due to defects in either liver function or hepatic circulation or both.

Drugs of high intrinsic clearance such as lidocaine (182), meperidine (101a,
202a), and propranolol (17a) show reduced clearance in patients with chronic
liver disease due to an accompanying decrease in total hepatic blood flow
and/or shunting of blood, whether intrahepatic or extrahepatic. Drugs with
low intrinsic clearance such as antipyrine (17a) may also be removed less
efficiently in chronic liver disease due to changes in the metabolic capacity of
the diseased organ.

Surgical anesthesia tends to decrease hepatic blood flow (46a). For example, cyclopropane (143a) increases sympathetic tone and decreases hepatic
blood flow by one-third, halothane (55b) causes a comparable reduction
associated with its hypotensive potential, and methoxyflurane (143b) causes
an even larger decrease in splanchnic blood flow. Similarly, hypotension due
to high levels of spinal anesthesia can also result in reduced splanchnic blood
flow—the higher the level of spinal block, the more marked are the effects
(46a). There is little information on whether morphine, thiopental, pentobarbital, or nitrous oxide affect splanchnic blood flow. However, if hypercarbia
occurs almost all anesthetics will decrease hepatic blood flow. In any of these
situations, it is obvious that the clearance of accompanying drugs with high
intrinsic clearance such as propranolol, meperidine, and lidocaine could be
reduced, leading to their accumulation to potentially toxic levels.

The already discussed drug clearance observed in diseases which markedly
alter the circulation of the liver can be further compromised by severe
circulatory disorders (13a). Both hepatic flow and metabolism are affected by
cardiac insufficiency. Blood flow to the liver is reduced in proportion to the

reduction in cardiac index (177a). Cardiac disease can influence the metabolic capacity of the liver either by hepatocellular damage resulting from hypoperfusion or by hypoxia with impaired microsomal drug action (48a). As above, substances with high intrinsic clearance such as lidocaine, aldosterone, and indocyanine green have their total clearance reduced as hepatic blood flow is reduced in congestive heart failure or circulatory shock (13a, 182). This is the basis for the greater toxicity of lidocaine in usual doses in patients with congestive heart disease. In addition, Tokola and associates (183a) have demonstrated reduced hepatic drug metabolizing activity in biopsy samples from patients with chronic cardiac insufficiency.

The influence of hepatic disease on drug metabolism and therapeutic effect can be illustrated by comparing the effects of prednisolone and prednisone in the presence of chronic hepatitis (46). Better therapeutic effects were obtained in these patients with prednisolone than with prednisone, a finding accountable as follows: In normal subjects the administration of either prednisone or prednisolone in equal amounts results in similar plasma levels of prednisolone, the biologically active form of the steroid. In patients with moderate to severe hepatic damage, however, the biotransformation of prednisone to prednisolone and the subsequent reduction and conjugation of prednisolone are all impaired. Lower plasma levels of prednisolone result from the impaired conversion of prednisone, but once present in the plasma, prednisolone concentrations will persist, since its further conversion is also impaired. Consequently, much higher levels of prednisolone are reached after prednisolone administration than after comparable doses of prednisone.

Although the kinetics of diphenylhydantoin metabolism were unchanged during acute viral hepatitis, the percentage of unbound drug in the plasma was found to be increased by almost one-third. The same was observed with another highly protein-bound and slowly cleared drug tolbutamide (84, 202c) which has been recently demonstrated to be responsible for the apparent increase of clearance, based on total concentration, in patients with acute viral hepatitis. The higher plasma levels of unbound drug may explain the diphenylhydantoin toxicity observed in patients with hepatitis in whom the total plasma concentrations were below or within the usual range (16).

The metabolism of lidocaine was also studied in patients with liver disease (183). This drug has been shown to undergo considerable "first-pass " effects; more than 60% of the lidocaine transiting the liver at any time is metabolized (182). In patients with liver disease the volume of distribution was found normal, but the plasma half-life was increased and the clearance correspondingly reduced. After a change in the rate of drug infusion into normal individuals 6 to 10 hr were required to reach a new steady-state plasma concentration; in subjects with liver disease, the time required was much longer. Similar data were obtained with diazepam in patients with cirrhosis, acute and chronic active hepatitis (Table 9-5) (101).

The rate of elimination of highly lipid-soluble drugs is usually not reduced

Table 9-5. Correlation between hepatic function tests and $t_{1/2}(\beta)$ of diazepam in the same patients at time of acute viral hepatitis and at recovery[a]

Subject	$t_{1/2}(\beta)$ (hr)	Bilirubin (mg/100 ml)	LDH (IU/ml)	SGOT (Karmen U/ml)	Alkaline phosphatase (IU/ml)	Albumin (g/100 ml)
B.S.[b]	70.8	5.3	250	1,100	123	3.9
B.S.[c]	47.4	0.9	169	43	79	5.0
M.C.[b]	94.2	15.4	525	1,290	97	4.5
M.C.[c]	62.2	1.0	150	30	42	4.2
J.H.[b]	59.6	7.3	407	1,620	91	4.3
J.H.[c]	57.2	0.8	133	30	37	5.6
J.H.[d]	36.0	0.3	167	12	40	5.6
J.M.[b]	64.8	3.8	330	1,020	85	4.0
J.M.[c]	53.3	1.1	138	25	65	4.8
J.J.[b]	89.3	13.3	n.a.[e]	2,350	68	2.6
J.J.[c]	49.0	1.1	156	44	41	4.2
Upper limit of normal	45.4	1.0	200	40	85	

[a]After Klotz *et al.* (101).

[b]Acute stage.

[c]Recovery stage.

[d]Four months after all hepatic function tests had returned to normal.

[e]n.a., not available.

in cases of impaired renal function. However, the rate of elimination of their more polar metabolites (e.g., hydroxyamobarbital) may be decreased. These predictions were also verified with diphenylhydantoin and chloramphenicol (54). In other words, the administration of lipid-soluble drugs to patients with impaired renal function may lead to accumulation of their polar metabolites. Should a metabolite retain significant pharmacologic activity, overdosage may result.

Severe renal impairment does prolong the serum half-life of highly water-soluble drugs that are usually eliminated almost unchanged in the urine. The resulting accumulation of such compounds (e.g., aminoglycoside antibiotics) may cause severe intoxication. This potential hazard can be minimized by using serum creatinine determinations as a guide to monitor drug administration in the presence of impaired renal function. Indeed, several relatively simple nomograms have been designed to calculate loading and maintenance doses of drugs such as digoxin, kanamycin, and gentamycin. Dettli and associates (54) have compiled a fairly extensive list of rate constants of drug elimination in anuric patients (k_m) and in patients with normal kidney function (k_n), which, together with the patient's sex, age, weight, and plasma creatinine levels can be used in a simplified method for determining dosage schedules. With their help it is possible to keep plasma concentrations within the therapeutic range.

The microsomal oxidations of tolbutamide, phenobarbital, pentobarbital, and phenacetin, and the glucuronide conjugation of chloramphenicol occur at normal rates in uremic patients (148a) even though they are lower in plasma protein binding values. The half-life of diphenylhydantoin is decreased in uremic patients, possibly because this drug binds poorly to plasma proteins (149). Certain nonmicrosomal pathways of drug metabolism are known to be hindered in patients with impaired renal function. These include the acetylation of sulfisoxazole (148) and isoniazid and the hydrolyses of insulin and procaine (147). The decrease in the rate of procaine hydrolysis by plasma pseudocholinesterase was proportional to the patient's blood urea nitrogen level. The significantly lower than normal plasma cholinesterase activity encountered in patients with end-stage kidney disease was increased with hemodialysis (205).

Not only therapeutic agents, but many other chemicals may induce or inhibit the activity of microsomal enzymes. Halogenated hydrocarbons such as DDT, lindane, and chlordane have been shown to decrease the rate of disappearance of phenylbutazone and increase the excretion of 6β-hydroxycortisol in factory workers exposed to these chemicals (102, 143). Since these individuals had serum and fat levels of DDT which were 20- to 30-fold higher than those of control subjects, it was not possible to ascertain the minimum concentrations of DDT required to produce induction in man. Using animals, induction is also readily produced with xanthines, flavones, volatile oils of terpene, and food preservatives. Conversely, enzyme inhibition has been demonstrated in animals with organophosphorus insecticides, which also have anticholinesterase activity; methylenedioxyphenyl-type synergists of pesticides, such as piperonyl butoxide (173); and carbon tetrachloride. However, following the administration of 50 mg of piperonyl butoxide by mouth to 9 human volunteers, no changes were observed in antipyrine metabolism.

Cigarette smoking has been found to increase the activity of benzpyrene hydroxylase, an enzyme system inducible by polycyclic hydrocarbons, in *in vitro* preparations of lungs, livers, intestines, and placentas of rats exposed to cigarette smoke and also the placentas of human cigarette smokers (199, 200). The benzypyrene hydroxylase system was studied because cigarette smoke contains various amounts of polycyclic hydrocarbons such as 3,4-benzpyrene, 1,2,5,6-dibenzanthracene, anthracene, chrysene, etc. All these stimulate the cytochrome *P*-448 microsomal system and increase the production of the metabolite hydroxybenzpyrene. Other studies showed that cigarette smoking enhances the metabolism of nicotine *in vivo*, which may account for the nicotine tolerance of smokers (13). It may also explain the larger doses of pentazocine required to achieve analgesia in smokers (100). Interestingly, neither nicotine nor phenobarbital stimulates placental benzpyrene hydroxylase activity in rats.

Several clinical studies have demonstrated induction of hepatic microsomal enzymes in cigarette smokers. In one study, plasma levels of phenacetin were markedly lower in smokers than in nonsmokers (137). At the same time, the ratio of the plasma concentration of the main metabolite N-acetyl-p-aminophenol to that of the parent compound phenacetin increased severalfold in smokers, confirming the induction of phenacetin metabolism. The source of the stimulated enzyme has not been established, although some evidence in animals indicates that benzpyrene hydroxylase activity in the intestinal mucosa (but probably not in the liver) is involved (136). This would explain the lack of change in the apparent plasma half-life of phenacetin, while the rate of overall metabolic degradation increased. Another study, comparing methemoglobin formation in normal subjects and phenacetin abusers who developed nephropathy, clearly showed significantly higher methemoglobin levels in the latter group (64). Nephropathies due to other causes do not increase methemoglobin levels. These findings have been interpreted to indicate, first, that the heavy consumption of phenacetin altered its own metabolism by increasing its rate of deacetylation, N-hydroxylation, and aromatic ring hydroxylation, and, second, that these changes resulted in increased production of toxic metabolites which facilitate methemoglobin formation. Such findings lead to the speculation, as yet unsupported by experimental data, that the enzyme induction caused by cigarette smoking may also accelerate the formation of other toxic metabolites which can cause methemoglobinemia. Stimulation of microsomal enzymes by cigarette smoking should accelerate the metabolism of many other drugs, especially if the cytochrome P-448 system is involved in the process (e.g., with chlorpromazine) (198). This postulate is supported by a recent clinical study with imipramine, which clearly showed induction of the aromatic hydroxylating system in cigarette smokers. The effect was more marked in male than in female patients (141).

Epoxides have been implicated as intermediate products in the microsomal hydroxylation of olefins and aromatic substrates in mammals (90). An epoxide is a highly reactive intermediate formed when an oxygen atom is attached to two adjacent carbons of olefins such as trichloroethylene or in an aromatic ring; the latter also is referred to as an arene oxide. Depending on the stability of the arene oxide and its reactivity with various enzyme systems, it may (a) be transformed within a fraction of a second into the corresponding phenol, (b) react covalently with proteins, or (c) be further metabolized either to a dihydrodiol by the action of a microsomal hepatic epoxide hydrase or to a glutathione derivative by the action of a soluble hepatic glutathione-epoxide conjugate (Figure 9-3). The two latter secondary products are then further metabolized to catechols and premercapturic acids, respectively. These reactions of arene oxides with tissue constituents may well be responsi-

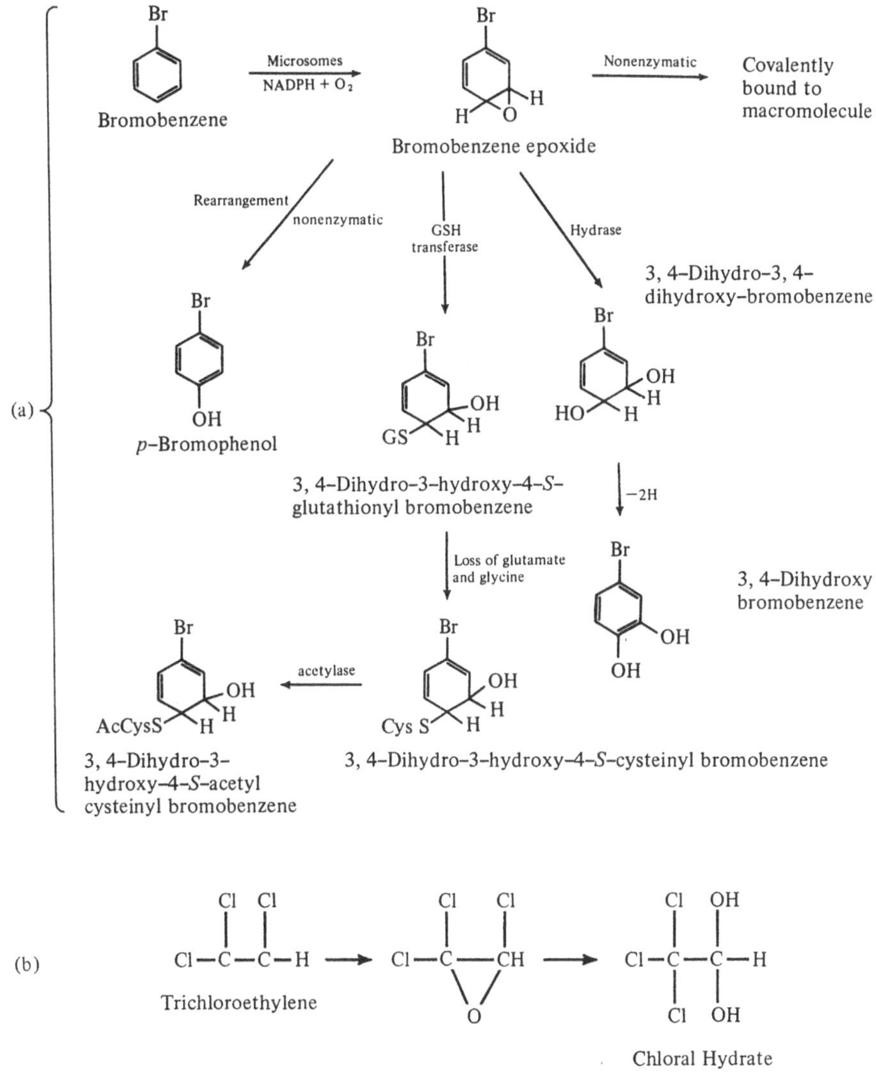

Figure 9-3. Epoxide-diol pathway in the metabolism of bromobenzene (a) and trichloroethylene (b).

ble for the toxicity and/or carcinogenicity of aromatic compounds. Brodie and his associates (25) have shown that phenobarbital induction in rats increases the hepatic toxicity of a number of relatively stable substances such as carbon tetrachloride and halogenated aromatic hydrocarbons (e.g., bromobenzene); meanwhile a covalently bonded compound, presumably resulting from an active intermediate, is fixed firmly in the tissue. Biochemical evidence indicates that the active intermediate of bromobenzene metabolism is an epoxide which reacts with glutathione in hepatocytes to yield a conjugate. Furthermore, when liver glutathione has been sufficiently depleted by this chemical reaction the epoxide then reacts with tissue macromolecules, including proteins. Preliminary evidence indicates that the bromobenzene is covalently bound to sulfur-containing amino acids. A similar epoxide intermediate has also been implicated in the metabolism of trichloroethylene to chloral hydrate by rat liver microsomes (35, 109).

Studies of the microsomal epoxide hydrase revealed it to be inducible with 3-methylcholanthrene and phenobarbital. Partial purification of the enzyme demonstrated the presence of at least two isozymes. The purified hydrase is competitively inhibited by 1,1,1-trichloropropene oxide and stimulated by metyrapone.

These observations may help to clarify the toxic effects of halogenated aromatic and olefinic compounds upon the liver. Humans show a wide range of interindividual variation in activity of their hepatic microsomal enzymes. Subjects whose enzymes metabolize drugs more rapidly are more susceptible to liver damage caused by halogenated agents. Obviously, patients receiving long-term drug therapy are likely to exhibit stimulation of their microsomal oxidases, rendering them potentially more susceptible to the hazard of liver damage from inhalation anesthetic agents. Furthermore, such "stimulated" patients, who had already received one anesthetic, are likely to be at greater risk if they are scheduled to receive a second anesthetic within a few days. It has been shown in animal experiments that some of the first anesthetic can remain bound in liver cells for several days without causing significant necrosis. A toxic reaction, however, is possible if the patient is again exposed to the same or a similar substance within a short period of time.

References

1. Ackermann, E., and Heinrich, I., Die Aktivitat der N-und O-Demethylase in der Leber des Menschen. *Biochem. Pharmacol.* **19**, 327 (1970).
2. Alexanderson, B., Price-Evans, D. A., and Sjöqvist, F., Steady-state levels of nortriptyline in twins: Influence of genetic factors and drug therapy. *Br. Med. J.* **4**, 764 (1969).
3. Altman, P. L., and Dittmer, D. S., eds., "Biological Handbooks: Growth," p. 33. Fed. Am. Soc. Exp. Biol., Washington, D.C.
4. Alvares, A. P., Kapelner, S., Sassa, S., and Kappas, A., Drug metabolism in normal children, lead-poisoned children, and normal adults. *Clin. Pharmacol. Ther.* **17**, 179 (1975).
5. Alvares, A. P., Parli, C. J., and Mannering, G. J., Induction of drug metabolism. VI. Effects of phenobarbital and 3-methycholanthrene administration on N-demethylating enzyme systems of rough and smooth hepatic microsomes. *Biochem. Pharmacol.* **22**, 1037 (1973).
6. Alvares, A. P., Schilling, G., Levin, W., and Kuntzman, R., Alteration of the microsomal hemoprotein by 3-methylcholanthrene: Effects of ethionine and actinomycin D. *J. Pharmacol. Exp. Ther.* **163**, 417 (1968).

7. Alvares, A. P., Schilling, G., Levin, W., Kuntzman, R., Brand, L., and Mark, L. C., Cytochromes P-450 and b_5 in human liver microsomes. *Clin. Pharmacol. Ther.* **10**, 655 (1969).

8. Anders, M. W., Enhancement and inhibition of drug metabolism. *Annu. Rev. Pharmacol.* **11**, 37 (1971).

9. Axelrod, J., Biochemical factors in the activation and inactivation of drugs. *Naunyn-Schmiedebergs Arch. Exp. Pathol. Pharmak.* **238**, 24 (1960).

10. Baekeland, F., and Greene, N. M., Effect of diethyl ether on tissue distribution and metabolism of pentobarbital in rats. *Anesthesiology* **19**, 724 (1958).

11. Ballou, D. P., Veeger, C., Van der Hoeven, T. A., and Coon, M. J., Properties of partially purified liver microsomal P-450: Acceptance of two electrons during anaerobic titration. *FEBS Lett.* **38**, 337 (1974).

12. Baron, J., and Tephly, T. R., Further studies on the relationship of the stimulatory effects of phenobarbital and 3,4-benzpyrene on hepatic synthesis to their effects on hepatic microsomal drug oxidations. *Arch. Biochem. Biophys.* **139**, 410 (1970).

13. Beckett, A. H., and Triggs, E. J., Enzyme induction in man caused by smoking. *Nature (London)* **216**, 587 (1967).

13a. Benowitz, N. L., and Meister, W., Pharmacokinetics in patients with cardiac failure. *Clin. Pharmacokinet.* **1**, 389 (1976).

14. Berman, M. L., and Bochantin, J. E., Nonspecific stimulation of drug metabolism in rats by methoxyflurane. *Anesthesiology* **32**, 500 (1970).

15. Berman, M. L., Lowe, H. J., Bochantin, J. E., and Hagler, K., Uptake and elimination of methoxyflurane as influenced by enzyme induction in the rat. *Anesthesiology* **38**, 352 (1973).

16. Blaschke, T. F., Meffin, P. J., Melmon, K. L., and Rowland, M., Influence of acute viral hepatitis on phenytoin kinetics and protein binding. *Clin. Pharmacol. Ther.* **17**, 685 (1975).

17. Bock, K. W., Krauss, E., and Frohling, W., Regulation of α-aminolevulinic acid synthetase by drugs and steroids in vivo and in perfused rat liver. *Eur. J. Biochem.* **23**, 366 (1971).

17a. Branch, R. A., James, J. A., and Read, A. E., The clearance of antipyrine and indocyanine green in normal subjects and in patients with chronic liver disease. *Clin. Pharmacol. Ther.* **20**, 81 (1976).

18. Brand, L., Mark, L. C., Heiber, S., and Perel, J. M., Interaction between diazepam and diphenylhydantoin. *Abstr. Sci. Pap. Annu. Meet. Am. Soc. Anesthesiol.*, p. 251 (1973).

19. Breckenridge, A., Clinical implications of enzyme induction. *Basic Life Sci.* **6**, 273 (1975).

20. Breckenridge, A., and Orme, M., Clinical implications of enzyme induction. *Ann. N.Y. Acad. Sci.* **179**, 421 (1971).

21. Breckenridge, A., Orme, L. E., Davies, L., Thorgeirsson, S. S., and Davies, D. S., Dose-dependent enzyme induction. *Clin. Pharmacol. Ther.* **14**, 514 (1973).

22. Breimer, D. D., "Pharmacokinetics of Hypnotic Drugs." Drukkerij-Uitgeverij Brakkenstein, Nijmegen, The Netherlands, 1974.

23. Brodie, B. B., Possible mechanisms of drug-induced tissue lesions. *Chem.-Biol. Interact.* **3**, 247 (1971).

24. Brodie, B. B., Gillette, J. R., and LaDu, B. N., Enzymatic mechanisms of drugs and other foreign compounds. *Annu. Rev. Biochem.* **27**, 427 (1958).

24a. Brodie, B. B., Papper, E. M., and Mark, L. C., Fate of procaine in man and properties of its metabolite diethylaminoethanol. *Anesth. Analg. (Cleveland)* **29**, 29 (1950).

25. Brodie, B. B., Reid, W. D., Cho, A. K., Sipes, G., Krishna, G., and Gillette, J. R., Possible mechanism of liver necrosis caused by aromatic organic compounds. *Proc. Natl. Acad. Sci. U.S.A.* **68**, 160 (1971).

26. Brodie, B. B., Weiner, M., Burns, J. J., Simon, G., and Yale, E. K. Y., The physiological disposition of ethyl biscoumacetate (tromexan) in man and a method of its estimation in biological material. *J. Pharmacol. Exp. Ther.* **106**, 453 (1952).

27. Brown, B. R., Miller, J. A., and Miller, E. C., The metabolism of methylated aminoazo dyes. IV. Dietary factors enhancing demethylation in vitro. *J. Biol. Chem.* **209**, 211 (1954).

28. Brown, B. R., Jr., and Sagalyn, A. M., Hepatic microsomal enzyme induction by inhalation anesthetics: Mechanism in the rat. *Anesthesiology* **40**, 152 (1974).

29. Brown, B. R., Jr., Sipes, I. G., and Sagalyn, A. M., Mechanism of acute hepatic toxicity: Chloroform, halothane and glutathionie. *Anesthesiology* **41**, 554 (1974).

30. Brown, B. R., Jr., and Vandam, L. D., A review of current advances in metabolism of inhalation anesthetics. *Ann. N.Y. Acad. Sci.* **179**, 235 (1971).

31. Buchthal, F., Svensmark, O., and Schiller, P. J., Clinical and electroencephalographic correlation with serum levels of diphenylhydantoin. *Arch. Neurol. (Chicago)* **2**, 624 (1960).

32. Burns, J. J., Rose, R. K., Chenkin, T., Goldman, A., Schulert, A., and Brodie, B. B., The physiological disposition of phenylbutazone (butazalodin) in man and a method for its estimation in biological material. *J. Pharmacol. Exp. Ther.* **109**, 346 (1953).

33. Busfield, D., Child, K. J., Atkinson, R. M., and Tomich, E. G., An effect of phenobarbitone on blood levels of griseofulvin in man. *Lancet* **2**, 1042 (1963).

34. Bush, M. T., and Weller, W. L., Metabolic fate of hexobarbital (HB). *Drug Metab. Rev.* **1**, 249 (1972).

35. Byington, K. H., and Leibman, K. C., Metabolism of trichloroethylene in liver microsomes. II. Identification of the reaction product as chloral hydrate. *Mol. Pharmacol.* **1**, 247 (1965).

36. Cascorbi, H. H., Blake, D. A., and Helrich, M., Halothane biotransformation in mice and man. *In* "Second Symposium on Cellular Toxicity of Anesthetics" (B. R. Fink, ed.), p. 197. Williams & Wilkins, Baltimore, Maryland, 1972.

37. Cascorbi, H. F., and Singh-Amaranth, A. V., Fluroxene toxicity. *Anesthesiology* **37**, 480 (1972).

38. Cascorbi, H. F., Vesell, E. S., Blake, D. A., and Helrich, M., Genetic and environmental influence on halothane metabolism in twins. *Clin. Pharmacol. Ther.* **12**, 50 (1971).

39. Cascorbi, H. F., Vesell, E. S., Blake, D. A., and Helrich, M., Halothane biotransformation in man. *Ann. N.Y. Acad. Sci.* **179**, 244 (1971).

40. Castleden, C. M., Kaye, C. M., and Parsons, R. L., The effect of age on plasma levels of propanolol and practolol in man. *Br. J. Clin. Pharmacol.* **2**, 303 (1975).

41. Chen, W., Vrindten. P. A., Dayton, P. G., and Burns, J. J., Accelerated aminopyrine metabolism in human subjects pretreated with phenylbutazone. *Life Sci.* **2**, 35 (1962).

42. Cohen, E. N., Trudell, J. R., Edmunds, H. N., and Watson, E., Urinary metabolites of halothane in man. *Anesthesiology* **43**, 392 (1975).

43. Committee on Problems of Drug Safety of the Drug Research Board, National Academy of Sciences—National Research Council, Report: Application of metabolic data to the evaluation of drugs. *Clin. Pharmacol. Ther.* **10**, 607 (1969).

44. Conney, A. H., Pharmacological implications of microsomal enzyme induction. *Pharmacol. Rev.* **19**, 317 (1967).

45. Conney, A. H., Levin, W., Jacobson, M., and Kuntzman, R., Effects of drugs and environmental chemicals on steroid metabolism. *Clin. Pharmacol. Ther.* **14**, 727 (1974).

46. Cooksley, W. G. E., and Powell, L. W., Drug metabolism and interaction with particular reference to the liver. *Drugs* **2**, 177 (1971).

46a. Cooperman, L. H., Effects of anaesthetics on the splanchnic circulation. *Br. J. Anaesth.* **44**, 967 (1972).

47. Cucinell, S. A., Conney, A. H., Sansur, M. S., and Burns, J. J., Drug interactions in man. I. Lowering effect of phenobarbital on plasma levels of bishydroxycoumarin (dicumarol) and diphenylhydantoin (dilantin). *Clin. Pharmacol. Ther.* **6**, 420 (1965).

48. Cucinell, S. A., Odessky, L., Weiss, M., and Dayton, P. G., The effect of chloral hydrate on bishydroxycoumarin metabolism. *J. Am. Med. Assoc.* **197**, 366 (1966).

48a. Cumming, J. F., and Mannering, G. J., Effect of phenobarbital administration on the oxygen requirement for hexobarbital metabolism in the isolated, perfused rat liver and in the intact rat. *Biochem. Pharmacol.* **19**, 973 (1970).

49. Dasberg, H. H., Van der Kleijn, E., Guelen, P. J. R., and Van Praag, H. M., Plasma concentrations of diazepam and of its metabolite N-desmethyldiazepam in relation to anxiolytic effect. *Clin. Pharmacol. Ther.* **15**, 473 (1974).

50. Davies, D. S., Gigon, P. L., and Gillette, J. R., Species and sex differences in electron transport systems in liver microsomes and their relationship to ethylmorphine demethylation. *Life Sci.* **8**, Part 2, 85 (1969).

51. Davis, D. C., Schroeder, D. H., Gram, T. E., Reagan, R. L., and Gillette, J. R., A comparison of the effects of halothane and CCl$_4$ on the hepatic drug-metabolizing system. *J. Pharmacol. Exp. Ther.* **177**, 556 (1971).

52. Dayton, P. G., and Perel, J. M., Physiological and physicochemical bases of drug interactions in man. *Ann. N.Y. Acad. Sci.* **179**, 67 (1971).

53. Dayton, P. G., Pruitt, A. W., Faraj, B. A., and Israili, Z. H., Metabolism and disposition of diazoxide. *Drug Metab. Dispos.* **3**, 226 (1975).

54. Dettli, L., Spring, P., and Ryter, S., Multiple dose kinetics and drug dosage in patients with kidney disease. *Acta Pharmacol. Toxicol.* **29**, Suppl. 3, 211 (1971).

55. DiFazio, C. A., and Brown, R. E., Lidocaine metabolism in normal and phenobarbital-pretreated dogs. *Anesthesiology* **36**, 238 (1972).

55a. Dutton, G. F., Glucuronide synthesis in fetal liver and other tissues. *Biochem. J.* **71**, 141 (1959).

55b. Epstein, R. M., Deutsch, S., Cooperman, L. H., Clement, A. J., and Price, H. L., Splanchnic circulation during halothane anesthesia and hypercapnia in normal man. *Anesthesiology* **27**, 654 (1966).

56. Estabrook, R. W., Cooper, D. Y., and Rosenthal, O., The light reversible monoxide inhibition of the steroid C^{21}-hydroxylase system of the adrenal cortex. *Biochem. J.* **338**, 741 (1963).

57. Fiserova-Bergerova, V., Changes of fluoride content in bone: An index of drug defluorination in vivo. *Anesthesiology* **38**, 345 (1973).

58. Fish, F., and Wilson, W. D. C., Excretion of cocaine and its metabolites in man. *J. Pharm. Pharmacol.* **21**, Suppl. 135S (1969).

59. Fleischner, G., and Arias, I. M., Recent advances in bilirubin formation, transport, metabolism and excretion. *Am. J. Med.* **49**, 576 (1970).

60. Forrest, F. M., Forrest, I. S., and Serra, M. T., Modification of chlorpromazine metabolism by some other drugs frequently administered to psychiatric patients. *Biol. Psychiatry* **2**, 53 (1970).

61. Forsman, A., and Ohman, R., On the pharmacokinetics of haloperidol. *Nord. Psykiatr. Tidsskr.* **28**, 441 (1974).

62. Freudenthal, R. I., and Carroll, F. I., Metabolism of certain commonly used barbiturates. *Drug Metab. Rev.* **2**, 265 (1973).

63. Garattini, S., Drug-blood levels in human newborns. *Abstr. Pharmacol.-Toxicol. Program Symp., NIGMS, 1971* (1971).

64. Gault, M. H., Shahidi, N. T., and Barber, V. E., Methemoglobin formation in analgesic nephropathy. *Clin. Pharmacol. Ther.* **15**, 521 (1974).

65. Geddes, I. C., Metabolism of volatile anesthetics. *Br. J. Anaesth.* **44**, 953 (1972).

66. Gigon, P. L., Gram, T. E., and Gillette, J. R., Effect of drug substrates on the reduction of hepatic microsomal cytochrome P-450 by NADPH. *Biochem. Biophys. Res. Commun.* **31**, 558 (1968).

67. Gigon, P. L., Gram, T. E., and Gillette, J. R., Studies on the rate of reduction of hepatic microsomal cytochrome P-450 by reduced nicotinamide adenine dinucleotide phosphate: Effect of drug substrates. *Mol. Pharmacol.* **5**, 109 (1969).

68. Gillette, J. R., Biochemistry of drug oxidation and reduction by enzymes in hepatic endoplasmic reticulum. *Adv. Pharmacol.* **4**, 219 (1966).

69. Gillette, J. R., Effects of various inducers on electron transport system associated with

drug metabolism by liver microsomes. *Metab., Clin. Exp.* **20**, 215 (1971).

70. Gillette, J. R., Davis, D. C., and Sasame, H. A., Cytochrome P-450 and its role in drug metabolism. *Annu. Rev. Pharmacol.* **12**, 57 (1972).

71. Gillette, J. R., and Stripp, B., Pre- and postnatal enzyme capacity for drug metabolite production. *Fed. Proc., Fed. Am. Soc. Exp. Biol.* **34**, 172 (1975).

72. Gion, H., Yoshimura, N., Holaday, D. A., Bergerova, F. V., and Chase, R. E., Biotransformation of fluorexene in man. *Anesthesiology* **40**, 553 (1974).

73. Govier, W. C., Reticuloendothelial cells as the site of sulfanilamide acetylation in the rabbit. *J. Pharmacol. Exp. Ther.* **150**, 305 (1965).

74. Gram, L. F., Christiansen, J., and Overø, K. F., Pharmacokinetic interaction between neuroleptics and tricyclic antidepressants in the rat. *Acta Pharmacol. Toxicol.* **35**, 223 (1974).

75. Gram, L. F., and Overø, K. F., Drug interaction: Inhibitory effect of neuroleptics on metabolism of tricyclic antidepressants in man. *Br. Med. J.* **1**, 463 (1972).

76. Gram, L. F., Overø, K. F., and Kirk, L., Influence of neuroleptics and benzodiazepines on metabolism of tricyclic antidepressants in man. *Am. J. Psychiatry* **131**, 863 (1974).

77. Gram, T. E., Guarino, A. M., Schroeder, D. H., and Gillette, J. R., Changes in certain kinetic properties of hepatic microsomal aniline hydroxylase and ethylmorphine demethylase associated with postnatal development and maturation in male rats. *Biochem. J.* **113**, 681 (1969).

78. Greenblatt, D. J., and Shader, R. I., "Benzodiazepines in Clinical Practice." Raven Press, New York, 1974.

79. Greenblatt, D. J., Shader, R. I., and Koch-Weser, J., Flurazepam hydrochloride. *Clin. Pharmacol. Ther.* **17**, 1 (1975).

80. Greene, N. M., The metabolism of drugs employed in anesthesia. Part II. *Anesthesiology* **29**, 327 (1968).

81. Hammer, W., Martens, S., and Sjöqvist, F., A comparative study of the rate of metabolism of desmethylimipramine, nortriptyline and oxyphenylbutazone in man. *Clin. Pharmacol. Ther.* **10**, 44 (1969).

82. Hayes, M. J., Langman, M. J. S., and Short, A. H., Changes in drug metabolism with increasing age. 2. Phenytoin clearance and protein binding. *Br. J. Clin. Pharmacol.* **2**, 73 (1975).

83. Heinonen, J., Influence of some drugs on toxicity and rate of metabolism of lidocaine and mepivacaine. *Ann. Med. Exp. Biol. Fenn.* **44**, Suppl. 3, 1 (1966).

84. Held, V. H., Eisert, R., and Oldershausen, H. F., Pharmakokinetik von Glymidine (Glycodiazin) und Tolbutamid bei akuten und chronischen Leber-Schaden. *Arzneim.-Forsch.* **23**, 1801 (1973).

85. Hildebrandt, A., Remmer, H., and Estabrook, R. W., Cytochrome P-450 of liver microsomes—one pigment or many. *Biochem. Biophys. Res. Commun.* **30**, 607 (1968).

86. Hillestad, L., Hansen, T., and Melson, H., Diazepam metabolism in normal man. II. Serum concentration and clinical effect after oral administration. *Clin. Pharmacol. Ther.* **16**, 485 (1974).

87. Holcomb, R. R., Gerber, N., and Bush, M. T., The metabolic fate of hexobarbital in the rat. *J. Pharmacol. Exp. Ther.* **188**, 15 (1974).

88. Holtzman, J. L., Gram, T. E., Gigon, P. L., and Gillette, J. R., The distribution of the components of mixed-function oxidase between the rough and the smooth endoplasmic reticulum of liver cells. *Biochem. J.* **110**, 407 (1968).

89. Holtzman, J. L., and Rumack, H., The kinetics of ethylmorphine activation of the NADPH-cytochrome P-450 reductase activity of hepatic microsomes from male and female rats. *Chem.-Biol. Interact.* **3**, 279 (1971).

90. Hucker, H. B., Intermediates in drug metabolism reactions. *Drug Metab. Rev.* **2**, 33 (1973).

91. Hunter, J., Maxwell, J. D., Carrella, M., Stewart, D. A., and Williams, R., Urinary

D-glucaric acid excretion as a test for hapatic enzyme induction in man. *Lancet* **1**, 572 (1971).

92. Imai, Y., and Sato, R., Substrate interaction with hydroxylase system in liver microsomes. *Biochem. Biophys. Res. Commun.* **22**, 620 (1966).

93. Ioannides, C., and Parke, D. V., The effect of ethanol administration on drug oxidations and possible mechanism of ethnol-barbiturate interactions. *Biochem. Soc. Trans.* **1**, 716 (1973).

94. Irvine, R. E., Grove, J., Toseland, P. A., and Trounce, J. R., The effect of age on the hydroxylation of amylobarbitone sodium in man. *Br. J. Clin. Pharmacol.* **1**, 41 (1974).

95. Kadar, D., Inaba, T., Endrenyi, L., Johnson, G. E., and Kalow, W., Comparative drug elimination capacity in man-glutethimide, amobarbital, antipyrine and sulfinpyrazone. *Clin. Pharmacol. Ther.* **14**, 552 (1973).

96. Kappas, A., and Song, C. S., Enzyme induction in the liver. *Gastroenterology* **55**, 731 (1968).

97. Kater, R. M. H., Roggin, G., Tobon, F., Zieve, P., and Iber, F. L., Increased rate of clearance of drugs from the circulation of alcoholics. *Am. J. Med. Sci.* **258**; 35 (1969).

98. Kato, R., Sex-related differences in drug metabolism. *Drug Metab. Rev.* **3**, 1 (1974).

99. Keberle, H., Hoffmann, K., and Bernhard, K., The metabolism of glutethimide (Doriden). *Experientia* **18**, 105 (1963).

100. Keeri-Szanto, M., and Pomeroy, J. R., Atmospheric pollution and pentazocine metabolism. *Lancet* **1**, 947 (1971).

101. Klotz, U., Avant, G. R., Hoyumpa, A., Schenker, S., and Wilkinson, G. R., The effects of age and liver disease on the disposition and elimination of diazepam in adult man. *J. Clin. Invest.* **35**, 347 (1975).

101a. Klotz, U., McHorse, T. S., Wilkinson, G. R., and Schenker, S., The effect of cirrhosis on the disposition and elimination of meperidine in man. *Clin. Pharmacol. Ther.* **16**, 667 (1974).

102. Kolmodin-Hedman, B., Decreased plasma half-life in workers exposed to chlorinated pesticides. *Eur. J. Clin. Pharmacol.* **5**, 195 (1973).

103. Kristensen, M., Hansen, J. M., and Skovsted L., The influence of phenobarbital on the half-life of diphenylhydantoin in man. *Acta Med. Scand.* **185**, 347 (1969).

104. Kupfer, D., and Rosenfeld, J., A sensitive radioactive assay for hexobarbital hydroxylase in hepatic microsomes. *Drug Metab. Disp.* **1**, 760 (1973).

105. Kutt, H., Interactions of antiepileptic drugs. *Epilepsia* **16**, 393 (1975).

106. Kutt, H., Winters, W., Kokenge, R., and McDowell, F., Diphenylhydantoin, metabolism, blood levels and toxicity. *Arch. Neurol. (Chicago)* **11**, 642 (1964).

107. Kutt, H., Wolk, M., Scherman, R., and McDowell, F., Insufficient parahydroxylation as a cause of diphenylhydantoin toxicity. *Neurology* **14**, 542 (1964).

108. Lee Son, S., Colella, J. J., Jr., and Brown, B. R., Jr., The effect of phenobarbitone on the metabolism of methoxyflurane to oxalic acid in the rat. *Br. J. Anaesth.* **44**, 1224 (1972).

109. Leibman, K. C., Metabolism of trichloroethylene in liver microsomes 1. Characteristics of the reaction. *Mol. Pharmacol.* **1**, 239 (1965).

110. Levi, A. J., Sherlock, S., and Walker, D., Phenylbutazone and isoniazid metabolism in patients with liver disease in relation to previous drug therapy. *Lancet* **1**, 1275 (1968).

111. Levin, W., Sernatinger, E., Jacobson, M., and Kuntzman, R., Destruction of cytochrome P-450 by secobarbital and other barbiturates containing allyl groups. *Science* **176**, 1341 (1972).

112. Levy, G., Kinetics of drug action in man. *Acta Pharmacol. Toxicol.* **29**, Suppl. 3, 203 (1971).

113. Lieber, C. S., and DeCarli, L. M., Hepatic microsomes: A new site for ethanol oxidation. *J. Clin. Invest.* **47**, 62a, (1969).

114. Lieber, C. S., and DeCarli, L. M., Effect of drug administration on the activity of the hepatic microsomal ethanol oxidizing system. *Life Sci.* **9**, 267 (1970).

115. Linde, H. W., and Berman, M. L., Non-specific stimulation of drug-metabolizing enzymes by inhalation anesthetic agents. *Anesth. Analg. (Cleveland)* **50**, 656 (1971).

116. Ludwig, B. J., Douglas, J. F., Powell, L. S., Meyer, M., and Berger, F. M., Structures of the major metabolites of meprobamate. *J. Med.-Pharm. Chem.* **3**, 53 (1961).

117. MacDonald, M. G., and Robinson, D. S., Clinical observations of possible barbiturate interference with anticoagulants. *J. Am. Med. Assoc.* **204**, 97 (1968).

118. Manian, A. A., Efron, D. H., and Harris, S. R., Appearance of monohydroxylated chlorphromazine metabolities in the central nervous system. *Life Sci.* **10**, 679 (1971).

119. Mannering, G. J., Properties of cytochrome *P*-450 as affected by environmental factors: Qualitative changes due to administration of polycyclic hydrocarbons. *Metab. Clin. Exp.* **20**, 228 (1971).

120. Mark, L. C., Metabolism of barbiturates in man. *Clin. Pharmacol. Ther.* **4**, 504 (1963).

121. Mark, L. C., Brand, L., Heiber, S., and Perel, J. M., Effects of deuteration and optical isomerism on activity of pentobarbitol. *Fed. Proc., Fed. Am. Soc. Exp. Biol.* **30**, 442 (1971).

122. Mark, L. C., Brand, L., Heiber, S., Smith, D., and Carroll, F. I., Pharmacologic activity and biotransformation of R^+ and S^- barbiturate enantioners in mouse and man. *Fed. Proc. Fed. Am. Soc. Exp. Biol.* **32**, 681 (1973).

123. Marniemi, J., Aitio, A., and Vainio, H., Ethanol induced alteration of microsomal membrane bound enzymes of rat liver in vitro. *Acta Pharmacol. Toxicol.* **37**, 222 (1975).

124. Maurer, H. M., Wolff, J. A., Finster, M., Poppers, P. J., Pantuck, E., and Conney, A. H., Reduction in concentration of total serum-bilirubin in offspring of women treated with phenobarbitone during pregnancy. *Lancet* **2**, 122 (1968).

125. Mazze, R. I., Trudell, J. R., and Cousins, M. J., Methoxyflurane metabolism and renal dysfunction: Clinical correlation in man. *Anesthesiology* **35**, 247 (1971).

126. McMahon, R. E., and Sullivan, H. R., The microsomal hydroxylation of ethylbenzene: Stereochemical induction and isotopic studies. *In* "Microsomes and Drug Oxidations" C. J. R. Gillette *et al.*, eds. Academic Press, p. 239. New York, 1969.

127. Meffin, P., Long, G. J., and Thomas, J., Clearance and metabolism of mepivacaine in the human neonate. *Clin. Pharmacol. Ther.* **14**, 218 (1973).

128. Melville, K. I., Jordon, G. E., and Douglas, D., Toxic and depressant effects of alcohol given orally in combination with glutethimide or secobarbital. *Toxicol. Appl. Pharmacol.* **9**, 363 (1966).

129. Miller, R. P., Roberts, R. J., and Fischer, L. J., Acetoaminophen elimination kinetics in neonates, children and adults. *Clin. Pharmacol. Ther.* **19**, 284 (1976).

130. Moore, M. R., Battistini, V., Beattie, A. D., and Goldberg, A., The effects of certain barbiturates on the hepatic porphyrin metabolism of rats. *Biochem. Pharmacol.* **19**, 751 (1970).

131. Morselli, P., Rizzo, M., and Garattini, S., Interaction between phenobarbital and diphenyl-hydantoin in animals and in epileptic patients. *Ann. N. Y. Acad. Sci.* **179**, 88 (1971).

132. Mueller, G. C., and Miller, J. H., The metabolism of methylated aminoazo dyes. II. Oxidative demethylation by rat liver homogenates. *J. Biol. Chem.* **202**, 579 (1953).

133. Munson, E. S., Malagodi, M. H., Shields, R. P., Tham, M. K., Bergerova, F. V., Holaday, D. A., Perry, J. C., and Embro, W. J., Fluroxene toxicity induced by phenobarbital. *Clin. Pharmacol. Ther.* **18**, 687 (1975).

133a. Nies, A. S., Shand, D. G., and Wilkinson, G. R., Altered hepatic blood flow and drug disposition. *Clin. Pharmacokinet.* **1**, 135 (1976).

134. O'Malley, K., Crooks, J., Duke, E., and Stevenson, I. H., Effect of age and sex on human drug metabolism. *Br. Med. J.* **3**, 607 (1971).

135. Orrenius, S., and Ernster, L., Phenobarbital-induced synthesis of the oxidative demethylating enzymes of rat liver microsomes. *Biochem. Biophys. Res. Commun.* **16**, 60 (1964).

136. Pantuck, E. J., Hsiao, K. -C., Kaplan, S. A., Kuntzman, R., and Conney, A. H., Effects of enzyme induction on intestinal phenacetin metabolism in the rat. *J. Pharmacol. Ther.* **191**, 45 (1974).

137. Pantuck, E. J., Hsiao, K. -C., Maggio, A., Nakamura, K., Kuntzman, R., and Conney, A. H., Effect of cigarette smoking on phenacetin metabolism. Clin. Pharmacol. Ther. 15, 9 (1974).

138. Pelkonen, O., Kaltiala, E. H., Larmi, T. K. I., and Karki, N. T., Comparison of activities of drug-metabolizing enzymes in human fetal and adult livers. Clin. Pharmacol. Ther. 14, 840 (1973).

139. Perel, J. M., Brand, L., Heiber, S., and Mark, L. C., Effect of methylphenidate on rate of disappearance of thiopental from plasma. Clin. Res. 20, 411 (1972).

140. Perel, J. M., O'Brien, L., Black, N. B., Bellward, G. D., and Dayton, P. G., Imipramine and chlorpromazine in hepatic microsomal systems. Adv. Biochem. Psychopharmacol. 9, 201 (1974).

141. Perel, J. M., Shostak, M., Gann, E., Kantor, S. J., and Glassman, A. H., Pharmacodynamics of imipramine and clinical outcome in depressed patients. In "Pharmacokinetics of Psychoactive Drugs" (L. Gottschalk and S. Marlies, eds.), p. 229. Spectrum, New York, 1975.

142. Peterson, J. A., Camphor binding by pseudomonas putida cytochrome P-450. Arch. Biochem. Biophys. 144, 678 (1971).

143. Poland, A., Smith, D., Kuntzman, R., Jacobson, M., and Conney, A. H., Effect of intensive occupational exposure to DDT on phenylbutazone and cortisol metabolism in humans. Clin. Pharmacol. Ther. 11, 725 (1970).

143a. Price, H. L., Deutsch, S., Cooperman, L. H., Clement, A. J., and Epstein, R. M., Splanchnic circulation during cyclopropane anesthesia in normal man. Anesthesiology 26, 312 (1965).

143b. Price, H. L., and Pauca, A. L., Effects of anesthesia on the peripheral circulation. Clin. Anesthesiol. 3, 73 (1969).

144. Quinn, G. P., Axelrod, J., and Brodie, B. B., Species and sex differences in metabolism and duration of action of hexobarbital (Evipan). Fed. Proc., Fed. Am. Soc. Exp. Biol. 13, 396 (1954).

145. Quinn, G. P., Axelrod, J., and Brodie, B. B., Species, strain and sex differences in metabolism of hexobarbitone, amidopyrine, antipyrine and aniline. Biochem. Pharmacol. 1, 152 (1958).

146. Rane, A., Garle, M., Borga, O., and Sjöqvist, F., Plasma disappearance of transplacentally transferred diphenyl hydantoin in the newborn studied by mass fragmentography. Clin. Pharmacol. Ther. 15, 39 (1974).

147. Reidenberg, M. M., James, M., and Dring, L. G., The rate of procaine hydrolysis in serum of normal subjects and diseased patients. Clin. Pharmacol. Ther. 13, 279 (1972).

148. Reidenberg, M. M., Kostenbauder, H., and Adams, W., Rate of drug metabolism in obese volunteers before and during starvation and in azotemic patients. Metab. Clin. Exp. 18, 209 (1969).

148a. Reidenberg, M. M., Lowenthal, D. T., Briggs, W., and Gasparo, M., Pentobarbital elimination in patients with poor renal function. Clin. Pharmacol. Ther. 20, 67 (1976).

149. Reidenberg, M. M., Odar-Cederlof, I., von Bahr, C., Borga, O., and Sjöqvist, F., Protein binding of diphenyl hydantoin and desmethylimipramine in plasma from patients with poor renal function. N. Eng. J. Med. 285, 264 (1971).

150. Remmer, H., Die Beschleunigung des Evipanabhause unter der Wirkung von Barbituraten. Naturwissenschaften 45, 189 (1958).

151. Remmer, H., The role of the liver in drug metabolism. Am. J. Med. 49, 617 (1970).

152. Remmer, H., Induction of drug metabolizing enzyme system in the liver. Eur. J. Clin. Pharmacol. 5, 116 (1972).

153. Remmer, H., and Merker, H. J., Effect of drugs on the formation of smooth endoplasmic reticulum and drug-metabolizing enzymes. Ann. N. Y. Acad. Sci. 123, 79 (1965).

154. Remmer, H., Schenkman, J. B., Estabrook, R. W., Sasame, H.A., Gillette, J., Narasimhulu, S., Cooper, D. Y., and Rosenthal, O., Drug interaction with hepatic microsomal cytochrome. Mol. Pharmacol. 2, 187 (1966).

155. Robinson, D. S., and Amidon, E. L., Interaction of benzodiazepines with warfarin in man. *In* "The Benzodiazepines" (S. Garattini, E. Mussini, and L. Randall, eds.) p. 641. Raven Press, New York, 1973.

156. Robinson, D. S., and Sylwester, D., Interaction of commonly prescribed drugs with warfarin. *Ann. Intern. Med.* **72**, 853 (1970).

157. Rubin, E., Bacchin, P., Gang, H., and Lieber, C. S., Induction and inhibition of hepatic microsomal and mitochondrial enzymes by ethanol. *Lab. Invest.* **22**, 569 (1970).

158. Rubin, E., and Lieber, C. S., Hepatic microsomal enzymes in man and rat: Induction and inhibition by ethanol. *Science* **162**, 690 (1968).

159. Rubin, E., Lieber, C. S., Alvares, A. P., Levin, W., and Kuntzman, R., Ethanol binding to hepatic microsomes—its increase by ethanol consumption. *Biochem. Pharmacol.* **20**, 229 (1971).

160. Rubin, A., Tephly, T. R., and Mannering, G. J., Kinetics of drug metabolism by hepatic microsomes. *Biochem. Pharmacol.* **13**, 1007 (1964).

161. Sasame, H. A., Castro, J. A., and Gillette, J. R., Studies on the destruction of liver microsomal cytochrome P-450 by carbon tetrachloride administration. *Biochem. Pharmacol.* **17**, 1759 (1968).

162. Sasame, H. A., and Gillette, J. R., Studies on the inhibitory effects of various substances on drug metabolism by liver microsomes: The effect of nicotinamide in altering the apparent mechanism of inhibition. *Biochem. Pharmacol.* **19**, 1025 (1970).

163. Sawyer, D. C., Eger, E. I., Bahlman, S. H., Cullen, B. F., and Impelman, D., Concentration dependence of hepatic halothane metabolism. *Anesthesiology* **34**, 230 (1971).

164. Schenkman, J. B., Cinti, D. L., Moldeus, P. W., and Orrenius, S., Newer aspects of substrate binding to cytochrome P-450. *Drug Metab. Dispos.* **1**, 111 (1973).

165. Schenkman, J. B., Frey, I., Remmer, H., and Estabrook, R. W., Sex differences in drug metabolism by rat liver microsomes. *Mol. Pharmacol.* **3**, 516 (1967).

166. Sellers, E. M., and Koch-Weser, J., Kinetics and clinical importance of displacement of warfarin from albumin by acidic drugs. *Ann. N. Y. Acad. Sci.* **179**, 213 (1971).

167. Sellman, R., Kanto, J., Raijola, E., and Pekkarinen, A., Human and animal study on elimination from plasma and metabolism of diazepam after chronic alcohol intake. *Acta Pharmacol. Toxicol.* **36**, 33 (1975).

168. Shahidi, N. T., Acetophenotidin sensitivity. *Am. J. Dis. Child.* **113**, 81 (1967).

169. Shahidi, N. T., Acetophenetidin induced methehemoglobinemia *Ann. N. Y. Acad. Sci.* **151**, 822 (1968).

170. Shideman, F. E., Kelly, A. R., and Adams, B. J., The role of the liver in the detoxication of thiopental (Pentothal) and two other thiobarbiturates. *J. Pharmacol. Exp. Ther.* **91**, 331 (1947).

171. Shoeman, D. W., Chaplin, D. W., and Mannering, G. J., Induction of drug metabolism. III. Further evidence for the formation of a new P-450 hemoprotein after treatment of rats with 3-methylcholanthrene. *Mol. Pharmacol.* **5**, 412 (1969).

172. Sjöqvist, F., and von Bahr, C., Interindividual differences in drug oxidation: Clinical importance. *Drug Metab. Dispos.* **1**, 469 (1973).

173. Skrinjaric-Spoljar, M., Mathews, H. B., Engel, J. L., and Casida J. E., Response of hepatic microsomal mixed function oxidase to various types of insecticide chemical synergists administered to mice. *Biochem. Pharmacol.* **20**, 1607 (1971).

174. Sladek, N. E., and Mannering, G. J., Induction of drug metabolism. I. Differences in the mechanisms by which polycyclic hydrocarbons and phenobarbital produce their inductive effects on microsomal N-demethylating systems. *Mol. Pharmacol.* **5**, 174 (1969).

175. Slater, T. F., Necrogenic action of carbon tetrachloride in the rat: A speculative mechanism based on activation. *Nature (London)* **209**, 36 (1966).

176. Smuckler, E. A., Iseri, O. A., and Benditt, E. P., An intracellular defect in protein synthesis induced by carbon tetrachloride. *J. Exp. Med.* **116**, 55 (1962).

177. Stenger, R. J., and Johnson, E. A., Effects of phenobarbital pretreatment on the response of rat liver to halothane administration. *Proc. Soc. Exp. Biol. Med.* **140**, 1319 (1972).

177a. Stenson, R. E., Constantino, R. E., and Harrison, D. C., Interrelationships of hepatic blood flow, cardiac output and blood levels of lidocaine in man. *Circulation* **43**, 205 (1971).

178. Stevenson, I. H., O'Malley, K., Turnbull, M. J., and Ballinger, B. R., The effect of chlorpromazine on drug metabolism. *J. Pharm. Pharmacol.* **24**, 577 (1972).

179. Stier, A., Kunz, H. W., Walli, A. K., and Schimassek, H., Effects on growth and metabolism of rat liver by halothane and its metabolite trifluoracetate. *Biochem. Pharmacol.* **21**, 2181 (1972).

180. Strong, J. M., Mayfield, D. E., Atkinson, A. J., Burris, B. C., Raymon F., and Webster, L. T., Jr., Pharmacological activity, metabolism and pharmacokinetics of glycinexylidide. *Clin. Pharmacol. Ther.* **17**, 184 (1975).

181. Taves, D. R., Fry, B., Freeman, R. B., and Gillies, A. J., Toxicity following methoxyflurane anesthesia. II. Fluoride concentration in nephrotoxicity. *J. Am. Med. Assoc.* **214**, 91 (1970).

182. Thomson, P. D., Melmon, K. L., Richardson, J. A., Cohn, K., Steinbounn, W., Cudihee, R., and Rowland, M., Lidocaine pharmacokinetics in advanced heart failure, liver disease, and renal failure in humans. *Ann. Intern. Med.* **78**, 499 (1973).

183. Thomson, P. D., Rowland, M., and Melmon K. L., The influence of heart failure, liver disease and renal failure on the disposition of lidocaine in man. *Am. Heart J.* **82**, 417 (1971).

183a. Tokola, O., Pelkonen, O., Karki, N. T., Luoma, P., Kaltiala, E. H., and Larmi, T. K. I., Hepatic drug-oxidizing enzyme systems and urinary d-glucaric acid excretion in patients with congestive heart failure. *Br. J. Clin. Pharmacol.* **2**, 429 (1975).

184. Triggs, E. J., and Nation, R. L., Pharmacokinetics in the aged: A review. *J. Pharmacokinet. Biopharm.* **3**, 387 (1975).

185. Trolle, D., Phenobarbitone for low-birth-weight babies. *Lancet* **2**, 1123 (1968).

186. Uehleke, H., Extrahepatic microsomal drug metabolism. *Proc. Eur. Soc. Study Drug Toxic.* **10**, 94 (1968).

187. Vajda, F. J. E., Prineas, R. J., and Lovell, R. R. H., Interaction between phenytoin and the benzodiazepines. *Br. Med. J.O,* 346 (1971).

188. Van Dam, F. E., and Gribnau-Overkamp, M. J. H., The effect of some sedatives (phenobarbital, glutethimide, chlordiazepoxide, chloral hydrate) on the rate of disappearance of ethyl biscoumacetate from the plasma. *Folia Med. Neerl.* **10**, 141 (1967).

189. Van der Kleijn, E., Van Rossum, J. M., Muskens, E. T. J. M., and Rijntjes, N. V. M., Pharmocokinetics of diazepam in dogs, mice and humans. *Acta Pharmacol. Toxicol.* **29**, Suppl. 3, 109 (1971).

190. Van Dyke, R. A., Chenoweth, M. B., and Van Poznak, A., Metabolism of volatile anesthetics I. Conversion in vivo of several anesthetics to $^{14}CO_2$ and chloride. *Biochem. Pharmacol.* **13**, 1239 (1964).

191. Van Dyke, R. A., and Wood, C. L., Binding of radioactivity from ^{14}C-labeled halothane in isolated perfused rat livers. *Anesthesiology* **38**, 328 (1973).

192. Vesell, E. S., Advances in pharmacogenetics. *Prog. Med. Genet.* **9**, 291 (1973).

193. Vesell, E. S., Factors causing interindividual variations of drug concentrations in blood. *Clin. Pharmacol. Ther.* **16**, 135 (1974).

194. Vesell, E. S., Application of pharmacokinetic principles to the elucidation of polygenically controlled differences in drug response. *In* "Pharmacology and Pharmcokinetics" T. Teorell, R. L. Dedrick, and P. G. Condliffe, eds., p. 261. Plenum, New York, 1975.

195. Vesell, E. S., and Page, J. G., Genetic control of phenobarbital induced shortening of plasma antipyrine half-lives in man. *J. Clin. Invest.* **48**, 2202 (1969).

196. Vesell, E. S., and Passananti, G. T., Inhibition of drug metabolism in man. *Drug Metab. Dispos.* **1**, 402 (1973).

197. Vest, M. F., and Streiff, R. R., Studies on glucuronide formation in newborn infants and older children. *J. Dis. Child.* **98**, 688 (1959).

198. Wattenberg, L. W., and Leony, J. L., Effects of phenothiazines on protective systems

against polycyclic hydrocarbons. *Cancer Res.* **25**, 365 (1965).

199. Welch, R. M., Cavallito, C. J., and Loh, A., Effect of exposure to cigarette smoke on the metabolism of benzopyrene and acetophenetidin by lung and intestine of rats. *Toxicol. Appl. Pharmacol.* **23**, 749 (1972).

200. Welch, R. M., Harrison, Y. E., Gommi, B. W., Poppers, P. J., Finster, M., and Conney, A. H., Stimulatory effect of cigarette smoking on the hydroxylation of 3.4-benzpyrene and n-demethylation of 3-methyl-4-monomethylaminoazobenzene by enzymes in human placenta. *Clin. Pharmacol. Ther.* **10**, 100 (1969).

201. Werk, E. E., Sholiton, L. J., and Olinger, C. P., *Int. Congr. Proc. Horm. Steroids 2nd, 1966 Excerpta Med. Found. Ser. No 132* Vol. 3, 301 (1967).

202. Whittaker, J. A., and Price Evans, D. A., Genetic control of phenylbutazone metabolism in man. *Br. Med. J.* **4**, 323 (1970).

202a. Wilkinson, G. R., and Schenker, S., Letter to the editor. Pharmacokinetics of meperidine in man. *Clin. Pharmacol. Ther.* **19**, 486 (1976).

202b. Wilkinson, G. R., and Shand, D. G., Commentary. A physiological approach to hepatic drug clearance. *Clin. Pharmacol. Ther.* **18**, 377 (1975).

202c. Williams, R. L., Blaschke, T. F., Meffin, P. J., Melmon, K. L., and Rowland, M., Influence of acute viral hepatitis on disposition and plasma binding of tolbutamide. *Clin. Pharmacol. Ther.* **21**, 301 (1977).

203. Williams, R. T., "Detoxification mechanisms." Wiley, New York, 1959.

204. Zamboni, L., Electron microscopic studies of blood embryogenesis in humans. I. The ultrastructure of the fetal liver. *J. Ultrastruct. Res.* **12**, 509 (1965).

205. Zarday, Z., Deery, A., Tellis, I., Soberman, R., and Foldes, F. F., Plasma and red cell cholinesterase activity in uremic patients. Effects of hemodialysis and renal transplantation. *J. Med.* **6**, 337 (1975).

Chapter 10

Enzymes of Synthesis and Degradation of Catecholamines and 5-Hydroxytryptamine

S. H. Ngai

Introduction

Catecholamines are biogenic amines with two hydroxyl groups on the phenyl ring at the 3- and 4-positions, and a side chain of ethanolamine or ethylamine. The well-known naturally occurring catecholamines are epinephrine, norepinephrine, and dopamine. Epinephrine found in the adrenal medulla is considered a "hormone," because it is carried by the circulation to other parts of the body to exert its physiologic action. Norepinephrine and dopamine are present in adrenergic neurons in both the central and peripheral nervous systems. It is now well established that norepinephrine is the transmitter of the sympathetic nervous system, regulating the function of most if not all organs and tissues [for reviews, see, for example, Acheson (1), Blaschko and Muscholl (12), and Iversen (52)]. There is evidence that certain functions of the central nervous system are regulated by these biogenic amines, such as the state of consciousness (56), vasomotor control, and thermoregulation. In the basal ganglia, dopamine has been demonstrated to be an important transmitter of extrapyramidal modulation of the somatic motor system (50). The range of physiologic roles of catecholamines in the central nervous system has yet to be completely explored. But, what is known or has been assumed concerning the function of the catecholamine system has already led to therapeutic applications. Many modern therapeutic agents for cardiovascular diseases, neurologic, and psychiatric disorders have their primary action on the adrenergic system, as agonists or antagonists. There are also numerous drugs which interfere with the biosynthesis, storage, release, or catabolism of catecholamines. It is therefore essential for anesthesiologists to have a working knowledge of the biosynthesis and degradation of catecholamines and the enzymes catalyzing these reactions. Equally important is the understanding of action of drugs on this system.

5-Hydroxytryptamine (serotonin, 5-HT) is another biogenic amine, having an indole nucleus with a hydroxyl group at the 5-position and ethylamine as the side chain. The role of 5-hydroxytryptamine as a peripheral neurotransmitter is not well defined. However, its presence in tissues and platelets and its effects on the function of many organs and systems bespeak of its possible or probable physiologic importance. In the central nervous system, 5-hydroxytryptamine is considered a putative transmitter, believed to play a role in the state of consciousness, behavior, and regulation of certain visceral functions. Additionally, there is evidence suggesting that interaction between norepinephrine and 5-hydroxytryptamine is important in central regulatory mechanisms.

Both the adrenergic and serotoninergic systems may be affected by the state of anesthesia, anesthetics, narcotics, and other anesthetic adjuvants, resulting in changes in visceral functions. In addition, many patients undergoing anesthesia and surgical procedures are being treated with drugs that act primarily on these biogenic amine systems. Knowledge of drug actions and interactions is important for the safe conduct of anesthesia.

Progress in studies on catecholamines and 5-hydroxytryptamine during the past two decades is attested by the numerous reports and frequent symposia, dealing with all aspects of the subject. Continued intensive research in this area lends promise of new discoveries in terms of concepts and pharmacologic agents which might have therapeutic implications.

There are many extensive reviews, monographs, and symposium proceedings (1, 9, 11–13, 18, 20, 37, 50, 52, 99, 100, 104, 114) which summarize the biochemical, physiologic, pharmacologic, and certain clinical significance of these biogenic amines and their enzymes. All are the work of experts who are principally occupied with the intricacies on these subjects. This author does not claim to have the same degree of expertise, being at best an amateur with interest in this area from the point of view of an anesthetist. This chapter deals therefore with only the fundamentals. Those readers wishing to delve deeper into the subject are respectfully referred to the preceding citations.

Biosynthesis and Degradation of Catecholamines

Before discussing those enzyme systems involved in the synthesis and degradation of catecholamines, it may be desirable to briefly outline the biochemical pathways and currently accepted concepts concerning their storage and release.

Figure 10-1 shows the reactions involved in the biosynthesis of catecholamines. L-Tyrosine, an amino acid derived from food, or from hydroxylation of phenylalanine at the 4-position on the phenyl ring, is hydroxylated at the 3-position to become dihydroxyphenylalanine (dopa). This step is catalyzed

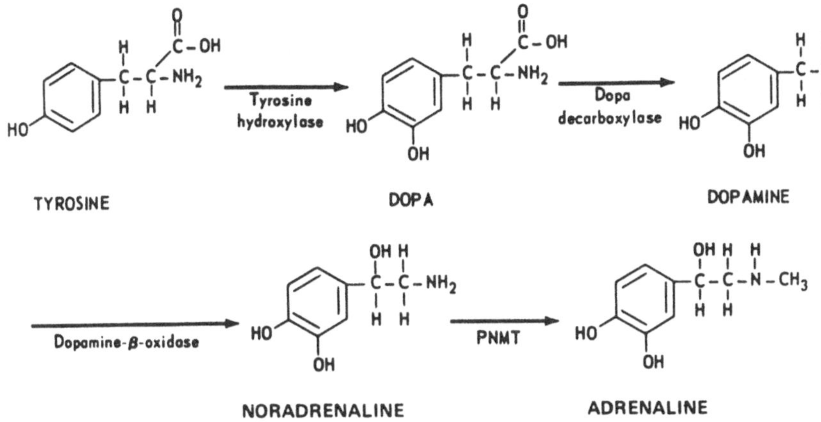

Figure 10-1. Pathways for the biosynthesis of norepinephrine and epinephrine. [From Axelrod (11), with kind permission of the author and Academic Press.]

by the enzyme tyrosine hydroxylase, which requires tetrahydropteridine as the cofactor, Fe^{2+}, and molecular oxygen. Dopa, which is also an amino acid, is then decarboxylated to become dopamine, a catecholamine. Dopamine is then hydroxylated at the beta carbon to norepinephrine by the action of the enzyme dopamine-β-hydroxylase. Finally, norepinephrine can be converted to epinephrine by a methylating enzyme, phenylethanolamine-N-methyltransferase. This reaction occurs principally in the adrenal medulla although there is evidence that it may also take place in the central nervous system (47).

Catecholamines are localized in adrenergic cells in the central nervous system and in the peripheral sympathetic system. Biosynthesis of dopamine and norepinephrine takes place within the neuronal cell body and nerve terminals where these amines act as neurotransmitters. The various specific enzymes and cofactors therefore are present within these cells.

Tyrosine is actively transported into the cell from blood. Conversion of tyrosine through dopa to dopamine proceeds in the cytoplasm, although there is still some question concerning the exact site of the enzyme tyrosine hydroxylase. In neurons containing dopamine as the neurotransmitter, it is then taken up by the amine storage granules and stored without further change until released. In norepinephrine-containing neurons, dopamine-β-hydroxylase, in or on the membrane of the storage granules, converts dopamine to norepinephrine.

Catecholamines stored in the granules are bound to ATP. The ratio of catecholamine and ATP has been estimated to be 4 moles of amines to 1 mole of ATP. The membrane of storage granules has the unique ability to concentrate these amines. The concentration of amines within the granules is said to be more than 200 times greater than that in the cytoplasm. This mechanism serves the important function of protecting catecholamines from catabolism by the enzyme monoamine oxidase.

Upon arrival of nerve impulses at the nerve endings, catecholamine in the storage granules is released through a Ca^{2+} dependent process of exocytosis together with ATP and a portion of dopamine-β-hydroxylase. The released catecholamine binds with effector cells to exert its specific action. Catecholamine also binds with nonspecific sites without known physiologic consequence. However, most of the released catecholamines (with the exception of epinephrine released from the adrenal medulla) are taken back into the nerve terminals and reenter the storage granules. The process of reuptake is believed to be the principal mechanism in terminating the physiolgic effect of the sympathetic transmitter. Reuptake of catecholamines involves active transport across the cell membrane. The process obeys saturation kinetics of the Michaelis–Menten type. Thus, under certain circumstances, as during severe stress or high frequency stimulation of the sympathetic nerve, massive release of catecholamines could saturate or even inhibit the reuptake process. Only then does the released catecholamine overflow into the blood stream. It is relevant to mention here that exogenously administered catecholamines, in

Figure 10-2.   Metabolic pathways of norepinephrine and epinephrine. COMT, catechol-O-methyltransferase; MAO, monoamine oxidase. [From Axelrod (11), with kind permission of the author and Academic Press.]

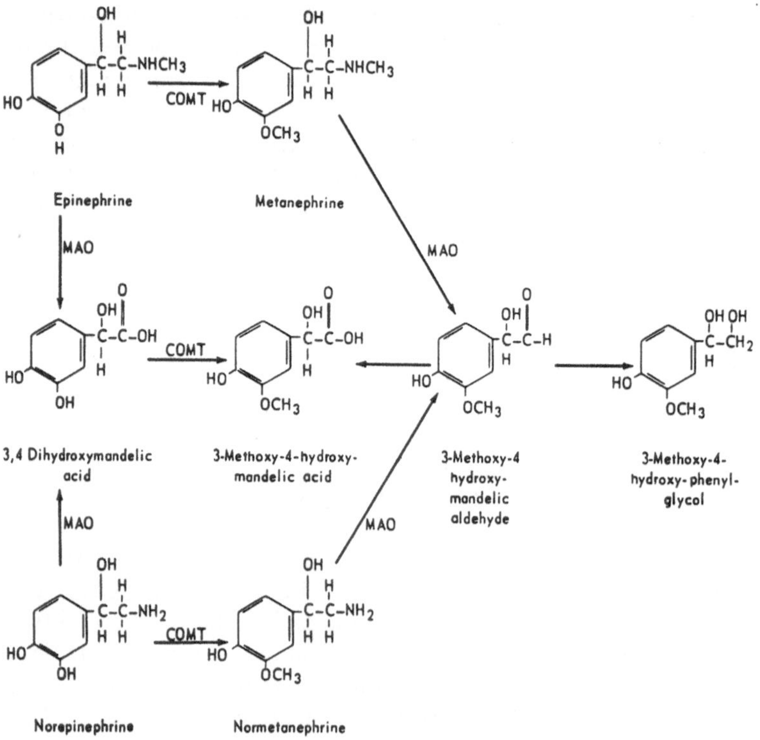

addition to being catabolized enzymatically, are also taken up by sympathetic nerve terminals. Epinephrine released from the adrenal medulla is carried by blood to act on effector cells remote from the site of release until inactivated by catabolizing enzymes or taken up by sympathetic nerve terminals.

Figure 10-2 shows the many pathways of catecholamine catabolism. The principal reactions are oxidative deamination catalyzed by monoamine oxidases, and O-methylation (methylation of the hydroxyl group at the 3-position on the phenyl ring by the action of catechol-O-methyltransferase). Within the neuron and nerve terminals, free catecholamines exist in the cytosol in equilibrium with those in the storage granules. Mitochondrial monoamine oxidases degradate free catecholamines to give acid and alcohol metabolites. O-Methylation of catecholamines takes place extraneuronally. Norepinephrine and epinephrine are converted to amine metabolites, normetanephrine and metanephrine, respectively. The deaminated products also undergo O-methylation. Conversely, O-methylated amines are oxidatively deaminated. The final products of catabolism of norepinephrine and epinephrine are 3-methyl-4-hydroxymandelic acid (VMA) and 3-methyl-4-hydroxyphenylglycol. These acid metabolites, together with amine metabolites, are found in the urine. Metanephrine and normetanephrine are excreted free or conjugated with sulfate or glucoronide.

In the case of dopamine, oxidative deamination yields 3,4-dihydroxyphenylacetic acid (DOPAC), O-methylation gives 3-methoxytyramine. The final metabolite is homovanillic acid (HVA) (Figure 10-3).

Figure 10-3. Biosynthesis of catecholamines from L-tyrosine and L-dopa, and metabolism of dopamine. Steps 1 and 2 as in Figure 10-1. Step 3 is catalyzed by catechol-O-methyltransferase; step 4 by monoamine oxidase, and step 5 by dopamine-β-hydroxylase. [Modified from Hornykiewicz (50).]

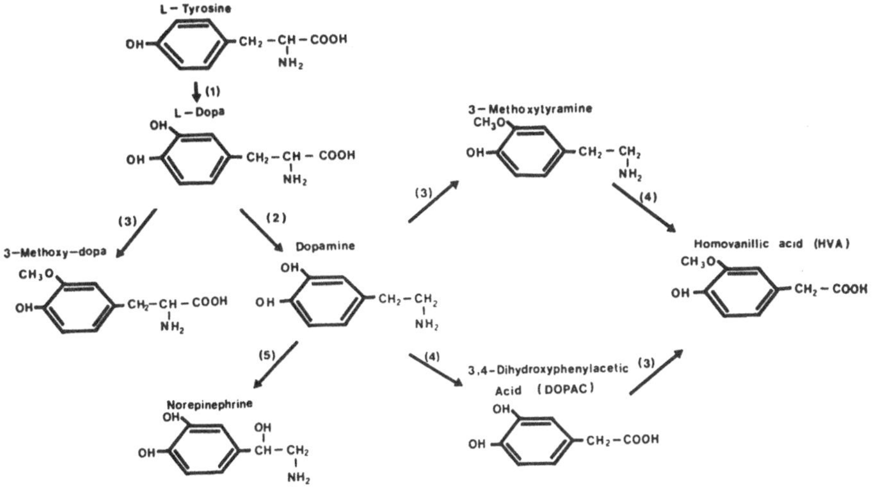

Effects of Anesthetics on the Catecholamine System

Commonly used anesthetics have varying effects on the circulation. Some agents, such as diethyl ether, cyclopropane, and ketamine (Ketalar) appear to activate the sympathoadrenal system, whereas others, halothane (Fluothane), methoxylurane (Penthrane), and enflurane (Ethrane) lack this action or depress circulation in clinically used concentrations. The basis for this could be the action of anesthetics on the central vasomotor control mechanisms. For example, it has been reported that cyclopropane increases efferent sympathetic discharge (84) and halothane depresses it (92). While there is general agreement on the central depressant action of halothane (34), controversy still exists on the site of action of cyclopropane. A recent study in the cat showed that cyclopropane, like halothane, also decreases the preganglionic sympathetic discharge frequency as measured along the splanchnic nerve (35).

Circulatory changes induced by anesthetics could have their basis in the periphery if the release, reuptake, biosynthesis, or degradation of catecholamines were influenced by anesthetics. Price *et al.* (83) reported in 1959 that in man cyclopropane increases plasma norepinephrine concentrations, whereas halothane has no significant effect. More recently, Roizen *et al.*, using a much more sensitive radiometric assay, have reported that in rats, plasma total catecholamine and norepinephrine concentrations decrease during both halothane and cyclopropane anesthesia (85a,b).

The availability of isotopic norepinephrine and its precursors made it easier to study this problem. Using ^{3}H-norepinephrine as a tracer, it has been found *in vivo* and *in vitro* that cyclopropane and halothane do not significantly affect the release and reuptake of norepinephrine by adrenergic nerve terminals in the heart (16, 75, 79). In isolated, perfused rat heart, ketamine appears to inhibit the uptake of ^{3}H-norepinephrine (69).

Using ^{14}C-labeled tyrosine it was found that in rats neither cyclopropane nor halothane has a significant effect on the rate of norepinephrine synthesis in the brain and in the heart. This implies that these anesthetics probably do not influence the enzyme activity involved in the synthesis or the rate of utilization (metabolism) of norepinephrine (80).

Catecholamines, as neurotransmitters, act on the cyclic 3′,5′-adenosine monophosphate (cyclic AMP) system by stimulating the enzyme adenylate cyclase. Cyclic AMP, designated as the second messenger in effector cells, alters the function of target organs, in the case of circulatory system, heart, and vascular smooth muscles. It is possible that anesthetics may alter receptor sensitivity to catecholamines. This has been demonstrated recently (96, 97). Halothane and isoflurane (Forane) relax vascular smooth muscles through their action on the cyclic AMP system. The formation of cyclic AMP from prelabeled ATP in the rat aorta is increased by these agents (97). In contrast, cyclopropane increases the contractile response of rat aortic strips to phenyl-ephrine and decreases cyclic AMP formation (96).

Enzymes for the Biosynthesis and Catabolism of Catecholamines

Enzymes for the synthesis of catecholamine, namely tyrosine hydroxylase (TH), aromatic L-amino acid decarboxylase (AADC), and dopamine-β-hydroxylase (DβH), are synthesized in the cell body and transported distally along the axon to the nerve terminals. So are the amine storage granules. These enzymes are catabolized, and in the case of dopamine-β-hydroxylase, partially released together with the amine. Thus, the loss is replenished through proximodistal axoplasmic transport. The rate of axoplasmic transport of these substances is considered rapid as compared to that of certain other cellular components. The rate of transport is fastest with dopamine-β-hydroxylase. It has been estimated that in the rat sciatic nerve dopamine-β-hydroxylase travels proximodistally at a rate of 2–3 mm per hr. Aromatic L-amino acid decarboxylase and tyrosine hydroxylase are transported at a slower rate, 0.6–0.8 mm per hr (24, 81). One recent report on the axoplasmic transport of dopamine-β-hydroxylase along the sural nerve in man gives a rate of 2 mm per hr (14).

The effect of anesthetics on axoplasmic transport has been studied recently. Halothane in clinically used concentrations does not appear to affect proximodistal transport of radioactive material (^{3}H-leucine) along the rabbit vagus nerve *in vitro* (57). Lidocaine (Xylocaine) inhibits axoplasmic transport in the rabbit vagus nerve *in vitro* in a dose-dependent manner (33). However, *in vivo* studies in guinea pigs indicate that application of lidocaine or etidocaine (Duranest) to the sciatic nerve to produce analgesia does not slow the axoplasmic transport of catecholamine-synthesizing enzymes (78a).

Although catecholamines are synthesized in the cell body and transported to the nerve terminals in the storage granules (22), it is safe to assume that most of the amines present at the nerve terminals are synthesized at the nerve terminals. The amine is constantly synthesized to replenish the losses from release and degradation. The rate of synthesis varies in different tissues; it is controlled by enzyme activity which in turn is regulated by various mechanisms. One of these is the relative level of neuronal activity. The intricate regulatory mechanisms serve to keep the tissue stores of amines relatively constant in the face of fluctuating sympathetic activity under most physiologic conditions. It is only during extreme conditions, for example, cold stress and vigorous exercise, when changes in tissue amine concentrations may be observed. However, tissue amine concentration can be drastically changed by pharmacologic agents which interfere with the uptake, release, and storage of the transmitters or drugs which inhibit enzymes of synthesis and degradation of these amines.

Tyrosine Hydroxylase (TH, EC 1.1.4.3)

Tyrosine hydroxylase was first isolated and characterized by Nagatsu *et al.* in 1964 (72, 73). It is present in the adrenal medulla, heart, brain, and all adrenergic tissues. The enzyme extracted from the beef adrenal medulla has been purified 400- to 500-fold. The measured K_m of purified tyrosine hydroxylase is approximately 5×10^{-5} M, and its V_{max} is 150 (51,103). The consensus is that tyrosine hydroxylase is a soluble enzyme present in the cytosol of the adrenergic neuron (108).

To convert tyrosine to dopa, tyrosine hydroxylase needs tetrahydropteridine as the cofactor. For maximal activity Fe^{2+} and molecular oxygen are also required (103). Hypoxia decreases tyrosine hydroxylase activity but this effect is not as marked as with tryptophan hydroxylase (17). Although the enzyme is capable of hydroxylating phenylalanine to tyrosine, its activity is rather specific in that D-tyrosine, tyramine, and tryptophan are not substrates of this enzyme.

Under usual conditions, the plasma and tissue concentrations of tyrosine are sufficient to saturate the enzyme tyrosine hydroxylase (103). As the subsequent steps of catecholamine biosynthesis can proceed at much faster rates, hydroxylation of tyrosine is considered the rate-limiting step of the whole biosynthetic process of catecholamines. It is also the step at which the rate of synthesis of catecholamines is regulated.

In addition to the adrenal medulla, the blood vessels (e.g., mesenteric arteries and veins) are rich in tyrosine hydroxylase, reflecting their relatively high norepinephrine concentrations. It is believed that the enzymatic component in blood vessels is a major factor in maintaining functional levels of the neurotransmitter and in regulating its disposition, hence, homeostatis of circulation (95).

The activity of tyrosine hydroxylase is subject to end-product inhibition, that is, intraneuronal concentrations of norepinephrine (or other catechols) modulate the rate at which tyrosine is converted to dopa. Tyrosine hydroxylase activity can be accelerated acutely in a matter of seconds without a change in its tissue concentration. Sympathetic activation releases norepinephrine. This is associated with an increase in the rate of norepinephrine synthesis, occurring at the rate-limiting step. It has been shown that stimulation of hypogastric nerve increases the synthesis of ^{14}C-norepinephrine from ^{14}C-tyrosine in the vas deferens (3). Similarly, stimulation of the cervical sympathetic nerve increased the formation of ^{14}C-norepinephrine from ^{14}C-tyrosine in the salivary gland (89).

On the other hand, elevation of catecholamine concentrations in adrenergic tissues, such as that caused by monoamine oxidase inhibition or by exogenously administered norepinephrine decreases norepinephrine synthesis at the step of hydroxylation of tyrosine. The formation of norepinephrine from labeled tyrosine, but not from dopa, is decreased (94). By using dopa as the precursor, the rate-limiting step is bypassed.

Therefore, it has been concluded that the intraneuronal concentration of catecholamine serves to modulate the activity of tyrosine hydroxylase, through a negative feedback mechanism on tyrosine hydroxylase. Norepinephrine as well as other catechols are inhibitors of tyrosine hydroxylase, being competitors of the pteridine cofactor. During increased nerve activity, norepinephrine is released from the nerve, resulting in a reduced concentration of norepinephrine associated with the inhibitory binding site on tyrosine hydroxylase. Consequently, tyrosine hydroxylase activity is increased through the release from end-product inhibition. Additionally, it has been suggested that sympathetic nerve stimulation may, in some unknown manner, increase tyrosine hydroxylase activity (108).

Another regulatory mechanism of tyrosine hydroxylase activity is a slower process, apparently occurring in hours or days. Chronic sympathetic activation can "induce" or increase the tyrosine hydroxylase levels in the neuron (11, 82). Stimulation of the hypothalamus, electric shock, cold exposure, and other stresses all result in an increase in enzyme concentrations. In animals reserpine treatment induces tyrosine hydroxylase, apparently owing to the loss of tissue catecholamines and to the compensatory increase in sympathetic activity (101). In rats, adrenal demedullation also increases the rate of norepinephrine synthesis from labeled tyrosine, again, in compensation for the loss of catecholamine normally derived from the adrenal medulla (77). Insulin increases tyrosine hydroxylase (109) and dopamine-β-hydroxylase activity in the adrenal medulla, probably through sympathetic activation.

The induction of tyrosine hydroxylase (and dopamine-β-hydroxylase) during chronic sympathetic excitation occurs transsynaptically. If the adrenal medulla or sympathetic ganglion is denervated (decentralized), then the tissue enyzme levels do not increase (82).

As is the case with acute increase in enzyme activity, tyrosine hydroxylase levels can be reduced in situations where the tissue concentrations of catecholamines are increased over a prolonged period of time. An example is chronic treatment with dopa. Dopa treatment leads to an increase in the total concentration of catechols. This is associated with a progressive decrease in tyrosine hydroxylase levels in the adrenal medulla, as well as other tissues such as the heart and blood vessels (23). In spontaneous hypertensive rats, tyrosine hydroxylase activity in the mesenteric artery is about one-third that of normotensive rats of the same strain (95). It appears that in this case compensatory changes in the enzyme activity in vascular tissues contribute to the circulatory homeostasis.

Synthesis of tyrosine hydroxylase, together with dopamine-β-hydroxylase and phenylethanolamine-N-methyltransferase, is also under the influence of the adrenocorticotrophic hormone (ACTH). Hypophysectomy leads to decreases in the adrenal medullary levels of these enzymes which can be restored partly by ACTH (71, 110).

Tyrosine Hydroxylase Inhibitors

Inhibitors of tyrosine hydroxylase have been used extensively as pharmacologic tools for the study of catecholamines. The potent inhibitors are α-methyl-p-tyrosine, some of the 3- or 3,5-halogenated tyrosines. Their K_1 values range from 5×10^{-5} to about 2×10^{-7} M (103). Certain catechol derivatives are also potent inhibitors, the most potent of these is 3,4-dihydroxyphenylpropylacetamide (Hässle 22/54). Its K_1 is 2×10^{-5} M. It has been mentioned earlier that norepinephrine itself is a tyrosine hydroxylase inhibitor, serving as an important modulator of its own synthesis. However, compared with the above-mentioned tyrosine and catechol derivatives, norepinephrine is less potent, with a K_1 of approximately 1×10^{-3} M (106).

The action of these enzyme inhibitors is competitive and reversible. Since tyrosine hydroxylase catalyzes the rate-limiting step in catecholamine biosynthesis, these inhibitors can reduce tissue catecholamine concentrations. Complete enzyme inhibition with α-methyl-p-tyrosine or Hässle 22/54 has been used to estimate the turnover rate of catecholamines in various tissues *in vivo*. According to steady-state kinetics, the rate of decline of tissue catecholamine concentrations upon inhibition of tyrosine hydroxylase is taken as the rate of biosynthesis (or utilization) of amines, provided that at the time of enzyme inhibition (0 time) the tissue amine concentration is not changing (steady state) (15). Presumably, turnover rates of catecholamine reflect adrenergic activity. In normal rats, the turnover rate of norepinephrine is approximately 0.25 μg/g/hr in the brain (77), 0.07 in the heart, 0.25 in the mesenteric artery, and 0.31 in the mesenteric vein (95).

Depletion of tissue catecholamine stores by enzyme inhibition is associated with physiologic changes, to some extent similar to those caused by drugs such as reserpine (Serpasil). Depletion of norepinephrine in the brain by α-methyl-p-tyrosine results in sedation. The pressor response to the indirectly acting amine, tyramine, is reduced as expected but not abolished as in the case of complete depletion of catecholamines by reserpine (93).

Tyrosine hydroxylase inhibitors have also been studied in patients suffering from pheochromocytoma and essential hypertension (90). α-Methyl-p-tyrosine decreases the urinary excretion of catecholamine metabolites by more than 50%. In patients with pheochromocytoma, the arterial pressure fell significantly. But this effect was not clearly noted in patients with essential hypertension. Potentially tyrosine hydroxylase inhibitors could be useful in preparing patients with pheochromocytoma for surgery, by stopping the synthesis of catecholamines in the tumor.

Aromatic L-Amino Acid Decarboxylase (AADC, EC 4.1.1.26)

This enzyme decarboxylates dopa to form dopamine. It was discovered in 1938 by Holtz *et al.* originally designated as dopa decarboxylase (48). Later evidence indicates that several other aromatic amino acids, including 5-hydroxytryptophan and tyrosine, are also substrates of this enzyme. The name aromatic-L-amino acid decarboxylase (AADC) was suggested by Lovenberg *et al.* (68).

More recent studies further support the view that AADC is the one enzyme concerned with biosynthesis of both catecholamines and 5-hydroxytryptamine. Purified enzyme from bovine adrenal medulla and hog kidney yields one broad band on disc gel electrophoresis and the pattern of enzyme activity is similar using either dopa or 5-hydroxytryptophan as the substrate (19, 42). This enzyme is ubiquitous in its distribution. In human brain, AADC is found in all parts containing either catecholamines or 5-hydroxytryptamine (65).

AADC purified from hog kidney has a K_m of 1.9×10^{-4} M and a V_{max} of 8900 with dopa as the substrate (19). AADC requires pyridoxal phosphate as the cofactor. In addition to adrenal medulla (chromaffin cells), catecholamine and 5-hydroxytryptamine containing neurons, it is present in large amounts in the liver and kidney. The biologic significance of hepatic and renal AADC is not clear. But in the treatment of Parkinson's disease with dopa the abundance of AADC and catechol-O-methyltransferase (COMT) in the peripheral tissue is of considerable importance.

Inhibitors of Aromatic L-Amino Acid Decarboxylase

α-Methyldopa (Aldomet) is an inhibitor of AADC. This action was thought to be the basis of the antihypertensive effect of this drug through a reduction in the synthesis of catecholamines. However, the antihypertensive action does not coincide temporally with the inhibition of AADC. Also, other AADC inhibitors do not have this hypotensive effect. Subsequently it was shown that α-methyldopa is converted to α-methyldopamine and α-methylnorepinephrine, the latter considered to be a "false transmitter." In other words, α-methylnorepinephrine so synthesized is stored in the adrenergic nerve terminals and released in the same manner as norepinephrine, but it is not as potent as norepinephrine. This view has been challenged again recently since it was found that α-methylnorepinephrine is as potent a transmitter as norepinephrine. It is now believed that α-methyldopa acts on the central nervous system after its conversion to α-methylnorepinephrine [for review, see Laverty (63)].

Other potent inhibitors of AADC studied extensively are hydrazine derivatives, hydrazine analogs of α-methyldopa (carbidopa, MK 486), N-(DL-seryl)-N'-(2, 3, 4-hydroxybenzyl)hydrazine (Ro 4-4602), and N-M = hydroxybenzyl-N-methylhydrazine (NSD-1034). MK 486 and Ro 4-4602 have been used in the treatment of parkinsonism in conjunction with L-dopa [see Yahr (114)].

The Use of L-Dopa and AADC Inhibitors in the Treatment of Parkinsonism

Parkinsonism is characterized by akinesia, rigidity, and tremor at rest. The basic defect is considered to be within the basal ganglia which are deficient in dopamine (50). L-Dopa has been used as a replacement therapy with considerable success, the rationale being that L-dopa is the precursor of dopamine,. and, as an amino acid, it can enter the central nervous system. Dopamine does not pass the blood–brain barrier. To be effective, rather large doses of L-dopa are required, up to 8–10 gm daily, administered in divided doses. This is so because L-dopa has a short biologic half-life (less than 1 hr in man). It is rapidly converted to dopamine by AADC and to 3-methoxytyramine by COMT in the periphery. Only a small fraction of the administered L-dopa finds its way into the central nervous system. Formation of dopamine from L-dopa in the periphery could explain some of the cardiovascular side effects of L-dopa therapy, such as cardiac arrhythmias observed in some patients soon after ingestion of the drug. Limited experience indicates that concurrent L-dopa therapy poses no significant hazards during anesthesia. It is suggested that L-dopa therapy should be continued up to the evening before anesthesia and resumed as soon as feasible postoperatively (78).

The combined use of a peripheral AADC inhibitor and L-dopa considerably reduces the dose (as much as 65–80 percent) of the latter as more of it becomes available to the central nervous system. The two AADC inhibitors currently used therapeutically are carbidopa and Ro 4-4602. This is because their action is principally in the periphery although larger doses of Ro 4-4602 could also inhibit enzyme activity in the central nervous system. Furthermore, it takes less time (1 week or so) to establish an effective therapeutic regimen. Gastrointestinal side effects as well as incidence of cardiac arrhythmias and orthostatic hypotension are reduced with the combined therapy.

The long-term effects of L-dopa therapy with or without AADC inhibitors remain to be defined from clinical experience. It has been observed in experimental animals that L-dopa reduces tyrosine hydroxylase levels in the adrenal medulla and the heart, most likely by end-product inhibition (see Tyrosine Hydroxylase). Chronic administration of L-dopa also decreases AADC activity in the liver but not in the heart, kidney, brain, and adrenal medulla (23).

It is relevant to note here the role of pyridoxal phosphate, the cofactor of AADC, in the context of L-dopa therapy. Therapeutic effects of L-dopa are antagonized by high intake of pyridoxine, the vitamin analog of the cofactor. On the other hand, with the combined use of L-dopa and peripheral inhibitors of AADC, pyridoxine seems to enhance the action of L-dopa, probably through facilitated conversion of L-dopa to dopamine in the basal ganglia.

For more detailed discussion the reader is referred to several monographs and reviews on this subject (50, 99, 100, 114).

Dopamine-β-Hydroxylase (DβH, EC 1.14.17.1)

This enzyme catalyzes the hydroxylation of the β-carbon of dopamine to form norepinephrine, the last step in the biosynthesis of catecholamines, except in the adrenal medulla and in dopaminergic neurons. In the adrenal medulla, norepinephrine is further converted into epinephrine by phenylethanolamine-N-methyltransferase. In the dopaminergic neurons no further conversion takes place. Dopamine-β-hydroxylase is nonspecific. A number of phenylethanolamines structurally related to dopamine are substrates of this enzyme (42). It contains Cu^{2+}, requires ascorbic acid and molecular oxygen and is stimulated by fumarate and catalase. The molecular weight of this enzyme is approximately 290,000. Purified enzyme from beef adrenal medulla has a K_m of 5.8×10^{-3} M, and a V_{max} of 50,000. Formation of norepinephrine from dopamine in perfused guinea pig heart is directly related to the substrate concentration up to 10^{-4} M (10, 103).

The enzyme in the adrenal medulla exists half in bound and half in soluble form. It is concentrated in the norepinephrine storage granules. In the central nervous system the highest concentrations of DβH are present in the hypothalamus and the brain stem, the lowest concentrations are found in the striatum which contains mostly dopaminergic nerve terminals. Using immunofluorescence technics to mark the presence of DβH, the basal ganglia appear to be devoid of any fluorescent fibers or nerve terminals (45). Thus the distribution of this enzyme parallels that of norepinephrine.

Part of DβH in the adrenal medulla and nerve terminals is released together with norepinephrine, epinephrine, ATP, and chromagranin proteins. Employing a variety of technics, evidence has been gathered indicating that the soluble contents of storage granules are released directly, by exocytosis, without first entering the cytosol (10, 58).

Exocytosis requires calcium ions and appears to depend on the integrity of microtubules and microfilaments. Colchicine and vinblastin disaggregate microtubules and cytochlasin B disrupts microfilaments and thereby inhibit the release of catecholamines and DβH (10).

DβH is present in the serum of a number of mammals, including man. The released enzyme presumably enters the circulating blood directly or indirectly. The level of DβH in the serum can be measured enzymatically (74, 111) or by radioimmunoassay using antibodies from rabbits immunized with purified enzyme from the beef adrenal medulla (86), or from human pheochromotocytoma (29). Serum $D\beta H$ activity in man varies widely, with a 300- to 400-fold range among healthy subjects (29, 74, 111). This variation of $D\beta H$ activity is believed to be genetically controlled (113).

Serum $D\beta H$ increases during chronic stress (10). It is still controversial, however, whether acute changes in sympathetic activity are reflected by similar changes of serum enzyme levels (10, 29, 42, 85). In patients with familial dysautonomia the serum enzyme levels are low (111). According to Geffen *et al.* (38) serum $D\beta H$ levels in patients with pheochromocytoma are not different from those with essential hypertension. However, the mean serum $D\beta H$ levels are higher in patients with labile and essential hypertension (39, 88). This disagreement may be due to the wide range of serum $D\beta H$ levels in apparently normal subjects and to the variable interpretation of clinical observations. In some but not all patients with neuroblastoma the serum $D\beta H$ activity is elevated (41).

Similar to tyrosine hydroxylase, tissue levels of $D\beta H$ are influenced by sympathetic activity and drugs. Various types of stress in animals, such as immobilization, cold, and insulin shock, chronic administration of reserpine and phenoxybenzamine (Regitine) increase $D\beta H$ activity in the adrenal gland and presumably also in norepinephrine-containing neurons (11, 70). These drugs lower the arterial pressure and increase the firing of sympathetic nerves. Increased sympathetic nervous activity induces enzyme activity trans-synaptically. The increase in enzyme activity could result from conformational changes of existing enzyme or from new enzyme synthesis. Evidence for the latter has been reported (44). Denervation of the adrenal gland prevents enzyme induction. Inhibition of protein synthesis by cycloheximide also prevents drug-induced increase in enzyme levels [see Axelrod (11)].

Hormonal control of $D\beta H$ by ACTH has been mentioned above under Tyrosine Hydroxylase.

Dopamine-β-Hydroxylase Inhibitors

As $D\beta H$ contains Cu^{2+}, compounds binding enzymatic Cu^{2++} inhibit its activity. Disulfiram (tetraethylthiurea, Antabuse), in addition to being an aldehyde dehydrogenase inhibitor, is a potent $D\beta H$ inhibitor. It reduces the enzyme activity to less than 10% at a concentration of $10^{-6} M$ (40).

Disulfiram and a number of other $D\beta H$ inhibitors (including fusaric acid, FLA-63, NSD-1024, Sch 10595, U-10157) and other aromatic or alkyl thiourea derivatives* have been used as pharmacologic tools in the study of norepinephrine synthesis. In experimental animals each of these compounds reduces the brain norepinephrine concentration with an initial increase in dopamine concentration (40, 55, 62, 98). However, depletion of brain

---

*FLA-63 is bis-(4-methyl-1-homopiperazinyl-thiocarbonyl) disulfide; NSD-1024, *m*-hydroxybenzyloxyamine; Sch-10595, 5(*n*-butyl)-picolinamide; U-10157, 1, 1-dimethyl-3-phenyl-2-thiourea.

norepinephrine is never complete even with rather large doses of these compounds. In DOCA-induced hypertensive rats, disulfiram and Sch 10595 have been found to lower the arterial pressure. This appears to be related to the inhibition of DβH (40, 62). Clinical application of DβH inhibitors awaits further studies.

Phenylethanolamine-*N*-Methyltransferase (PNMT, EC 2.1.1.—)

This enzyme is highly localized in the adrenal medulla, mainly in the soluble fraction of the chromaffin cells. Recently the presence of PNMT in the brain stem has been reported (47). PNMT converts norepinephrine to epinephrine by transferring a methyl group to the amine radical. The methyl group comes from methionine and *S*-adenosylmethionine. A number of phenylethanolamines such as normetanephrine, octopamine, synephrine, epinephrine, and metanephrine, are substrates of PNMT. Purified enzyme from bovine adrenal medulla has an approximate molecular weight of 40,000. Isoenzymes with higher molecular weight (80,000 and 160,000) have been isolated (42).

PNMT is inhibited both by its substrate norepinephrine and by its product epinephrine [see Axelrod (9)]. The inhibition of epinephrine occurs at concentrations normally present in the adrenal medulla, probably representing end-product inhibition to regulate epinephrine synthesis. Like tyrosine hydroxylase and dopamine-β-hydroxylase, PNMT activity can be increased (induced) trans-synaptically by stress and drugs such as reserpine and phenoxybenzamine. Hypophysectomy lowers PNMT activity.

Recently, purified PNMT has been used in the assay of norepinephrine and dopamine-β-hydroxylase. With ^{14}C-*S*-adenosylmethionine as the methyl donor, radioactivity of enzymatically formed product, i.e., ^{14}C-epinephrine, gives a measure of the norepinephrine concentration (53). This new technic has a sensitivity of 0.5 ng for norepinephrine and should allow its accurate determination in smaller volumes of serum than those necessary for spectrofluorometric methods.

The same principle has been used for the assay of dopamine-β-hydroxylase activity (112). Tyramine is hydroxylated to octopamine by serum or tissue DβH. Octopamine is then quantitatively converted to ^{14}C-synephrine by PNMT with ^{14}C-*S*-adenosylmethionine as the methyl donor. The amount of ^{14}C-synephrine formed is determined by scintillation counting, reflecting DβH activity.

Monoamine Oxidases (MAO, EC 1.4.3.4)

Monoamine oxidase (monoamine: oxygen oxidoreductase (deaminating)) was first identified by Hare in 1928 using tyramine as the substrate (43). Subsequently, it was found that this enzyme also deaminates catecholamines and indolealkylamines (5-hydroxytryptamine) and a number of other mono-amines. The term MAO was then used to distinguish it from diamine oxidase (116).

MAO first deaminates monoamines to form an aldehyde. Further oxidation gives an acid. Alternatively, reduction of the aldehyde yields an alcohol metabolite (see Figure 10-2). The cofactor of MAO is flavine–adenine dinucleotide (FAD).

The molecular weight of MAO purified from various tissues varies over a wide range. The minimum is approximately 100,000 with some preparations having molecular weights of over 1,000,000 (102). This variation could be the result of different degrees of polymerification.

MAO contains flavin, Fe^{2+}, Cu^{2+}, and phospholipid. But the removal of Cu^{2+} or phospholipid does not seem to influence enzyme activity significantly. MAO is present in most organs of the body. Liver, kidney, stomach, and intestine are rich in this enzyme [see Tipton (102)]. All catecholamines and 5-hydroxytryptamine-containing neurons contain MAO. The enzyme is principally present on the outer membrane of mitochrondria. It serves an important function in the intraneuronal catabolism of catecholamines (and 5-hydroxytryptamine). Catecholamines present in the cytosol are subject to the enzymatic action of MAO. As mentioned in previous sections the concentration of norepinephrine or epinephrine in the cytoplasm of neurons regulates the synthesis of these amines through a feedback inhibition of tyrosine hydroxylase.

For some time it was thought that catabolism of catecholamines by MAO occurs primarily within the neuron and nerve terminals. More recent evidence suggests that extraneuronal MAO also has a role in the degradation of released norepinephrine (49).

In the cardiovascular tissues, the resistance vessels (mesenteric artery) seem to have higher MAO activity than the heart and aorta (95). Together with higher tyrosine hydroxylase activity and a rather fast turnover rate of norepinephrine (see above) this finding may have some functional significance on the maintenance of circulatory homeostasis.

Multiple forms of MAO (isoenzymes) have been demonstrated using different technics of purification, solubilization, and separation by electrophoresis. These processes involve rather severe and drastic treatment of tissue homogenates and extracts. Some serious questions have been raised as to the authenticity of these isoenzymes, and to their existence *in vivo*. However, experiments with MAO inhibitors *in vivo* provided evidence supporting the concept of multiple forms of MAO. Isoenzymes have different activities toward various substrates, differ in their tissue distribution and in their susceptability to inhibitors (87, 115).

As many laboratories are studying the multiple forms of MAO, there are certain degrees of confusion in terminology. Using gel electrophoresis, isoenzymes as distinct bands have been named MAO_1, MAO_2, etc. according to their relative movement toward the anode. One isoenzyme moves toward the cathode and is named MAO_R (87). Others named isoenzymes isolated from the rat brain as enzymes A and B, according to substrate specificity and susceptability to the MAO inhibitor clorgyline. Isoenzyme A oxidatively deaminates tyramine, 5-hydroxytryptamine, norepinephrine, and normetanephrine. It is inhibited by clorgyline. Isoenzyme B metabolizes tyramine but not 5-hydroxytryptamine and is relatively insensitive to clorgyline [see Neff and Goridis (76)]. Ninety percent of MAO present in the superior cervical ganglion is Type A isoenzyme. On the other hand, 90% of MAO in the rat pineal gland is Type B. Presumably it is isoenzyme A present in sympathetic nerves which plays an important role in the catabolism of norepinephrine.

MAO_R isolated from human or rat brain is unique in that it not only moves toward the cathode on gel electrophoresis but is highly specific for the oxidation of dopamine. The highest MAO_R activity is found in the basal ganglia of the human brain. For these reasons MAO_R has been named "dopamine monoamine oxidase" (87).

Monoamine Oxidase Inhibitors (MAOI)

A large number of compounds of different structures inhibit MAO activity. Many of these inhibitors have been used clinically as antidepressants or antihypertensive drugs. The earliest clinical use of an MAOI was iproniazid (Marsilid) as an antidepressant. The inhibitory action of different drugs on MAO is quite selective. For example, harmaline is effective in inhibiting the oxidation of 5-hydroxytryptamine by rat brain MAO, but only 1/20,000 as active in blocking phenethylamine oxidation, and 1/5000 as active when tyramine is the substrate. Tranylcypromine (Parnate) and pargyline (Eutonyl) both have the greatest inhibitory effect on MAO with phenethylamine and tyramine as substrates (36). The preferential inhibitory action of clorgyline on Type A isoenzyme has been mentioned above.

The explanation for the selective blockade of different substrates by various MAOI is that different isoenzymes may be responsible for the oxidation of these substrates. Or, an inhibitor may preferentially combine with a particular isoenzyme.

Administration of appropriate MAOI results in accumulation of monoamines (catecholamines and 5-hydroxytryptamine) in tissues, intraneurally and extraneurally. Increase in intraneural concentration of norepinephrine is associated with a decrease in the rate of its synthesis through inhibition of tyrosine hydroxylase.

Presumably, the increase in neuronal norepinephrine in the brain is responsible for changes in behavior and elevation of mood. In addition to the accumulation of norepinephrine in the cytoplasm of adrenergic nerve terminals, tyramine, originating from food or formed by decarboxylation of tyrosine, is also protected from oxidative deamination. Tyramine is further converted to octopamine through β-hydroxylation. It has been suggested that octopamine formed under this condition enters into storage granules and acts as "false transmitter (61)." Neural impulses arriving at the nerve terminals would release a mixture of norepinephrine and octopamine, the latter is a less potent neurotransmitter than the former. The hypotensive action of certain MAOI has been explained on this basis although this concept has been questioned recently.

On the other hand, MAOI-induced increase in norepinephrine concentration in sympathetic nerve terminals and inability of the organism to catabolize tyramine is an important consideration during MAOI therapy. Ingestion of food rich in tyramine or administration of indirectly acting sympathomimetic amines may result in massive release of norepinephrine and hypertensive crisis (59).

MAOI therapy is also associated with other possibilities of drug interaction with potential hazards. During or within weeks after MAOI therapy, administration of tricyclic antidepressants may result in hyperpyrexia, coma, and death (91). Supposedly, both norepinephrine and 5-hydroxytryptamine are concerned with the hypothalamic mechanisms for thermoregulation, but the exact mode of action of these drugs in causing hyperpyrexia has not been elucidated. Patients being treated with MAOI, such as pargyline and tranylcypromine, are more liable to develop hypotension when given narcotics. Studies in dogs showed that MAOI may interfere with the degradation of histamine which is released by narcotics (S. Markee, personal communication).

The therapeutic use of MAOI in psychiatry has been replaced by other drugs in recent years. This trend may be considered fortunate because of the many undesirable side effects and drug interactions just mentioned. However, if a patient is being treated with MAOI these factors should be considered in the conduct of anesthesia to avoid hypertensive crises or profound hypotension.

Regeneration of MAO after complete irreversible inhibition may require a considerable length of time. The half-life of MAO in the peripheral tissues (superior cervical ganglion and submaxillary gland) is estimated to be 4 days and that in the brain, about 11 days (76).

Catechol-O-Methyltransferase (COMT, EC 2.1.1.6)

O-Methylation of catecholamines was first demonstrated by Axelrod in 1957 (7). The enzyme is named catechol-O-methyltransferase because all catechols are suitable substrates. These include epinephrine, norepinephrine, dopamine, dopa, and their deaminated metabolites. Thus, epinephrine and norepinephrine can be catabolized by MAO first to their respective acids or alcohols and then O-methylated, or vice versa (see Figure 10-2). *In vivo*, O-methylation occurs primarily at the meta (3-) position on the phenyl ring; but *in vitro*, O-methylation of monophenols and polyphenols has been described (8, 9).

COMT has been purified approximately 450-fold from the rat liver (5, 6). The purified enzyme has a molecular weight of approximately 24,000 and is labile. It has a sufhydryl group in the region of its active sites, as dichloromercuribenzoate inhibits its activity. A methyl group donor, S-adenosylmethionine, and a divalent cation, Mg^{2+}, are required [see Axelrod (9)].

COMT is distributed widely in mammalian tissues with the highest activity in the liver and kidney. COMT is primarily found in the soluble fraction of cells outside the adrenergic neurons, in contrast to MAO which is present within these neurons.

Functionally, COMT is important for the degradation of released catecholamines and for the inactivation of exogenously administered catechols. Almost all of the norepinephrine and epinephrine formed in the body is catabolized and excreted as O-methylated, deaminated products. The daily urinary output of VMA ranges from 2 to 4 mg and that of metanephrine and normetanephrine, only about 200–300 μg (in free or conjugated form) (8).

In man, more than 80 percent of intravenously administered labeled epinephrine is recovered in the urine in the form of O-methylated metabolites (VMA, 40 percent; metanephrine, 40 percent; and 3-methoxy-4-hydroxyphenylglycol, 7 percent). In animals (mice and cats), exogenous epinephrine or norepinephrine is rapidly O-methylated. Within 5 min about 50 percent of administered norepinephrine is metabolized, almost all of it to normetanephrine. Apparently, this process of inactivation occurs in all peripheral tissues but the liver is probably the major site [see Axelrod (8)].

However, O-methylation, together with oxidative deamination, does not appear to be essential for the termination of action of endogenously released norepinephrine. The major process of inactivation is reuptake of the neurotransmitter into the nerve terminals. Inhibition of COMT prolongs but does not potentiate the responses to intravenously administered catecholamines (21).

Biosynthesis and Degradation of 5-Hydroxytryptamine

Pathways for the synthesis and degradation of 5-hydroxytryptamine (5-HT) are shown in Figure 10-4. Tryptophan, an amino acid derived from food, is hydroxylated at the 5-position on the indole nucleus to 5-hydroxytryptophan. Decarboxylation of 5-hydroxytryptophan yields the amine, 5-HT. The primary catabolic process is through oxidative deamination, the final metabolite being 5-hydroxyindoleacetic acid (5-HIAA).

As mentioned in the introductory remarks, the physiologic function of

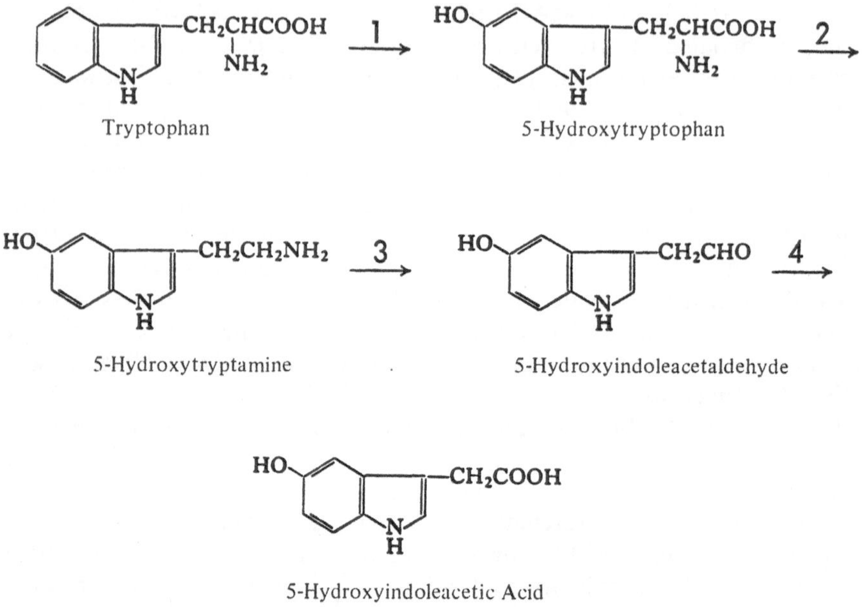

Tryptophan   5-Hydroxytryptophan

5-Hydroxytryptamine   5-Hydroxyindoleacetaldehyde

5-Hydroxyindoleacetic Acid

Figure 10-4. Biosynthesis of 5-hydroxytryptamine from tryptophan and principal metabolic pathway of 5-hydroxytryptamine. Step 1 is catalyzed by tryptophan hydroxylase; step 2 by aromatic L-amino acid decarboxylase; step 3 by monoamine oxidase and step 4 by aldehyde dehydrogenase.

5-HT is not as well defined as that of catecholamines in spite of intensive studies during the past 2 decades. In the periphery, 5-HT is present in abundant amounts in the gastrointestinal tract, the lung, the spleen, and platelets [see Garattini and Valzelli, (37) Appendix 1]. 5-HT causes smooth muscle contraction, such as bronchospasm and vasoconstriction. Some of the symptoms of carcinoid syndrome are believed to be mediated through 5-HT.

In the central nervous system, 5-HT is found in the brain stem (median raphé nuclei), the forebrain, and the spinal cord. 5-HT containing neurons send ascending fibers to the forebrain and descending tracts to the spinal cord. 5-HT neural pathways have been defined by histochemical and electrophysiologic technics (4, 46). Destruction of 5-HT neurons in the midbrain raphé nuclei results in degeneration of 5-HT terminals in the forebrain and selective reduction in forebrain 5-HT concentrations. Similarly, transection of the spinal cord is followed by a decrease in the cord 5-HT concentration after 9–14 days (4).

It has been suggested, based on experiments using drugs, precursors of 5-HT and enzyme inhibitors, that 5-HT is involved in sleep–waking cycle (56), temperature regulation, neuroendocrine function, seizure disorders, behavior, and sexuality (18). This multitude of physiologic and behavioral functions of 5-HT strongly indicate that 5-HT is probably one of the important neurotransmitters in the central nervous system. However, the problem is more complicated than it appears. At present definitive experimental evidence is lacking because of the limitations of experimental approach. Drugs used to modify the synthesis and degradation of 5-HT cannot be considered to act on the 5-HT system alone. It is not within the purview of this chapter to enter into the discussion on this subject.

The enzyme for biosynthesis and degradation of 5-HT are tryptophan hydroxylase, aromatic-L-amino acid decarboxylase, and monoamine oxidases. The latter two have been discussed under the section on Catecholamines. Drugs acting on aromatic-L-amino acid decarboxylase and monoamine oxidases have been used for studies of the 5-HT system.

Tryptophan Hydroxylase (EC 1.99.1.4)

Hydroxylation of tryptophan as the first step in 5-HT synthesis was established in 1956 by Udenfriend *et al.* (105). However, it was not until recently that the enzyme tryptophan hydroxylase was isolated and partially purified (67). Its molecular weight is about 50,000 and it requires reduced pteridine and oxygen. The activity of this enzyme is specific for tryptophan with an apparent K_m of 0.2–0.5 mM (using dimethyltetrahydropteridine as the cofactor). The distribution of tryptophan hydroxylase within the central nervous system correlates well with that of 5-HT. In the brain stem raphé nuclei, the enzyme exists in the soluble phase of the cell. In the forebrain tryptophan hydroxylase is found in the synaptosomes. Like tyrosine hydroxylase, the enzyme is synthesized in the neuronal cell bodies and transported to nerve terminals [see Lovenberg (66)].

5-Hydroxylation of tryptophan is considered the rate-limiting step in 5-HT synthesis. However, because of the relatively high K_m values this enzyme is not normally saturated with its substrate, tryptophan, in contrast to tyrosine hydroxylase. Thus, upon tryptophan loading the brain 5-HT concentration increases (17, 30). The increase is limited to areas of 5-HT containing neurons. Fernstrom and Wurtman (32) also showed that dietary intake can affect 5-HT concentrations in the brain. On the other hand, as decarboxylase is present in both 5-HT and catecholamine-containing neurons, 5-hydroxytryptophan loading results in 5-HT formation in both types of neurons. This is one of the reasons for the difficulties encountered in pharmacologic studies of the 5-HT system in the central nervous system.

Oxygen tension plays an important role in hydroxylation of tryptophan. As compared to tyrosine hydroxylase, tryptophan hydroxylase is more sensitive to hypoxia (17, 25). *In vivo*, the increase in the inspired oxygen concentration from 21 to 100% accelerated the rate of 5-HT synthesis, presumably through the increased activity of tryptophan hydroxylase (26). However, the physiologic significance of hyperoxia-induced increase in 5-HT biosynthesis is unclear.

Among clinically used anesthetics studied, only diethyl ether appears to increase the rate of 5-HT synthesis, suggesting that the state of general anesthesia is probably not associated with changes in 5-HT metabolism (27). Morphine, administered acutely or chronically, also increases the rate of 5-HT synthesis in rats. However, there are still controversies with respect to the role of central biogenic amines (catecholamines and 5-HT) in the analgesic action of narcotics and in the development of tolerance and physical dependence [see Chase and Murphy (18) and Way (107)].

The psychotropic agent, lysergic acid diethylamide (LSD) has been hypothesized to block or mimic the action of 5-HT. LSD increases the brain 5-HT concentration, and decreases its turnover rate (28, 64). Additionally, LSD in small doses (10 μg/kg) promptly inhibits neuronal discharge of 5-HT-containing neurons (2). A trans-synaptic feedback mechanism in the regulation of 5-HT synthesis has been suggested, although the problem is more complicated than it appears because of limitations of experimental approach (2).

Tryptophan Hydroxylase Inhibitor

Koe and Weissman (60) first reported in 1966 that p-chlorophenylalanine (PCP) is a potent depletor of brain 5-HT. It was soon concluded that PCP causes an apparent irreversible inactivation of brain tryptophan hydroxylase (54). Since then PCP has been used extensively as a pharmacologic tool in studies of 5-HT metabolism as related to physiologic changes associated with 5-HT depletion. In animals PCP treatment causes loss of both rapid eye movement (REM) and nonrapid eye movement (NREM) sleep with 80–90 percent depletion of brain 5-HT. Sleep disturbance has also been observed in monkeys and in man after PCP. As pointed out earlier, 5-HT is not considered to be the sole neurotransmitter assumed to regulate the sleep patterns; norepinephrine appears to have a role as well.

PCP has been used clinically in attempts to inhibit 5-HT synthesis in patients with carcinoid syndrome. The urinary excretion of 5-HIAA can be reduced by PCP by 72–88 percent. PCP treatment appears to relieve gastrointestinal symptoms but not flushing episodes (31).

5-HT is not totally depleted by PCP treatment. It is possible that the small amount of residual 5-HT is newly formed and can maintain activity of the 5-HT neuronal systems. It has also been reported that PCP may cause a small decrease of brain catecholamine concentrations and inhibition of phenylalanine hydroxylase, resulting in increases in the concentration of phenylalanine and that of its abnormal metabolites. PCP itself is decarboxylated in brain to p-chlorophenylethylamine, which may have some pharmacologic action of its own. Thus, interpretation of experimental results following inhibition of tryptophan hydroxylase by PCP is made difficult by these complicating factors [for review, see Chase and Murphy (18)].

Concluding Remarks

This chapter summarizes only a small fraction of current knowledge and concepts concerning biogenic amines and their metabolizing enzymes. If viewed from the point of experts, the disposition of the subject may be considered inadequate with numerous omissions. The field is vast, with new information being published at a rapid pace, each having its own implications in terms of probable or possible physiologic significance. However, an understanding of the systems discussed hopefully should be useful in the clinical practice of anesthesia, as we are dealing with drugs acting on these systems at all times.

Acknowledgment

I am grateful to Drs. Sydney Spector and Wallace Dairman of the Roche Institute of Molecular Biology for their suggestions and criticisms.

References

1. Acheson, G. H., Proceedings of the Second Symposium on Catecholamines. *Pharmacol. Rev.* **18**, 1-804 (1966).
2. Aghajanian, G. K., and Haigler, H. J., Studies on the physiological activity of 5-HT neurons. *Proc. Int. Congr. Pharmacol., 5th, 1972* Vol. 4, pp. 269–285 (1973).
3. Alousi, A., and Weiner, N., The regulation of norepinephrine synthesis in sympathetic nerves. Effect of nerve stimulation and catecholamine releasing agents. *Proc. Natl. Acad. Sci. U. S. A.* **56**, 1491 (1966).
4. Anderson, E. G., Bulbospinal serotonin-containing neurons and motor control. *Fed. Proc., Fed. Am. Soc. Exp. Biol.* **31**, 107 (1972).
5. Anderson, P. J., D'Iorio, A., Purification and properties of catechol-O-methyltransferase. *Biochem. Pharmacol.* **17**, 1943 (1968).
6. Assicot, M., and Bohuon, A. C., Purification and studies of catechol-O-methyltransferase of rat liver. *Eur. J. Biochem.* **12**, 490 (1970).
7. Axelrod, J., O-Methylation of catecholamines in vivo and in vitro. *Science* **126**, 400 (1957).
8. Axelrod, J., Methylation reactions in the formation and metabolism of catecholamines and other biogenic amines. *Pharmacol. Rev.* **18**, 95 (1966).
9. Axelrod, J., Biochemistry of catecholamines. *Annu. Rev. Biochem.* **40**, 465 (1971).
10. Axelrod, J., Dopamine-β-hydroxylase: Regulation of its synthesis and release from nerve terminals. *Pharmacol. Rev.* **24**, 233 (1972).
11. Axelrod, J., The fate of noradrenaline in the sympathetic neurons. *Harvey Lect.* **67**, 175 (1973).
12. Blaschko, H., and Muscholl, E., eds., "Catecholamines, Handbook of Experimental Pharmacology," Vol. 33. Springer-Verlag, Heidelberg, Berlin, New York, 1972.
13. Bloom, F. E., and Acheson, G. H., eds., "Brain Nerves and Synapses, Pharmacology and the Future of Man," Vol. 4, pp. 74–147 and 232–305. Karger, Basel, 1973.
14. Brimijoin, W. S., Capek, P., and Dyck, P. J., Axonal transport of dopamine-β-hydroxylase by human sural nerve in vitro. Fed. Proc. *Fed. Am. Soc. Exp. Biol.* **32**, 707 (abstr.) (1973).

15. Brodie, B. B., Costa, E., Dlabac, A., Neff, N. H., and Smookler, H. H., Application of steady-state kinetics to the estimation of synthesis rate and turnover time of tissue catecholamines. *J. Pharmacol. Exp. Ther.* **154**, 493 (1966).

16. Brown, B. R., Jr., Tatum, E. N., and Crout, J. R., The effects of inhalation anesthetics on the uptake and metabolism of L-^{3}H-norepinephrine in guinea pig atria. *Anesthesiology* **36**, 263 (1972).

17. Carlsson, A., Bédard, P., Davis, J. N., Kehr, W., Lindquist, M., and Magnusson, T., Physiological control of 5-HT synthesis and turnover in the brain. *Proc. Int. Congr. Pharmacol., 5th, 1972* Vol. 4, pp. 286–298 (1973).

18. Chase, T. N., and Murphy, D. L., Serotonin and central nervous system function. *Annu. Rev. Pharmacol.* **13**, 181 (1973).

19. Christenson, J. G., Dairman, W., and Udenfriend, S., Preparation and properties of a homogenous aromatic L-amino acid decarboxylase from hog kidney. *Arch. Biochem. Biophys.* **141**, 356 (1970).

20. Costa, E., and Sandler, M., eds., "Monoamine Oxidases—New Vistas," Adv. Biochem. Psychopharmacol., Vol. 5, Raven Press, New York, 1972.

21. Crout J. R., Effect of inhibiting both catechol-O-methyl-transferase and monoamine oxidase on cardiovascular responses to norepinephrine. *Proc. Soc. Exp. Biol. Med.* **108**, 482 (1961).

22. Dahlström, A., The effects of drugs on axonal transport of amine storage granules. *In* "New Aspects of Storage and Release Mechanisms of Catecholamines" (H. J. Schümann and G. Kronenberg, eds.), pp. 20–36. Springer-Verlag, Heidelberg, Berlin, New York, 1970.

23. Dairman, W., Christenson, J. G., and Udenfriend, S., Changes in tyrosine hydroxylase and dopa decarboxylase induced by pharmacological agents. *Pharmacol. Rev.* **24**, 269 (1972).

24. Dairman, W., Geffen, L., and Marchelle, M., Axoplasmic transport of aromatic L-amino acid decarboxylase (EC 4.1.1.26) and dopamine-β-hydroxylase (EC 1.14.2.1) in rat sciatic nerve. *J. Neurochem.* **20**, 1617 (1973).

25. Davis, J. N., and Carlsson, A., The effect of hypoxia on monoamine synthesis, levels and metabolism in rat brain. *J. Neurochem.* **21**, 783 (1973).

26. Diaz, P. M., Ngai, S. H., and Costa, E., Effect of oxygen on brain serotonin metabolism in rats. *Am. J. Physiol.* **214**, 591 (1968).

27. Diaz, P. M., Ngai, S. H., and Costa, E., The effects of cyclopropane, halothane and diethyl ether on the cerebral metabolism of serotonin in the rat. *Anesthesiology* **29**, 959 (1968).

28. Diaz, P. M., Ngai, S. H., and Costa, E., Factors modulating brain serotonin turnover. *Adv. Pharmacol.* **6b**, 75 (1968).

29. Ebstein, R. P., Park, D. H., Freedman, L. S., Levitz, S. M., Ohuchi, T., and Goldstein, M., A radioimmunoassay of human circulating dopamine-β-hydroxylase. *Life Sci.* **13**, 769 (1973).

30. Eccleston, D., A critical evaluation of techniques for studying synthesis and turnover of 5-HT in the brain in vivo. *Proc. Int. Congr. Pharmacol., 5th, 1972* Vol. 4, pp. 257–268 (1973).

31. Engelman, K., Lovenberg, W., and Sjoerdsma, A., Inhibition of serotonin synthesis by para-chlorophenylalanine in patients with cardinoid syndrome. *N. Engl. J. Med.* **277**, 1103 (1967).

32. Fernstrom, J. D., and Wurtman, R. J., Brain serotonin content: Physiological dependence on plasma tryptophan levels. *Science* **173**, 149 (1971).

33. Fink, B. R., Kennedy, R. D., Hendrickson, A. E., and Middaugh, M. E., Lidocaine inhibition of rapid axonal transport. *Anesthesiology* **36**, 422 (1972).

34. Fukunaga, A., and Epstein, R. M., Sympathetic excitation during nitrous oxide-halothane anesthesia in the cat. *Anesthesiology* **39**, 23 (1973).

35. Fukunaga, A., and Epstein, R. M., Effects of cyclopropane on the sympathetic nervous system and on the neural regulation of the circulation in the cat. *Anesthesiology* **40**, 323 (1974).

36. Fuller, R. W., Selective inhibition of monoamine oxidase. *Adv. Biochem. Psychopharmacol.* **5**, 339–354 (1972).

37. Garattini, S., and Valzelli, L., "Serotonin." Elsevier, Amsterdam, 1965.

38. Geffen, L. B., Rush, R. A., Louis, W. J., and Doyle, A. E., Plasma catecholamine and dopamine-β-hydroxylase amounts in phaeochromocytoma. *Clin. Sci.* **44**, 421 (1973).

39. Geffen, L. B., Rush, R. A., Louis, W. J., and Doyle, A. E., Plasma dopamine-β-hydroxylase and noradrenaline amounts in essential hypertension. *Clin. Sci.* **44**, 617 (1973).

40. Goldstein, M., Inhibition of norepinephrine biosynthesis at the dopamine-β-hydroxylase stage. *Pharmacol. Rev.* **18**, 77 (1968).

41. Goldstein, M., Freedman, L. S., Bohuon, A. C., and Guérinot, F., Serum dopamine-β-hydroxylase activity in neuroblastoma. *N. Engl. J. Med.* **286**, 1123 (1972).

42. Goldstein, M., Fuxe, K., and Hökfelt, T., Characterization and tissue localization of catecholamine synthesizing enzymes. *Pharmacol. Rev.* **24**, 293 (1972).

43. Hare, M. L. C., Tyramine oxidase. I. A new enzyme system in liver. *Biochem. J.* **22**, 968 (1928).

44. Hartman, B. K., and Udenfriend, S., The application of immunological techniques to the study of enzymes regulating catecholamine synthesis and degradation. *Pharmacol. Rev.* **24**, 311 (1972).

45. Hartman, B. K., Zide, D., and Udenfriend, S., The use of dopamine-β-hydroxylase as a marker for the central noradrenergic nervous system in rat brain. *Proc. Natl. Acad. Sci. U.S.A.* **69**, 2722 (1972).

46. Heller, A., Neuronal control of brain serotonin. *Fed. Proc., Fed. Am. Soc. Exp. Biol.* **31**, 81 (1972).

47. Hökfelt, T., Fuxe, K., Goldstein, M., and Johansson, O., Evidence for adrenaline neurons in the rat brain. *Acta Physiol. Scand.* **89**, 286 (1973).

48. Holtz, P., Heise, R., and Lüdtke, K., Fermentativer Abban von L-Dioxyphenyl-alanin (Dopa) durch Niere. *Arch. Exp. Pathol. Pharmakol.* **191**, 87 (1938).

49. Horita, A., and Lowe, M. C., On the extraneural nature of cardiac monoamine oxidase in the rat. *Adv. Biochem. Psychopharmacol.* **5**, 227-242 (1972).

50. Hornykiewicz, O., Dopamine (3-hydroxytyramine) and brain function. *Pharmacol. Rev.* **18**, 925 (1966).

51. Ikeda, M., Fahien, L. A., and Udenfriend, S., A kinetic study of bovine adrenal tyrosine hydroxylase. *J. Biol. Chem.* **241**, 4452 (1966).

52. Iversen, L. L., ed., Catecholamines. *Br. Med. Bull.* **29**, 91-178 (1973).

53. Iversen, L. L., and Jarrott, B., Modification of an enzyme radiochemical assay procedure for noradrenaline. *Biochem. Pharmacol.* **19**, 1841 (1970).

54. Jequier, E., Lovenberg, W., and Sjoerdsma, A., Tryptophan hydroxylase inhibition. The mechanism by which p-chlorophenylalanine depletes rat brain serotonin. *Mol. Pharmacol.* **3**, 274 (1967).

55. Johnson, G. A., Boukma, S. J., and Kim, E. G., Inhibition of dopamine-β-hydroxylase by aromatic and alkyl thioureas. *J. Pharmacol. Exp. Ther.* **168**, 229 (1969).

56. Jouvet, M., Monoaminergic regulation of the sleep-waking cycle. *Proc. Int. Congr. Pharmacol., 5th, 1972* Vol. 4, pp. 103-107 (1973).

57. Kennedy, R. D., Fink, B. R., and Byers, M. R., The effect of halothane on rapid axonal transport in the rabbit vagus. *Anesthesiology* **36**, 433 (1972).

58. Kirschner, N., and Viveros, O. H., Quantal aspects of the secretion of catecholamines and dopamine-β-hydroxylase from the adrenal medulla. *New Aspects Storage Release Mech. Catecholamines, Bayer-Symp., 2nd, 1969*, pp. 78-88 (1970).

59. Knoll, J., and Magyar, K., Some puzzling pharmacological effects of monoamine oxidase inhibitors. *Adv. Biochem. Psychopharmacol.* **5**, 393-408 (1972).

60. Koe, B. K., and Weissman, A., p-Chlorophenylalanine. A specific depletor of brain serotonin. *J. Pharmacol. Exp. Ther.* **154**, 499 (1966).

61. Kopin, I. J., Fischer, J. E., Musacchio, J. M., Horst, W. D., and Weise, V. K., "False

neurochemical transmitters" and the mechanism of sympathetic blockade by monoamine oxidase inhibitors. *J. Pharmacol. Exp. Ther.* **147**, 186 (1965).

62. Korduba, C. A., Veals, J., Wohl, A., Symchowicz, S., and Tabachnick, I. I. A., Sch 10595, an effective dopamine-β-hydroxylase inhibitor and a hypotensive agent. *J. Pharmacol. Exp. Ther.* **184**, 671 (1973).

63. Laverty, R., The mechanism of action of some antihypertensive drugs. *Br. Med. Bull.* **29**, 152 (1973).

64. Lin, R. C., Ngai, S. H., and Costa, E., Lysergic acid diethylamide: Role in conversion of plasma tryptophan to brain serotonin (5-hydroxytryptamine). *Science* **166**, 237 (1969).

65. Lloyd, K. G., and Hornykiewicz, O., Occurence and distribution of aromatic L-amino acid (L-dopa) decarboxylase in the human brain. *J. Neurochem.* **19**, 1549 (1972).

66. Lovenberg, W., Tryptophan hydroxylase and serotonin synthesis in the brain. *Proc. Int. Congr. Pharmacol., 5th, 1972* Vol. 4, pp. 232–244 (1973).

67. Lovenberg, W., Besselaar, G. H., Bensinger, R. E., and Jackson, R. L., Physiologic and drug-induced regulation of serotonin synthesis. *In* "Serotonin and Behavior" (J. D. Barchas and E. Usdin, eds.), pp. 49–54. Academic Press, New York, 1973.

68. Lovenberg, W., Weissbach, H., and Udenfriend, S., Aromatic L-amino acid decarboxylase. *J. Biol. Chem.* **237**, 89 (1962).

69. Miletich, D. J., Ivankovic, A. D., Albrecht, R. F., Zahed, B., and Ilahi, A. A., The effect of ketamine on catecholamine metabolism in the isolated perfused rat heart. *Anesthesiology* **39**, 271 (1973).

70. Molinoff, P. B., Brimijoin, W. S., Weinshilboum, R. M., and Axelrod, J., Neurally mediated increase in dopamine-β-hydroxylase activity. *Proc. Natl. Acad. Sci. U. S. A.* **66**, 453 (1970).

71. Mueller, R. A., Thoenen, H., and Axelrod, J., Effect of pituitary and ACTH on the maintenance of basal tyrosine hydroxylase activity in the rat adrenal gland. *Endocrinology* **86**, 751 (1970).

72. Nagatsu, T., Levitt, M., and Udenfriend, S., Conversion of L-tyrosine to 3,4-dihydroxy-phenylalanine by cell-free preparations of brain and sympathetically innervated tissues. *Biochem. Biophys. Res. Commun.* **14**, 543 (1964).

73. Nagatsu, T., Levitt, M., and Udenfriend, S., Tyrosine hydroxylase—the initial step in NE biosynthesis. *J. Biol. Chem.* **239**, 29 10 (1964).

74. Nagatsu, T., and Udenfriend, S., Photometric assay of dopamine-β-hydroxylase activity in human blood. *Clin. Chem.* **18**, 980 (1972).

75. Naito, H., and Gillis, C. N., Anesthetics and response of atria to sympathetic nerve stimulation. *Anesthesiology* **29**, 259 (1968).

76. Neff, N. H., and Goridis, C., Neuronal monoamine oxidase: Specific enzyme types and their rate of formation. *Adv. Biochem. Psychopharmacol.* **5**, 307–324 (1972).

77. Neff, N. H., Ngai, S. H., Wang, C. T., and Costa, E., Calculation of the rate of catecholamine synthesis from the rate of conversion of ^{14}C-tyrosine to catecholamines: Effect of adrenal demedullation on synthesis rate. *Mol. Pharmacol.* **5**, 90 (1969).

78. Ngai, S. H., Parkinsonism, levodopa and anesthesia. *Anesthesiology* **37**, 344 (1972).

78a. Ngai, S. H., Dairman, W., Marchelle, M., Effects of lidocaine and etidocaine on the exoplasmic transport of catecholamine-synthesizing enzymes. *Anesthesiology* **41**, 542 (1974).

79. Ngai, S. H., Diaz, P. M., and Ozer, S., The uptake and release of norepinephrine: Effects of cyclopropane and halothane. *Anesthesiology* **31**, 45 (1969).

80. Ngai, S. H., Neff, N. H., and Costa, E., The effects of cyclopropane and halothane on the biosynthesis of norepinephrine in vivo. *Anesthesiology* **31**, 53 (1969).

81. Oesch, F., Otten, U., and Thoenen, H., Relationship between the rate of axoplasmic transport and subcellular distribution of enzymes involved in the synthesis of norepineph-rine. *J. Neurochem.* **20**, 1691 (1973).

82. Pletscher, A., Regulation of catecholamine turnover by variation of enzyme levels. *Pharmacol. Rev.* **24**, 225 (1972).

83. Price, H. L., Linde, H. W., Jones, R. E., Black, G. W., and Price, M. L., Sympathoadrenal responses to general anesthesia in man and their relation to hemodynamics. *Anesthesiology* **20**, 563 (1959).

84. Price, H. L., Warden, J. C., Cooperman, L. H., and Millar, R. A., Central sympathetic excitation caused by cyclopropane. *Anesthesiology* **30**, 426 (1969).

85. Roffman, M., Freedman, L. S., and Goldstein, M., The effect of acute and chronic swim stress on dopamine-β-hydroxylase activity. *Life Sci.* **12**, 369 (1973).

85a. Roizen, M. F., Moss, J., Henry, D. P., and Kopin, I. J., Effects of halothane on plasma catecholamines. *Anesthesiology* **41**, 432–439 (1974).

85b. Roizen, M. F., Thoa, N. B., Moss, J., and Kopin, I. J., Inhibition by cyclopropane of release of norepinephrine, but not dopamine-β-hydroxylase, from the guinea-pig vas dererens. *Anesthesiology* **44**, 54–56 (1976).

86. Rush, R. A., and Geffen, L. B., Radioimmunoassay and clearance of circulating dopamine-β-hydroxylase. *Circ. Res.* **31**, 444 (1972).

87. Sandler, M., and Youdim, M. B. H., Multiple forms of monoamine oxidase: Functional significance. *Pharmacol. Rev.* **24**, 331 (1972).

88. Schanberg, M., Stone, R. A., Kirschner, N., Gunnells, J. C., and Robinson, R. R., Plasma dopamine-β-hydroxylase: A possible aid in the study and evaluation of hypertension. *Science* **183**, 523 (1974).

89. Sedvall, G. C., and Kopin, I. J., Acceleration of norepinephrine synthesis in the rat submaxillary gland in vivo during sympathetic nerve stimulation. *Life Sci.* **6**, 45 (1967).

90. Sjoerdsma, A., Catecholamine-drug interaction in man. *Pharmacol. Rev.* **18**, 673 (1966).

91. Sjöqvist, F., Psychotropic drugs (2). Interaction between monoamine oxidase (MAO) inhibitors and other substances. *Proc. R. Soc. Med.* **58**, 967 (1965).

92. Skovsted, T., Price, M. L., and Price, H. L., The effects of halothane on arterial pressure, pre-ganglionic sympathetic activity in basostatic reflexes. *Anesthesiology* **31**, 507 (1969).

93. Spector, S., Inhibitors of endogenous catecholamine biosynthesis. *Pharmacol. Rev.* **18**, 599 (1966).

94. Spector, S., Gordon, R., Sjoerdsma, A., and Udenfriend, S., End-product inhibition of tyrosine hydroxylase as a possible mechanism for regulation of norepinephrine synthesis. *Mol. Pharmacol.* **3**, 549 (1967).

95. Spector, S., Tarver, J., and Berkowitz, B., Effects of drugs and physiological factors in the disposition of catecholamines in blood vessels. *Pharmacol. Rev.* **24**, 191 (1972).

96. Sprague, D. H., and Ngai, S. H., Effect of cyclopropane on the contractility and the cyclic 3′,5′-adenosine monophosphate (cyclic AMP) system in the rat aorta. *Anesthesiology* **40**, 336 (1974).

97. Sprague, D. H., Yang, J. C., and Ngai, S. H., Effects of isoflurane (Forane) and halothane on the contractility and the cyclic 3′,5′-adenosine monophosphate (cyclic AMP) system in rat aorta. *Anesthesiology* **40**, 162 (1974).

98. Svensson, T. H., and Waldeck, S., On the significance of central noradrenaline for motor activity: Experiments with a new dopamine-β-hydroxylase inhibitor. *Eur. J. Pharmacol.* **7**, 278 (1969).

99. Symposium on Levodopa in Parkinson's Disease: Clinical and Pharmacological Aspects, *Clin. Pharmacol. Ther.* **12**, 317 (1971).

100. Symposium on Levodopa in Parkinson's Disease: Collaborative Studies *Neurology* **22**, Suppl., 1–102 (1972).

101. Thoenen, H., Mueller, R. A., and Axelrod, J., Trans-synaptic induction of adrenal tyrosine hydroxylase. *J. Pharmacol. Exp. Ther.* **169**, 249 (1969).

102. Tipton, K. F., Biochemical aspects of monoamine oxidase. *Br. Med. Bull.* **29**, 116 (1973).

103. Udenfriend, S., Tyrosine hydroxylase. *Pharmacol. Rev.* **18**, 43 (1966).

104. Udenfriend, S., and Spector, S., eds., Regulation of catecholamine metabolism in the sympathetic nervous system. *Pharmacol. Rev.* **24**, 163–449 (1972).

105. Udenfriend, S., Titus, E., Weissbach, H., and Peterson, R. E., Biogenesis and metabolism of 5-hydroxyindole compounds. *J. Biol. Chem.* **219**, 335 (1956).

106. Udenfriend, S., Zaltzman-Nirenberg, P., and Nagatsu, T., Inhibition of purified beef adrenal tyrosine hydroxylase. *Biochem. Pharmacol.* **14**, 837 (1965).
107. Way, E. L., Brain neurohormones in morphine tolerance and dependence. *Proc. Int. Congr. Pharmacol., 5th, 1972* Vol. 1, pp. 77–94 (1973).
108. Weiner, N., Cloutier, G., Bjur, R., and Pfeffer, R. I., Modification of norepinephrine synthesis in intact tissue by drugs and during short term adrenergic nerve stimulation. *Pharmacol. Rev.* **24**, 203 (1972).
109. Weiner, N., and Mosimann, W. F., The effect of insulin on the catecholamine content and tyrosine hydroxylase activity of cat adrenal glands. *Biochem. Pharmacol.* **19**, 1189 (1970).
110. Weinshilboum, R. M. and Axelrod, J., Dopamine-β-hydroxylase activity in the rat after hypophysectomy. *Endocrinology* **87**, 894 (1970).
111. Weinshilboum, R. M., and Axelrod, J., Reduced plasma dopamine-β-hydroxylase activity in familial dysautonomia. *N. Engl. J. Med.* **285**, 938 (1971).
112. Weinshilboum, R. M. and Axelrod, J., Serum dopamine-β-hydroxylase activity. *Circ. Res.* **28**, 307 (1971).
113. Weinshilboum, R. M., Raymond, F. A., Elveback, L. R., and Weidman, W. H., Serum dopamine-β-hydroxylase activity: Sibling–sibling correlation. *Science* **181**, 943 (1973).
114. Yahr, M. O., ed., "Advances in Neurology" Vol. 2. Raven Press, New York, 1973.
115. Youdim, M. B. H., Multiple forms of mitochondrial monoamine oxidase. *Br. Med. Bull.* **29**, 120 (1973).
116. Zeller, E. A., Oxidation of amines. *In* "The Enzymes: Chemistry and Mechanism of Action" (J. B. Sumner and K. Myrbäck, eds.), Vol. 2, Part 1, pp. 536–558. Academic Press, New York, 1951.

Chapter 11

The Cyclic AMP System

Lubos Triner

Introduction

Cyclic $3', 5'$-adenosine monophosphate (cAMP), a nucleotide present in small amounts in almost every type of mammalian cell, has a key role in the regulation of many processes in the body. Cyclic AMP was discovered by E. W. Sutherland almost 20 years ago in the course of studies of the hyperglycemic effect of epinephrine. Sutherland postulated that in order to increase hepatic glucose output epinephrine must increase the activity of the rate-limiting enzyme involved in the conversion of glycogen to glucose. This enzyme was identified as phosphorylase, which catalyzes the hydrolysis of glycogen to glucose 1-phosphate. Indeed, the addition of epinephrine to a liver slice or preparation of ruptured liver cells produced an increase in phosphorylase activity. However, when phosphorylase was partially purified, epinephrine did not increase enzyme activity, an observation that led to the assumption that the hormone does not activate phosphorylase directly but exerts its effect at a step prior to the activation of the enzyme. When fragments of cell membrane were exposed to epinephrine, boiled, and added to purified phosphorylase, its activity increased. Epinephrine evidently interacted with an enzyme present in the cell membrane producing a heat-stable compound that in turn activated phosphorylase. The heat-stable compound was identified by Sutherland as cAMP in 1957, and the enzyme with which epinephrine interacts was identified as adenylate cyclase in 1962.

Sutherland's discovery of the mechanism of the hyperglycemic action of epinephrine would not have had the impact it has had (including the award to Sutherland of the Nobel prize in physiology and medicine for 1971), despite its originality, were it limited to the understanding of this process. The discovery led to further studies which showed that cAMP mediates the action of catecholamines and other hormones regulating various cellular processes. Since this is a very wide and rapidly developing field, only those areas where significant progress has been made in explaining cAMP function and which are of particular interest to the anesthesiologist are reviewed. The interested reader is referred to recently published books which cover the subject *in extenso* (19, 40, 41).

The Cyclic AMP System

The following diagram represents the sequence of events involved in the synthesis and breakdown of cAMP:

$$\text{ATP} \xrightarrow[\text{Mg}^{2+}]{\text{Adenylate cyclase}} 3',5'\text{-cAMP} \xrightarrow[\text{Mg}^{2+}]{\text{cAMP-phosphodiesterase}} 5'\text{-AMP}$$

Adenylate cyclase, the enzyme catalyzing the formation of cAMP from ATP, is located mainly in the cell membrane. The enzyme is probably a lipoprotein, but its structure has not yet been identified and all attempts so far to purify and isolate it have been unsuccessful. Magnesium ions are necessary for the activity of the enzyme and the range of optimum pH is relatively broad, from 7.2 to 8.2. Adenylate cyclase retains its activity quite well for long periods of time at low temperatures ($-70°C$), but at higher temperatures loses activity rapidly.

The initial procedure for adenylate cyclase activity measurement used by Sutherland was a bioassay based on the measurement of phosphorylase activation by cAMP. Simpler methods have been developed based on the measurement of radioactive cAMP derived from radioactive ATP of known specific activity and chromatographic separation of the cyclic nucleotide (29). The hormones (catecholamines, glucagon, ACTH, LH, ICSH, ADH, parathyroid hormone, TSH) which act via cAMP interact with receptors which are closely related to or are part of adenylate cyclase (regulatory subunit or receptor site). It is assumed that adenylate cyclase consists of two subunits: the hormone receptor site or regulatory subunit which is oriented toward the outer surface of the membrane and the catalytic subunit which is oriented toward the inside of the cell. The result of the hormone–receptor interaction—the signal (stimulatory or inhibitory) to adenylate cyclase—is appropriately reflected by the catalytic subunit, and increased or decreased rate of cAMP formation instantaneously results. This dramatic change in the rate of cAMP formation is usually reflected in a corresponding change of cAMP level which occurs despite ongoing destruction by cAMP-phosphodiesterase.

Phosphodiesterase (cyclic 3', 5'-adenosine monophosphate-phosphodiesterase) is the enzyme catalyzing the hydrolysis of cAMP to 5'-AMP, the main physiologic mechanism of cAMP inactivation. Cyclic AMP-phosphodiesterase, like adenylate cyclase, has been detected in almost all mammalian tissues. In the cell it is partly soluble and partly structure-bound. The enzyme is stable, with a molecular weight varying over a wide range. It is probably composed of subunits, and different conditions may give rise to different subunit structures and thus to different molecular weights. Magnesium ions are required for the activity of the enzyme and optimum pH is between 7.5 and 8.0. Several methods are available for measurement of phosphodiesterase activity, based either on decrease in labeled substrate or on end-product measurement (8).

Since cAMP-phosphodiesterase by its action participates in the control of cAMP level in the cell, factors that affect its activity influence the extent and duration of cAMP action and thereby the hormonal effect mediated by this nucleotide. As the intracellular level of cAMP is a result of a balance between the rate of synthesis and hydrolysis, an inhibitor of cAMP-phosphodiesterase causes intracellular cAMP to rise and an activator causes cAMP to decrease. Many therapeutic agents, such as methylxanthines (aminophylline, caffeine), papaverine, phenothiazine derivatives, reserpine (24), puromycin (4), and some benzothiadiazine derivatives (44) inhibit cAMP-phosphodiesterase activity (7). Some endogenous factors that regulate cAMP-phosphodiesterase activity are ATP, pyrophosphate, cyclic guanosine monophosphate and other nucleotides and citrate, which are inhibitory, and certain protein factors and Ca^{2+}, which activate the enzyme.

In most tissues there seem to be at least two types of cAMP-phosphodiesterase, one with a low K_m and the other with a high K_m for cAMP. The two forms of the enzyme provide an additional dimension to the control of intracellular concentrations of cAMP by cAMP-phosphodiesterase. At normal physiologic concentrations of cAMP in the cell, primarily the low K_m enzyme is involved in the process of inactivation. The K_m of this enzyme is about one–two orders of magnitude lower than that of the high K_m cAMP-phosphodiesterase. When adenylate cyclase is activated, the concentration of cAMP increases rapidly, and at higher substrate concentrations the high K_m cAMP-phosphodiesterase becomes more operative. Thus the function of low K_m cAMP-phosphodiesterase may be to control basal levels of cAMP and that of the high K_m enzyme to protect the cell from exceedingly high concentrations of the cyclic nucleotide.

The structural formula of cAMP (Fig. 11-1) shows the ester bond (between the 3'-C of the ribose and the phosphate) which is essential for the biologic activity of the molecule; this is a high energy bond which is split by cAMP-phosphodiesterase. The intracellular concentration of cAMP in the absence of stimulation of adenylate cyclase is approximately 0.1 to 0.5 nmoles/gm of tissue or about 0.5 to 2.6 pmoles/mg protein. Assuming an even distribution in intracellular water, the concentration would be in the range of 0.1 μM (roughly 1000 times less than that of ATP). The actual

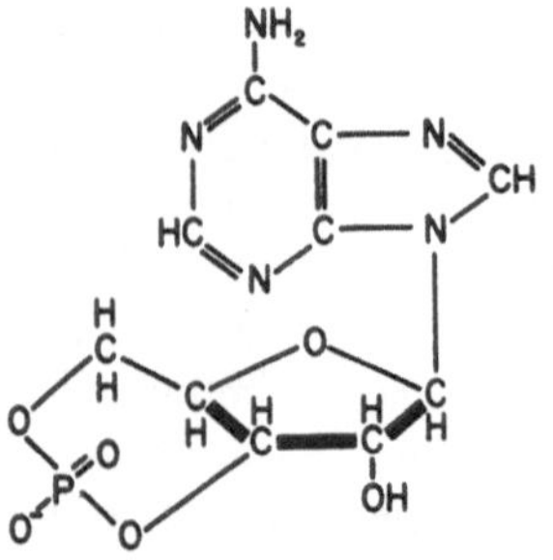

Figure 11-1. Structural formula of cyclic AMP.

concentration of free cAMP in the cell, however, is likely to be less, due to specific and nonspecific protein binding. On the other hand, in certain parts of the cell it may be much higher as a result of compartmentalization. There are a number of methods currently in use for the assay of cAMP. These methods are highly sophisticated and have made it possible to detect small amounts of intracellular cAMP. Most of these procedures use saturated enzyme systems and are based on radioisotope dilution or enzymatic cyclic procedures (21).

Cyclic AMP mediates the action of a number of hormones and regulates many biochemical processes in the cell. How does the cAMP system provide for the specificity of response of so many different hormones in so many different tissues if the intracellular mediator of the hormonal action is (in all these cases) cAMP? The basal rate of adenylate cyclase activity in intact cells is not high. However, when a proper stimulatory signal reaches the enzyme, the rate immediately increases manyfold, with a corresponding rise in intracellular cAMP concentration occurring within seconds. The cAMP, in turn, diffuses and is channeled by its affinity to certain proteins within the cell, some of which are its sites of action. This, in brief, is the sequence of events when a hormone reaches the target cell and interacts with a specific receptor site of the hormone-sensitive adenylate cyclase on the outside of the cell membrane. A hormone can be viewed as a carrier of information instructing a given organ or tissue to modify its specific functioning. The hormonal message has to be recognized by the cell (receptor), translated into a signal (cAMP in this case) which is transmitted within the cell to the target system where it acts, modifying the particular function. In addition, the hormone concentration is usually small in relation to the amplitude of the response it triggers. Thus the cAMP system also provides amplification of the original message. Sutherland named cAMP the "second messenger" in recognition of its role as intracellular mediator of the action of hormones, which act as the "first messenger."

The specificity is mainly determined by the configuration of the hormone and that of the receptor site on the target cell. Their interaction leads to a change in the activity of adenylate cyclase, as described above. The rate of cAMP formation in the cell is proportional to the quantum of hormone interacting with the receptor. This represents the first step in the process of amplification, since many cAMP molecules are produced for each hormone molecule interacting with the receptor. Thus, the same intracellular mediator, cAMP, is produced regardless of which hormone is interacting with the receptor, which is related to or is part of adenylate cyclase. Despite this, the final response varies from one type of cell to another. The nature of the response to cAMP in the target cell is not only determined by the hormone but also by the functional differentiation and effector systems of the cell. Thus, an epinephrine-induced rise in cAMP level increases lipolysis in adipose tissue, glycogenolysis in the liver, or force of contraction in the myocardium.

The mechanism of action of cAMP in the cell is not yet fully understood. Walsh and colleagues (59) described a cAMP-dependent protein kinase in skeletal muscle which catalyzes the phosphorylation of phosphorylase kinase. Subsequently, cAMP-dependent protein kinases were found in liver (32), nervous tissue (36), and other preparations (30) and the hypothesis was proposed that all the actions of cAMP are mediated through regulation of the activity of these specific protein kinases (31). Cyclic AMP, by interacting with its receptor (protein kinase), causes the dissociation of the receptor moiety from the protein kinase, thereby activating the enzyme. The activated kinase then catalyzes the transfer of phosphate from ATP to the specific protein substrates (e.g., phosphorylase kinase). This hypothesis also provides an explanation for the diverse effects the cAMP molecule has in different types of cells. The tissue-specific effects of cAMP would be determined by the nature of the protein kinase and of the particular substrate for this kinase in a given tissue.

Hormonal Effects Mediated by cAMP

The effects of catecholamines exerted through β-adrenergic receptors are mediated by increased cAMP concentrations caused by adenylate cyclase activation. This increase leads to accelerated glycogenolysis in the liver and muscle, lipolysis in adipose tissue, a positive inotropic effect in the heart, arteriolar and bronchial relaxation, and decreased intestinal and uterine motility. The relation between catecholamine action and the cAMP system in heart and smooth muscle is discussed in later sections of this chapter.

The mechanism of the mediation of the metabolic effects of catecholamines by cAMP, most thoroughly studied and understood in liver and muscle glycogenolysis, is shown schematically in Figure 11-2. The diagram shows the sequence of events in the liver cell leading to glycogenolysis after catecholamines stimulate adenylate cyclase. This cascade of reactions is amplified at

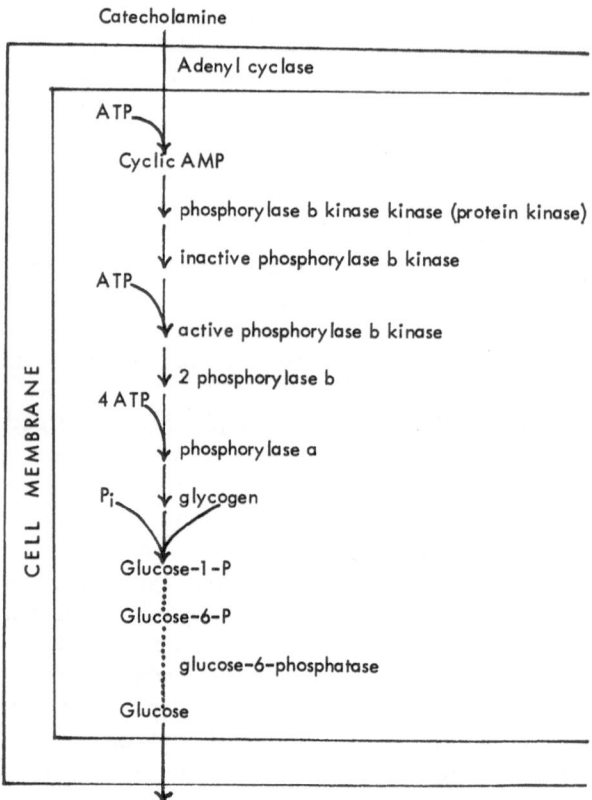

Figure 11-2. Diagram of the sequence of events in activation of glycogenolysis in the liver.

each of the six steps and, as a result, the quantum of glucose released far exceeds the quantum of catecholamines interacting with adenylate cyclase. Increased levels of cAMP in the liver lead, therefore, to an increased rate of glucose production and output. This is mainly due to the activation of phosphorylase, the decreased activity of glycogen synthetase and stimulation of gluconeogenesis, all under the control of cAMP. Glycogen synthetase I kinase, activated by cAMP, catalyzes the conversion of glycogen synthetase I, the physiologically active form, to the *d*, or inactive form. Therefore, the action of catecholamines, through this stimulatory effect of cAMP, results in a decreased glycogen synthesis. The combined changes in phosphorylase and glycogen synthetase activity in the liver lead to a net loss of glycogen and a net increase of glucose. In addition, as a result of increased muscle glycogenolysis and glycolysis, lactic acid is converted to glucose at higher rates in the liver. (In muscle, lactic acid leaves the cell as an end product of glycogenolysis, because of the absence of glucose 6-phosphatase.) Modification of other pathways by an epinephrine-induced increase in cAMP may also result in increased gluconeogenesis, either indirectly as a result of increased substrate levels or directly through a number of mechanisms (e.g., facilitation at the pyruvate and phosphoenol–pyruvate step, increased uptake of amino acids, induction of new enzymes). The effect of catecholamines on carbohydrate metabolism in muscle is mediated through a similar chain of reactions as in the liver. Increase in cAMP causes phosphorylase activation and a less pronounced decrease in glycogen synthetase activity. The effect of catecholamines in the pancreas should be mentioned here. The overall action of catecholamines on the beta cells of the islets of Langerhans results in a lower level of cAMP which, in turn, slows down the release of insulin that would be expected to occur in response to increasing blood glucose levels. The catecholamine effect, in this case, is exerted through α-adrenergic receptors which are apparently predominant in these cells. When the α-adrenergic receptors are blocked, epinephrine induces an increased rate of cAMP formation and increased insulin release.

These diverse metabolic effects of catecholamines may be interpreted as compensatory mechanisms designed to save and shunt glucose from the liver and muscle to the brain. The increased glucose in the blood, released from the liver, is not accompanied by a proportionally increased uptake of glucose by muscle because of the decreased plasma levels of insulin and because of the saturation of the initial steps of the glycolytic pathway as a result of increased muscle glycogenolysis. On the other hand, the uptake of glucose by the central nervous system is much less controlled by insulin and is mainly dependent on blood glucose levels.

Catecholamines increase lipolysis in adipose tissue by stimulating adenylate cyclase of the fat cell. Here again, cAMP, formed at higher rates, initiates the breakdown of stored triglycerides by activating protein kinase which in turn activates triglyceride lipase. The lipolytic response to catecholamines is facilitated by the simultaneous decrease of insulin release, described above, which would interfere with the catecholamine-induced increase in cAMP level in adipose tissue. A number of other hormones (e.g., ACTH, glucagon, thyroxine, growth hormone) stimulate adenylate cyclase in fat cells and thereby induce lipolysis.

ACTH and Steroidogenesis

In 1958, Haynes (22) established that in the adrenal cortex ACTH causes an increase in the level of cAMP and that this nucleotide substitutes for ACTH in inducing steroidogenesis. Subsequently, it was demonstrated that ACTH activates adenylate cyclase in the membrane of the adrenal cell. Two types of ACTH receptors have been identified (33): one with a high affinity but few in number, the other with a low affinity and greater in number. When interacting with the hormone, both types of receptors have the same effect on cAMP production. The high affinity sites probably fill rapidly but empty slowly, thus providing a more steady response. The low affinity receptors fill as hormone concentration rises sharply and empty promptly as the plasma level of ACTH falls, thereby providing a system by which sharp, brief elevations in plasma ACTH produce proportionately sharp, short-lived responses. The activated adenylate cyclase catalyzes the formation of cAMP and the nucleotide acts at the step of conversion of cholesterol to pregnenolone. Garren (17) identified the cAMP receptor in adrenal cortex as a specific protein kinase.

Antidiuretic Hormone (ADH)

In 1961, cAMP was proposed as an intracellular mediator of the action of ADH and related neurohypophyseal hormones. Further experimental evidence documented that cAMP plays a key role in the cellular mechanism of the posterior pituitary antidiuretic hormone action in the skin and urinary bladder of amphibia (which served as the experimental model) as well as in the mammalian kidney. The action of ADH on the collecting duct of mammalian nephrons may be described, according to present knowledge, as follows: ADH reaches the receptor on the basilar cell membrane ("the body side") via the blood stream and interstitial fluid, binds to it and stimulates adenylate cyclase; the increased amount of cAMP within the cell acts on the luminal cell membrane ("tubular side") to increase its permeability to water. In other words, the action of the hormone is initiated at one pole of the cell and results in an increased water permeability of the cell membrane at the opposite pole of the cell.

The ADH-sensitive adenylate cyclase is localized in the medullopapillary part of the kidney and ADH seems to be a quite specific stimulant since a number of other hormones known to stimulate adenylate cyclase in other tissues are inactive in this preparation.

Comparison of the stimulatory effect of neurohypophyseal hormones on adenylate cyclase with their antidiuretic activities determined *in vivo* reveals a good correlation between the hormone-dependent formation of cAMP and the functional response (14).

Nervous System

Considerable information is available on the factors regulating the activity of adenylate cyclase, cAMP-phosphodiesterase, and cAMP levels in nervous tissue. However, the role of cAMP in the function of the nervous system is far from understood. Brain contains high activities of adenylate cyclase and cAMP-phosphodiesterase and has relatively high concentrations of cAMP (7, 49). The enzymes and cAMP are not evenly distributed throughout the brain. The cAMP levels are the highest in the cerebellum and brain stem, intermediate in the hypothalamus and midbrain, and the lowest in the hippocampus and cortex (43). This correlates approximately with the ratios of adenylate cyclase and cAMP-phosphodiesterase in these areas (62). Cellular distribution studies show that adenylate cyclase, cAMP-phosphodiesterase (11), cAMP-dependent protein kinase (35), and protein substrate for cAMP-dependent protein kinase are located mainly in the region of the synaptic membrane (26).

Experimental evidence suggests that cAMP mediates the action of certain neurotransmitters (norepinephrine, dopamine, serotonin, histamine) in the nervous system. For the action of the neurotransmitters to be short lasting, the neurotransmitter itself (see chapter on catecholamines) and its "second messenger," cAMP, have to be inactivated or removed efficiently and rapidly. For some neurotransmitters, the mechanisms of inactivation or removal are known. Thus, for example, acetylcholine released at the synaptic junction is inactivated by acetylcholinesterase, and norepinephrine is removed primarily by uptake mechanisms. Both adenylate cyclase and cAMP-phosphodiesterase are capable of altering cAMP levels in nervous tissue very rapidly. Even though brain has a much greater cAMP-phosphodiesterase than adenylate cyclase activity, an increased activity of the latter is manifested by higher levels of cAMP. This suggests that in the brain the activity of cAMP-phosphodiesterase is controlled by intracellular components (see page 253) and/or cAMP is protected by compartmentation or channeled away from cAMP-phosphodiesterase by its preferential affinity to specific protein kinases or nonspecific proteins.

Numerous studies of factors regulating the activity of both adenylate cyclase and cAMP-phosphodiesterase and of the changes in cAMP level under various conditions in mammalian brain have been reported (10). Attempts have been made to correlate physiologic responses to changes in cAMP in nervous tissue. In a series of elegant experiments combining microtechniques, cytochemistry, neuropharmacology, and electrophysiology, Bloom and his group (5) demonstrated the presence of a central neuronal circuit utilizing norepinephrine as a transmitter, adenylate cyclase as its receptor and cAMP as the mediator of norepinephrine action. This pathway originates in the *locus coeruleus* and its norepinephrine-containing axons provide an inhibitory synaptic connection to cerebellar Purkinje cells (23). The description of this pathway led to the suggestion that central synaptic mechanisms may share some of the features of peripheral sympathetic junctions in utilizing cAMP to mediate the action of synaptically released norepinephrine. Evidence for cAMP mediation of synaptic transmission was provided by studies of the superior cervical ganglion. Greengard and his group (20) showed that cAMP mediates transmission at a dopaminergic synapse of the superior cervical ganglion and proposed the following hypothesis for the role of cAMP in the physiology of synaptic transmission in sympathetic ganglia: Activity in the presynaptic fiber leads to excitation of postganglionic neurons and generation of impulses in the postganglionic axons; simultaneously, the activity in the presynaptic fiber causes excitation of a dopamine-containing interneuron resulting in the release of dopamine, which in turn activates dopamine-sensitive adenylate cyclase in the membrane of the postganglionic neuron and thus causes an increased cAMP content. The rise in cAMP causes, via protein kinase activation leading to phosphorylation of some membrane proteins, an alteration in the movement of ions across the membrane, resulting in hyperpolarization of the cell. This hyperpolarization makes the postganglionic neurons less responsive to subsequent impulses from the presynaptic fibers, resulting in a change in the physiologic response to bursts of sympathetic activity. There is also some evidence that cAMP may modulate transmission at the neuromuscular junction (46).

Another approach to the study of the role of cAMP in the function of the central nervous system was prompted by the observation that urinary excretion of cAMP was increased in agitated and decreased in depressed patients (1, 38). It remains to be shown, however, that the concentration of cAMP in urine reflects the concentration of cAMP in brain. Studies in laboratory animals indicate that changes in cAMP level in specific areas of the brain may be associated with a variety of behavioral changes (18) and that these changes also influence the effects of some psychotropic drugs (55). Tranquilizers of the phenothiazine group, which alleviate certain symptoms of psychosis, prevent the rise of cAMP induced by various stimuli in cerebellum and brain stem. Trifluoperazine, followed by chlorpromazine, are the most effective. This effect of phenothiazine-type tranquilizers is due to their inhibitory action on adenylate cyclase. Promethazine, a weak ataractic agent, and trifluoperazine sulfoxide, a metabolite of trifluoperazine practically devoid of tranquilizing activity, both failed to prevent the increase of cAMP. These findings indicate that there is a correlation between the clinical activity of these drugs as antipsychotic agents and their effectiveness in blocking the rise in cAMP.

The effects of other nonphenothiazine-type potent tranquilizers (e.g., haloperidol, chlorprothixene) on cAMP formation in the brain are similar to those of the pharmacologically active phenothiazine derivatives. Haloperidol and chlorprothixene also inhibit adenylate cyclase activity. The pharmacologically inactive derivatives of these two compounds have little or no effect on cAMP formation. Lithium, which has been used for many years in the management of the manic episodes of manic–depressive disease, also inhibits adenylate cyclase activity in brain (13). It is possible that the therapeutic effect of this cation may be due to a reduction of cAMP level in brain.

Dopaminergic transmission in the caudate nucleus of mammalian brain is well established and an inadequate activation of the dopamine receptor in the caudate nucleus is believed responsible for the symptoms of Parkinson's disease (25). Subsequently, dopamine-sensitive adenylate cyclase was identified in the caudate nucleus (27), and its activation by dopamine was shown to be inhibited by antipsychotic compounds (fluphenazine) (9). It appears that the dopamine receptor in the caudate nucleus is the dopamine-binding portion of a dopamine-sensitive adenylate cyclase and that the Parkinsonian-like side effects of the antipsychotic drugs may be due to the inhibition of this enzyme in the caudate nucleus and the antischizophrenic effects of these compounds to the inhibition of dopamine-sensitive adenylate cyclase in the limbic region of the brain.

Hallucinogenic drugs (e.g., LSD, mescaline, psilocybin), which produce psychotic symptoms in man, were reported to increase the concentration of cAMP in the brain stem of experimental animals and trifluoperazine, to prevent it. Such a finding is in agreement with the observation that tranquilizers prevent electrophysiologic and behavioral responses to psychotomimetic drugs (2, 28). On the other hand, it has been suggested that hallucinogenic drugs act by inhibiting serotonin-sensitive adenylate cyclase in the brain (P. Greengard, personal communication).

Heart

The inotropic effect of catecholamines and glucagon is thought to be exerted through cAMP. These hormones stimulate adenylate cyclase and increase cAMP in the heart. The sequence of events by which cAMP mediates the positive inotropic response is not known. However, experimental evidence suggests that the inotropic action of epinephrine and glucagon may be due to an increase of sarcotubular calcium stores, an effect which is related to the activation of adenylate cyclase located in the sarcoplasmic reticulum. Although catecholamines and glucagon appear to act through a common mechanism, the cAMP system in the heart, each hormone interacts with a different receptor. This is demonstrated by the effect of propranolol, which inhibits the enzymatic and mechanical effects of catecholamines on the heart but does not alter the increase in adenylate cyclase activity nor the positive inotropic effects of glucagon (34). Furthermore, the positive inotropic effect, as well as the ability of glucagon to stimulate adenylate cyclase, is substantially diminished in chronic cardiac decompensation. In constrast, there is normal contractile and adenylate cyclase response to norepinephrine in chronic heart failure (16).

For further information on the role of cAMP in the modulation of cardiac contractility, the reader is referred to a recently published review article (15).

Smooth Muscle

The relaxing effects of catecholamines on the uterus, bronchi, vessels, and intestinal tract are associated with increased levels of cAMP. Catecholamines, through their β-adrenergic activity, stimulate adenylate cyclase in smooth muscle and, thus, induce a rise in intracellular cAMP content. Their relative potency to stimulate adenylate cyclase correlates with their relative potency to decrease smooth muscle tone. Both effects of catecholamines, biochemical and functional, are inhibited by β-adrenergic blocking compounds and potentiated by phosphodiesterase inhibitors. When exogenous cAMP is added to a smooth muscle preparation, the relaxing effect of catecholamines is mimicked. These data are consistent with the concept that the β-adrenergic receptor in the uterus, bronchus, arteries, and intestines is closely related to or part of adenylate cyclase, which represents one of the initial steps of the pathway through which the relaxing effect of catecholamines is exerted (3, 12, 37, 50, 56).

Further studies suggest that the role of the cAMP system in smooth muscle extends beyond the mediation of the relaxing effect of catecholamines. Increase in intracellular cAMP induced by other compounds (other than catecholamines) which either stimulate adenylate cyclase or inhibit cAMP-phosphodiesterase leads to a decrease in smooth muscle tone. Aminophylline and papaverine, compounds with a strong inhibitory effect on cAMP-phosphodiesterase, are thought to exert their relaxing effect on bronchus, uterus or vessels through cAMP, which increases in the cell as a result of the inhibition of cAMP-phosphodiesterase (39, 50, 56).

Synthetic β-adrenergic agents such as terbutaline and albuterol (which are potent bronchodilators but, in contrast to isoproterenol, have a weak effect on the heart) have shown similar differences in their stimulatory effect on adenylate cyclase in bronchus and heart (6, 57, 58). Our recent data show that this difference can be explained by the considerably lower intrinsic activity of terbutaline and albuterol to stimulate adenylate cyclase in the myocardium and in the heart conduction system. In fact, the activity of these compounds is so small that they act as partial agonists of isoproterenol on adenylate cyclase in the heart (57).

Nitroprusside, a potent vasodilator, was found to elevate the intracellular cAMP level in the artery; however, the mechanism of its action is not clear (Triner, unpublished observation).

Anesthetic Agents

At present, there are only a limited number of studies on the effect of anesthetic agents on the cAMP system. Brominated barbiturates were found to inhibit adenylate cyclase in guinea pig heart and lung and it was suggested that this is caused by a reaction with an essential sulfhydryl group of the cyclase (61). Diabutal in 0.1 mM concentration inhibits basal adenylate cyclase activity by 25% but does not modify the catecholamine stimulatory effect on the enzyme (Triner, unpublished observation).

Anesthetic concentrations of chloroform inhibit, noncompetitively, dog heart cAMP-phosphodiesterase and it was suggested that cAMP may play a part in the sensitizing action of chloroform to the arrhythmogenic effect of catecholamines (54). The chloroform-induced susceptibility to the arrhythmogenic effect of epinephrine may be related to the decreased rate of the removal of cAMP caused by the inhibition of cAMP-phosphodiesterase. This hypothesis would be compatible with the evidence of the participation of the β-adrenergic action in the arrhythmogenic effect of catecholamines.

The effect of halothane on the cAMP system and on smooth muscle contractility was first investigated in rat uterus (63). Halothane increased the activity of both adenylate cyclase and cAMP-phosphodiesterase in a dose-dependent fashion. The adenylate cyclase activity was increased to a greater extent than that of cAMP-phosphodiesterase, and tissue levels of cAMP in the uterus were elevated when exposed to halothane. Acetylcholine-induced contraction of rat uterus was inhibited by halothane in concentrations similar to those affecting the cAMP system. These results suggest that the relaxing effect of halothane on uterine muscle is related to an increased intracellular cAMP level caused by the stimulation of adenylate cyclase. In contrast to epinephrine, these effects of halothane were not antagonized by β-adrenergic blocking compounds, propranolol and sotalol (MJ 1999). Halothane, therefore, in its action on uterine smooth muscle, cannot be characterized as a β-adrenergic agonist.

Extending this observation, Sprague et al. (48) described an increased cAMP formation proportional to a direct relaxing effect of halothane and isoflurane in rat aortic strip. On the other hand, cyclopropane in concentrations of 10 to 30 percent, which increased contractility of the strip, caused a decrease in cAMP formation (47).

The bronchodilating effects of halothane and ether also appear to be mediated by cAMP. When compared in equipotent clinical concentrations, halothane appears to be, in in vitro preparations, a stronger bronchodilating agent than ether. Both agents increase the intracellular content of cAMP in this tissue in proportion to their relaxing effects. Both effects are pH-dependent and are not inhibited by a β-adrenergic blocking compound, propranolol (51). Studies in this direction may provide valuable information for the management of patients with bronchial asthma during anesthesia. Similar effects of enflurane and isoflurane on bronchial tone and cAMP were reported recently (52). Halothane has also been shown to increase adenylate cyclase in rat liver (42) and human platelets (60) and to increase cAMP content in cultured mouse neuroblastoma cells (45).

Further studies of the mechanism of halothane action on adenylate cyclase in preparations of rat uterus show an increased response of the enzyme to different agonists (catecholamines, prostaglandin E_1, sodium fluoride) as well as an increased activity of adenylate cyclase induced by halothane in clinically used concentrations. The significantly increased V_{max} of the ATP-dependent adenylate cyclase activity in the presence of halothane suggests that halothane exerts its action on the catalytic unit of adenylate cyclase through conformational changes of lipoprotein structures of the cell membrane, unmasking and rendering operative more catalytic sites of the enzyme (53). One can speculate that such a mechanism of action is not unique to halothane and adenylate cyclase and that other anesthetic agents can affect, through a similar mechanism, other biologically active membrane-bound structures (receptors or enzymes). Biochemical changes thus induced could then contribute to the spectrum of pharmacologic effects of anesthetic agents and, perhaps, even to their anesthetic action.

References

1. Abdulla, Y. H., and Hamadah, K., 3',5'Cyclic adenosine monophosphate in depression and mania. *Lancet* 1, 378 (1970).
2. Adey, W. R., Neurophysiological action of LSD. *In* "Neurophysiological and Behavioral Aspects of Psychotropic Drugs" (A. G. Karczmar and W. P. Koella, eds.), p. 5. Thomas, Springfield, Illinois, 1969.
3. Anderson, R. G. G., Cyclic AMP and calcium ions in mechanical and metabolic responses of smooth muscles; influence of some hormones and drugs. *Acta Physiol. Scand., Suppl.* 382, 1 (1972).
4. Appleman, M. M., and Kemp, R. G., Puromycin: A potent metabolic effect independent of protein synthesis. *Biochem. Biophys. Res. Commun.* 24, 564 (1966).
5. Bloom, F. E., Hoffer, B. J., and Siggins, G. R., Norepinephrine mediated cerebellar synapses: A model system for neuropsychopharmacology. *Biol. Psychiatry* 4, 157 (1972).
6. Burges, R. A., and Blackburn, K. J., Evidence for two kinds of adrenoreceptors. *Nature (London), New Biol.* 235, 249 (1972).
7. Butcher, R. W., and Sutherland, E. W., Adenosine 3',5'-phosphate in biological materials. I. Purification and properties of cyclic 3',5' nucleotide phosphodiesterase and use of this enzyme to characterize adenosine 3',5'-phosphate in human urine. *J. Biol. Chem.* 237, 1244 (1962).
8. Cheung, W. Y., Cyclic nucleotide phosphodiesterase. *Adv. Biochem. Psychopharmacol.* 3, 51 (1970).
9. Clement-Cormier, Y. C., Kebabian, J. W., Petzold, G. L., and Greengard, P., Dopamine-sensitive adenylate cyclase in mammalian brain: A possible site of action of antipsychotic drugs. *Proc. Natl. Acad. Sci. U.S.A.* 71, 1113 (1974).
10. Daly, J. W., Huang, M., and Shimizu, H., Regulation of cyclic AMP levels in brain tissue. *Adv. Cyclic Nucleotide Res.* 1, 411 (1972).
11. De Robertis, E., Rodrigues de Lores Arnaiz, G., Alberici, M., Butcher, R. W., and Sutherland, E. W., Subcellular distribution of adenyl cyclase and cyclic phosphodiesterase in rat brain cortex. *J. Biol. Chem.* 242, 3487 (1967).
12. Dobbs, J. W., and Robison, G. A., Functional biochemistry of beta-receptors in the uterus. *Fed. Proc., Fed. Am. Soc. Exp. Biol.* 27, 352 (1968).
13. Dousa, T., and Hechter, O., Lithium and brain adenyl cyclase. *Lancet* 1, 834 (1970).
14. Dousa, T., Sands, H., and Hechter, O., Cyclic 3',5'-AMP dependent phosphorylation of renal medullary plasma membranes. *Fed. Proc., Fed. Am. Soc. Exp. Biol.* 30, 200 (1971).
15. Entman, M. L., The role of cyclic AMP in the modulation of cardiac contractility. *Adv. Cyclic Nucleotide Res.* 4, 163 (1974).
16. Epstein, S. E., Levey, G. S., and Skelton, C. L., Adenyl cyclase and cyclic AMP. Biochemical links in the regulation of myocardial contractility. *Circulation* 43, 437 (1971).
17. Garren, L. D., Gill, G. N., and Walton, G. M., The isolation of a receptor for adenosine

3′,5′ cyclic monophosphate (cAMP) from the adrenal cortex: The role of the receptor in the mechanism of action of cAMP. *Ann. N.Y. Acad. Sci.* **185**, 210 (1971).

18. Gessa, G. L., Krishna, G., Forn, J., Tagliamonte, A., and Brodie, B. B., Behavioral and vegetative effects produced by dibutyryl cyclic AMP injected into different areas of the brain. *Adv. Biochem. Psychopharmacol.* **3**, 371 (1970).

19. Greengard, P., and Costa, E., eds., "Advances in Biochemical Psychopharmacology," Vol. 3. Raven Press, New York, 1970.

20. Greengard, P., McAfee, D. A., and Kebabian, J. W., On the mechanism of action of cyclic AMP and its role in synaptic transmission. *Adv. Cyclic Nucleotide Res.* **1**, 337 (1972).

21. Greengard, P., Robison, G. A., and Paoletti, R., eds., "Advances in Cyclic Nucleotide Research," Vol. 1. Raven Press, New York, 1972.

22. Haynes, R. C., Jr., The activation of adrenal phosphorylase by the adrenocorticotropic hormone. *J. Biol. Chem.* **233**, 1220 (1958).

23. Hoffer, B. J., Siggins, G. R., Oliver, A. P., and Bloom, F. E., Cyclic AMP-mediated adrenergic synapses to cerebellar Purkinje cells. *Adv. Cyclic Nucleotide Res.* **1**, 411 (1972).

24. Honda, F., and Imamura, H., Inhibition of cyclic 3′,5′-nucleotide phosphodiesterase by phenothiazine and reserpine derivatives. *Biochim. Biophys. Acta* **161**, 267 (1968).

25. Hornykiewicz, O., Neurochemical pathology and pharmacology of brain dopamine and acetylcholine: Rational basis for current drug treatment of Parkinsonism. *Contemp. Neurol.* **8**, 34 (1971).

26. Johnson, E. M., Maeno, H., and Greengard, P., Phosphorylation of endogenous protein of rat brain by cyclic adenosine 3′,5′-monophosphate-dependent protein kinase. *J. Biol. Chem.* **246**, 7731 (1971).

27. Kebabian, J. W., Petzold, G. L., and Greengard, P., Dopamine-sensitive adenylate cyclase in the caudate nucleus of the rat brain and its similarity to the "dopamine receptor." *Proc. Natl. Acad. Sci. U.S.A.* **69**, 2145 (1972).

28. Key, B. J., Effect of chlorpromazine and lysergic acid diethylamide on the rate of habituation of the arousal response. *Nature (London)* **190**, 275 (1961).

29. Krishna, G., Weiss, B., and Brodie, B. B., A simple sensitive method for the assay of adenyl cyclase. *J. Pharmacol. Exp. Ther.* **163**, 379 (1968).

30. Kuo, J. F., and Greengard, P., Cyclic nucleotide-dependent protein kinases. IV. Widespread occurrence of adenosine 3′,5′-monophosphate-dependent protein kinase in various tissues and phyla of the animal kingdom. *Proc. Natl. Acad. Sci. U. S. A.* **64**, 1349 (1969).

31. Kuo, J. F., and Greengard, P., An adenosine 3′,5′-monophosphate-dependent protein kinase from Escherichia coli. *J. Biol. Chem.* **244**, 3417 (1969).

32. Langan, T. A., Histone phosphorylation: Stimulation by adenosine 3′,5′-monophosphate. *Science* **162**, 579 (1968).

33. Lefkowitz, R. J., Roth, J., and Pastan, I., ACTH-receptor interaction in the adrenal: A model for the initial step in the action of hormones that stimulate adenyl cyclase. *Ann. N. Y. Acad. Sci.* **185**, 195 (1971).

34. Levey, G. S., and Epstein, S. E., Activation of adenyl cyclase by glucagon in cat and human heart. *Circ. Res.* **24**, 151 (1969).

35. Maeno, H., Johnson, E. M., and Greengard, P., Subcellular distribution of adenosine 3′,5′-monophosphate-dependent protein kinase in rat brain. *J. Biol. Chem.* **246**, 134 (1971).

36. Miyamoto, E., Kuo, J. F., and Greengard, P., Adenosine 3′,5′-monophosphate-dependent protein kinase from brain. *Science* **165**, 63 (1969).

37. Oppelt, W. W., and Rose, W. E., Effects of dibutyryl cyclic AMP on guinea pig bronchospasm in vivo. *Fed. Proc., Fed. Am. Soc. Exp. Biol.* **31**, 556 (1972).

38. Paul, M. I., Ditzion, B. R., Pauk, G. L., and Janowsky, D. S., Urinary adenosine 3′,5′-monophosphate excretion in affective disorders. *Am. J. Psychiatry* **126**, 1493 (1970).

39. Pöch, G., and Kukovetz, W. R., Studies on the possible role of cyclic AMP in drug-induced coronary vasodilatation. *Adv. Cyclic Nucleotide Res.* **1**, 195 (1972).

40. Robison, G. A., Butcher, R. W., and Sutherland, E. W., "Cyclic AMP." Academic Press, New York, 1971.

41. Robison, G. A., Nahas, G. G., and Triner, L., eds., "Cyclic AMP and Cell Function," Ann. N. Y. Acad. Sci., Vol. 185., N. Y. Acad. Sci., New York. 1971.

42. Rosenberg, H., and Pohl, S., Stimulation of rat liver adenylate cyclase by halothane. *Life Sci.* 17, 431 (1975).

43. Schmidt, M. J., Schmidt, D. E., and Robison, G. A., Cyclic adenosine monophosphate in brain areas: Microwave irradiation as a means of tissue fixation. *Science* 173, 1142 (1971).

44. Schultz, G., Senft, G., Losert, W., and Sitt, R., Biochemical basis of diazoxide hyperglycemia. *Naunyn-Schmiedebergs Arch. Exp. Pathol. Pharmakol.* 253, 372 (1966).

45. Seager, O. A., Effects of anesthetics on cyclic AMP levels in mouse neuroblastoma cells in culture. *Prog. Anesthesiol.* 1, 471 (1975).

46. Singer, J. J., and Goldberg, A. L., Cyclic AMP and transmission at the neuromuscular junction. *Adv. Biochem. Psychopharmacol.* 3, 335 (1970).

47. Sprague, D. H., and Ngai, S. H., Effects of cyclopropane on contractility and the cyclic 3′,5′-adenosine monophosphate system in the rat aorta. *Anesthesiology* 40, 336 (1974).

48. Sprague, D. H., Yang, J. C., and Ngai, S. H., Effects of isoflurane and halothane on contractility and the cyclic 3′,5′-adenosine monophosphate system in the rat aorta. *Anesthesiology* 40, 162 (1974).

49. Sutherland, E. W., Rall, T. W., and Menon, T., Adenyl cyclase. I. Distribution, preparation and properties. *J. Biol. Chem.* 237, 1220 (1962).

50. Triner, L., Nahas, G. G., Vulliemoz, Y., Overweg, N. I. A., Verosky, M., Habif, D. V., and Ngai, S. H., Cyclic AMP and smooth muscle function. *Ann. N.Y. Acad. Sci.* 185, 458 (1971).

51. Triner, L., Vulliemoz, Y., and Verosky, M., Role of cAMP in the bronchodilating effect of halothane and diethylether. *Abstr. Sci. Pap., Annu. Meet. Am. Soc. Anesthesiol.* p. 411 (1974).

52. Triner, L., Vulliemoz, Y., and Verosky, M., Effects of halothane, enflurane and isoflurane on bronchial tone and cAMP. *Fed. Proc., Fed. Am. Soc. Exp. Biol.* 34, 798 (1975).

53. Triner, L., Vulliemoz, Y., and Verosky, M., The action of halothane on adenylate cyclase. *Mol. Pharmacol.* 13, 976 (1977).

54. Ueda, I., and Okumura, F., Effects of chloroform, diethyl ether and a propiophenone derivative, 3-dimethylamino-2-methyl-2-phenoxypropiophenone hydrochloride, upon cyclic 3′,5′-nucleotide phosphodiesterase. *Biochem. Pharmacol.* 20, 1967 (1971).

55. Uzunov, P., and Weiss, B., Psychopharmacological agents and the cyclic AMP system of rat brain. *Adv. Cyclic Nucleotide Res.* 1, 435 (1972).

56. Vulliemoz, Y., Verosky, M., Nahas, G. G., and Triner, L., Adenyl cyclase-phosphodiesterase system in bronchial smooth muscle. *Pharmacologist* 13, 256 (1970).

57. Vulliemoz, Y., Verosky, M., and Triner, L., Effect of albuterol and terbutaline, synthetic beta adrenergic stimulants on the cyclic 3′,5′-adenosine monophosphate system in smooth muscle. *J. Pharmacol. Exp. Ther.* 195, 549 (1975).

58. Vulliemoz, Y., Verosky, M., Katz, R., and Triner, L., Effect of some synthetic beta adrenergic agonists on the cyclic AMP system in smooth muscle. *Fed. Proc., Fed. Am. Soc. Exp. Biol.* 32, 712 (1973).

59. Walsh, D. A., Perkins, J. P., and Krebs, E. G., An adenosine 3′,5′-monophosphate dependent protein kinase from rabbit skeletal muscle. *J. Biol. Chem.* 243, 3763 (1968).

60. Walter, F., Vulliemoz, Y., Verosky, M., and Triner, L., Halothane effect on adenylate cyclase and cAMP-phosphodiesterase in human platelets. (Manuscript in preparation.)

61. Weinryb, I., and Michel, I. M., Effects of barbiturate derivatives on adenyl cyclase from guinea pig heart and lung. *Fed. Proc., Fed. Am. Soc. Exp. Biol.* 30, 219 (1971).

62. Weiss, B., and Costa, E., Regional and subcellular distribution of adenyl cyclase and 3′,5′-cyclic nucleotide phosphodiesterase in brain and pineal gland. *Biochem. Pharmacol.* 17, 2107 (1968).

63. Yang, J. C., Triner, L., Vulliemoz, Y., Verosky, M., and Ngai, S. H., Effects of halothane on the cyclic 3′,5′-adenosine monophosphate (cyclic AMP) system in rat uterine muscle. *Anesthesiology* 38, 244 (1973).

Chapter 12

Effect of Anesthetic Drugs on Respiratory Enzymes

Richard W. Patterson and Stuart F. Sullivan

Introduction

A series of compounds with sequentially increasing oxidation–reduction potentials, the respiratory enzymes, provide for the passage of electrons from the electronegative respiratory substrates to the ultimate electron acceptor, oxygen. [Recent reviews are pertinent to the unique fitness of oxygen (18), and to the components (46, 47) and kinetics (27) of electron-transfer reactions.] This sequential multienzyme chain, together with its requisite components, a system for generating electron donors (the Kreb cycle), and a system to provide an energy output (oxidative phosphorylation) is contained within the mitochondrion, a cellular organelle.

The use of narcotics as experimental respiratory inhibitors can be traced back to 1925 when Keilin (26a) found that urethane and certain alcohols induced a block of electron transfer to cytochromes c and a_3. Subsequently, hypnotics, tranquilizers, and general anesthetics, drugs of dissimilar structure, have been shown to have similar blocking activity and in dose-dependent fashion to affect the other two components of energy transduction, thus providing stimulus for hypotheses and experiments involving the respiratory enzymes in the accelerating search for the elusive biochemical basis of anesthesia.

Metabolic Pathways

Reduced nicotinamide adenine dinucleotide (NADH), produced by the oxidation of Kreb cycle substrates with the exception of succinate, transfers reducing equivalents to the ferroflavoprotein, NADH dehydrogenase (FP_1).

Competing for the final common pathway to oxygen (Fig. 12-1), the NADH oxidase system and the succinate oxidase system converge at the level of ubiquinone (coenzyme Q, CoQ). Though thermodynamically unfavorable, NAD^+ can be reduced by succinate, when energy is made available either from adenosine triphosphate (ATP) or from the oxidation of succinate itself, demonstrating direct linkage between the two flavoproteins via CoQ.

Choline, acyl fatty acids, sarcosine, and L-α-glycerophosphate are also oxidized by enzyme systems involving direct linkage with the respiratory chain components. The exact level of this linkage is still uncertain, though it is generally accepted that CoQ is accessible to all substrates. The entrance of glycerophosphate, as shown in Figure 12-1, is based on the inhibitory effect of phenylurethane (thought to be an inhibitor of cytochrome b) on reactions involving succinate or choline as substrate and lack of inhibition when L-α-glycerophosphate is used. L-α-glycerophosphate is an essential part of the "shuttle" system permitting cytoplasmic or glycolytic NADH to enter the intact mitochondrial membrane for intramitochondrial reoxidation. In contradistinction to liver and heart mitochondria, brain mitochondria oxidize

Figure 12.1. Electron-transport scheme, showing respiratory substrate input via flavoproteins 1–5. Site of metabolic blockers denoted by × × for cyanide, × for antimycin. Locus of blocking activity of anesthetics is imposed between non-heme ion and CoQ as discussed in text. Energy-transfer coupling points are denoted by ◆, leading to ATP production or by ▲ to denote the other options of the disposition of the energized state generated by electron transfer, i.e., transhydrogenation, energized swelling, or ion translocation.

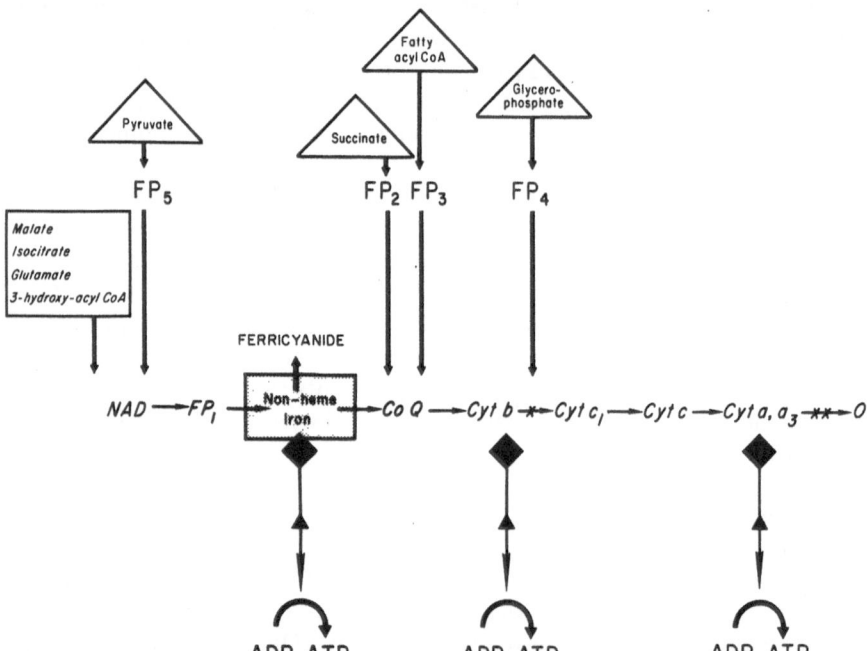

glycerophosphate at a rate of 50 percent greater than that of succinate, and 100 percent greater than that of NADH (39). Located at the beginning of the final pathway, cytochrome b accepts electrons from ubiquinone and transfers them to cytochrome c_1. Cytochrome c oxidase (cytochrome $a + a_3$, EC 1.9.3.1), the terminal enzyme of the respiratory chain, is the oxidant of ferrocytochrome c_1 and is itself oxidized by oxygen.

Specific and stoichiometric inhibitors of the respiratory chain are hydrogen cyanide, acting between cytochrome a, a_3 and oxygen, antimycin A, acting between cytochrome b and cytochrome c, and rotenone, blocking the NADH dehydrogenase to CoQ transfer.

Inhibition of Oxygen Transport by Anesthetics

Extensive investigation of the locus of this latter activity has led to the use of amytal (5-ethyl-5-isoamyl barbituric acid), also as a specific inhibitor of the NADH oxidase system (40). NADH oxidase activity measured in sub-mitochondrial particles (electron transport particles, ETP) during ^{14}C-rotenone binding studies demonstrated both specific and nonspecific binding sites; binding at the latter sites could be eliminated by treatment with bovine serum albumin. Prior treatment with amytal and piericidin A tends to abolish binding at the specific site when ^{14}C-roteonone is subsequently added. Therefore it was concluded that all three drugs compete for the same site and that this is the site responsible for the inhibition of NADH oxidase activity in the respiratory chain. Piericidin was found to be the most tightly bound at the specific site and it was surmised that the titer for specific binding of piericidin per milligram of protein should increase on cleaving the complete respiratory chain into shorter units. In progressing from ETP (26) (which contains the complete electron transport system) to the NADH-cytochrome reductase particle (22) (containing all of the components from flavoprotein to cytochrome c_1) and then to "Complex I" (23) (a particle free of cytochrome but containing flavoprotein, structural protein, and lipids including CoQ) the binding site titer for piericidin did increase. Subsequent demonstrations that succinate oxidation was not blocked, that NADH oxidation by succinate was blocked, and that amytal did not block the NAD-ferricyanide reaction established the site of block to be between the non-heme iron and CoQ.

Added to rat liver mitochondria, 1.8×10^{-3} M amytal completely inhibits the oxidation of NAD linked substrates, and partially inhibits oxidation of extramitochondrial NADH, but has no effect on the oxidation of succinate or of the phosphorylation coupled to it. Specificity of action, however, is obscured when high concentrations are used.

In a rat liver mitochondrial preparation with glutamate and malate as substrates, the following concentrations of amytal were required to inhibit by 50 percent (ID_{50}) various portions of the electron transport system: 1.2×10^{-4}

M, respiration; 3.2×10^{-4} M, NAD$^+$ reduction; 4.5×10^{-4} M, cytochrome b reduction; and 7×10^{-3} M, succinate oxidase activity (5). This relative sensitivity of the NAD-dependent system was again demonstrated in an inactivation study which required 1.7×10^{-4} M amytal for 50 percent inhibition of NADH oxidase; 6.4×10^{-4} M was required for 50 percent inhibition of choline oxidase (44). An amytal insensitive pathway of dihydronicotinamide adenine dinucleotide, NAD, metabolism has been demonstrated (13). Menadione (vitamin K$_3$) can bridge the amytal-inhibited site of the NADH oxidase system, restoring NADH oxidation and two sites of phosphorylation (14) (*vide infra*).

Analysis of the requirement for an amytal-like inhibition of NADH oxidase in beef heart submitochondrial particles determined that the essential structure consisted of a nonspecific hydrocarbon group attached to an amide, carbamide, or barbituric acid; compounds having in common the —CO—NH— group (10). Nonpolar organic solvents such as iso-octane cause a decrease in succinate oxidation and an increase in oxidation of NAD, whereas polar solvents such as diethyl ether cause a rapid, specific, and complete inhibition of NAD (11, 37).

Subsequently, a variety of structurally heterogenous hypnotic, analgesic, and anesthetic compounds have been shown to act on the sensitive NADH oxidase system with little effect on the succinate system: narcotic steroids and related compounds such as diethylstilbesterol and progesterone (48), demerol (24), phenothiazine (16), diethyl ether (11), methoxyflurane (2,2-dichloro-1, 1 difluorotheyl methyl ether) (21), and halothane (2-bromo-2-chloro-1 : 1 : 1-trifluoroethane) (21). Treatment with ether releases an NAD-menadione reductase (EC 1.6.99.2) which is inhibited by high levels of amytal (11). Additional ETP investigations demonstrated that diethyl ether, halothane, octanol, carbon tetrachloride, methoxyflurane, chloroform, and fluroxene (trifluoroethyl vinyl ether) individually inhibited NADH oxidation and NADH-cytochrome c reduction, but had little effect upon succinic oxidase or NADH ferricyanide reductase activities, demonstrating the block to be after the flavoprotein and non-heme iron but preceding CoQ (21).

The two ferroflavoproteins, succinic dehydrogenase and NADH dehydrogenase, are attached to the inner membrane of the mitochondria in a phospholipid environment apparently required for the stability of the components of the complex and for activation so that electron transport to ubiquinone can take place (1). The inhibition of NADH-cytochrome c oxidoreductase activity of 10 substituted barbiturates, including amytal, was highly correlated ($r = 0.95$) with their lipid solubility (42). Investigations with submitochondrial particles fail to reveal anesthetic effects on the properties of the respiratory chain which are dependent upon protein conformation, membrane relationship, and complex aggregation. For this reason, such studies are generally conducted with mitochondria suspended in an isotonic sucrose-mannitol medium, provided with oxygen, inorganic phosphate, ADP, NAD substrate or succinate, with exposure to the anesthetic at 0–4°C and with oxygen uptake polarographically measured at 25°C.

Factors Controlling Respiratory Rate

The energized state generated by electron transfer in the complex has to be transduced to supply energy for the synthesis of ATP by the union of ADP and orthophosphate. Intimately associated and synchronized with the electron transport system, this step is referred to as coupled phosphorylation. The rate of respiration depends on the supply and concentration of respiratory substrates (which independently have been shown to be affected by anesthetics) (35) and of oxygen, and on the concentrations of the reactants, adenosine 5'-diphosphate (ADP), inorganic phosphate (Pi), and the product adenosine 5'-triphosphate (ATP) of the coupled oxidative phosphorylation occurring in the respiratory chain. One test of tightly coupled intact mitochondria is the decrease in respiration, oxygen uptake, and electron transport, in the presence of an increased concentration of ATP. This mechanism of decreased respiratory enzyme activity may be operative during exposure to anesthetics as evidenced by the demonstration that both ether and halothane decrease the total content and the rate of utilization of ATP (2, 45). The respiratory rate of "uncoupled" mitochondria is little or not at all dependent on ADP. Though a decrease in oxygen consumption (V_{0_2}) has been found during exposure to anesthetics in preparations ranging from whole body, tissue, and cell culture to submitochondrial particles, evidence for the specificity in regard to the locus of this action is lacking due to the multiplicity of sites of demonstrated anesthetic action. Oxygen is but one of many controlled inputs in this output-driven metabolic system where a cellular "sink" of ATP-requiring activities "pulls" the process of ATP-yielding catabolism.

Chance and Williams have defined the conditions controlling the respiratory rate of intact mitochondria and the oxidation–reduction states of the components of the respiratory chain (3). State 1 is the condition in which both ADP and respiratory substrates are lacking and state 2 is the condition in which only respiratory substrates are lacking. State 3 is the condition in which all required components are present and the respiratory chain itself is the rate limiting factor; this state is also called the state of "active" respiration. State 4 is the condition in which only ADP is lacking: This is so-called controlled, or resting, or ADP-less state. State 5 is the condition in which only oxygen is lacking. The respiratory carriers are in their most oxidized state when the rate-limiting factor is the substrate and all other components are present in excess; the most reduced state occurs in the absence of oxygen. A block in the carrier system provides reduced substances on the substrate side of the block and oxidized substances on the oxygen side. In the aerobic steady state (state 3) there is an intermediate degree of reduction with a gradient of increasing degree of oxidation between pyridine nucleotides and oxygen. The addition of ADP during a state 4 condition produces state 3 and increases respiration. The rate of respiration (μM_{0_2} per sec) during state 3 divided by that during state 4, the controlled state, is defined as the respiratory control index or the acceptor control ratio. "Tightly coupled" intact mitochondria have high acceptor control ratios and high P:O ratios.

By measuring the respiratory rate under such conditions, dose-related reversible blockage of electron transport at the NADH dehydrogenase locus was demonstrated for the following volatile anesthetics: halothane, methoxyflurane, diethyl ether, enflurane (Ethrane, 2-chloro-1, 1, 2-trifluoroethyl difluoromethyl ether), its isomer isoflurane (Forane, 1-chlor-2,2,2,-trifluroethyl difluoromethyl ether), and fluroxene (6, 7, 9). The anesthetic concentrations required for 50% inhibition of state 3 respiration (ID_{50}) were calculated for all six agents and found to bear an inverse log–log relationship to lipid solubility. Previously, the relationship had been established between the minimum alveolar concentration (MAC) of an inhalation agent necessary to prevent gross movement in response to surgical incision in 50 percent of patients and its lipid solubility (12, 38). Therefore, potency *in vitro* (ID_{50}) could be related to potency *in vivo* (MAC). Depression of state 3 oxygen uptake is linearly depressed by increasing concentrations of halothane, methoxyflurane and enflurane up to a point beyond which there is no further effect, due to lack of inhibition of the amytal insensitive pathway of dihydronicotinamide adenine dinucleotide metabolism (8, 34). It was also shown that nitrous oxide is capable of suppressing state 3 oxygen uptake and if added to an equipotent concentration of halothane, the effect of the mixture on mitochondrial respiration is additive (33).

The ultimate benefit of transferring electrons through the respiratory chain is the conservation of some fraction of the energy of oxidation of the substrates in a form available for cellular work; i.e., synthesis of ATP, movement of inorganic ions and organic substrates, or the reduction of $NADP^+$ by NADH. The energy is released in discrete packets, as between NAD and FP_1 (NADH dehydrogenase) and some fraction of this is captured and utilized by an, as yet, unclarified mechanism. Thus, when electrons are transferred from NADH through the entire electron-transfer chain, three molecules of ATP are formed in energy-conserving reactions (oxidative phosphorylation). When succinate is oxidized by the flavin-containing enzyme succinic dehydrogenase (FP_2) electrons are transferred through the cytochrome chain but only two ATP molecules are formed.

Even at high concentrations of amytal, the increased reduction of pyridine nucleotide of phosphorylating mitochondria can be reversed by uncoupling agents, thus demonstrating that amytal combines an inhibition of electron transfer with an inhibition of energy transfer in tightly coupled mitochondria (4). Similar effects have been obtained with ethylene glycol and progesterone (5). In addition to blocking energy transfer at the first phosphorylating site, the second phosphorylating site can also be blocked in tightly coupled mitochondria. Thus, amytal in high concentrations is capable of additionally interfering with succinate oxidation (36). Similar to the responses obtained with amytal, dose-related depression of various components of the respiratory chain has been demonstrated with halothane (6–9, 17, 21, 30). With this agent clinical anesthesia results from an alveolar concentration of 0.7–1.0 percent. This corresponds to a whole tissue concentration of 1.0–1.5 mM in brain and liver. In these low concentrations, reversible inhibition of electron transport is nearly completely localized in the NADH dehydrogenase fraction. With higher concentrations than are used clinically, irreversibility and uncoupling of phosphorylation can be produced. It was observed by Miller and Hunter (30), using rat liver mitochondria and the NAD^+-dependent substrate, 3-hydroxybutyrate, that the ADP-stimulated rate of oxygen uptake (state 3 respiration) was inhibited by low concentrations of halothane (0.5–2.0% in the gas phase) in a concentration-dependent manner, with a maximal inhibition of 80%. Similar results were obtained by Harris (21), with comparable halothane concentration (<2 mM) as determined by gas liquid chromatography, and by Cohen et al. (6–8). At these concentrations succinate oxidation was not affected (21), the ADP/O ratios were not altered, and the addition of uncouplers did not antagonize the inhibition of state 3 respiration. The inhibition of NAD^+-dependent oxidation remains unchanged as the concentration is raised to 2–4% in the gas phase (7, 30) and is accompanied by a concentration-dependent slowing of succinate oxidation during state 3 respiration and an increase in state 2 respiration. With 3% halothane the state 2 and state 4 rates were more than doubled (30) implying an uncoupling of phosphorylation, which with time becomes irreversible (21). In addition to the apparent decrease in respiratory control, 4% halothane inhibits all response to the addition of ADP, at which time electron transfer, with succinate as the substrate, is inhibited 63%. Halothane, in concentrations between 5 and 10% alters membrane permeability and causes energy-independent swelling of mitochondria.

When the electron-transferring sequence is released from control by an acceptor, permitting respiration to proceed at an increased rate, oxygen uptake is increased, orthophosphate uptake is decreased, and "uncoupling" is considered to have occurred. According to such loosely defined characterization, the uncoupling of oxidation from phosphorylation has been demonstrated with exposure to anesthetic steroids, barbiturates, and ether. In a stricter sense, an uncoupler may be defined as an agent which disconnects the energy-transferring sequence from the electron-transferring sequence at the coupling site. The demonstration that the reaction has shunted energy conservation from ATP synthesis into another channel, as for example into ion transport, indicates that energy conservation has not been dislocated. On this basis Miller produced evidence that halothane is not a true uncoupler of oxidative phosphorylation and that any uncoupling by anesthetic concentrations is very limited (31). The rate and extent of Ca^{2+} uptake by mitochondria, a function extremely sensitive to true uncouplers, were measured in the presence of halothane concentrations as high as 4 percent. As the concentration increased, the rate of Ca^{2+} accumulation slowed but the amount accumulated at equilibrium was the same. This suggests that instead of uncoupling, other factors, for instance, inhibition of the rate of electron transfer, affect the kinetics of Ca^{2+} uptake.

Conformational Changes

When the anesthetics inhibit electron transport between flavoprotein and cytochrome b in intact mitochondrial preparations, a large positive entropy change occurs at the same time. This indicates the possibility of disruption of hydrophobic bond formation and of conformational changes. It has been proposed that anesthetic molecules combine with hydrophobic regions of proteins causing them to unfold and that this conformational change to less active or inactive forms is accompanied by enthalpy and entropy changes. This is manifested, among other changes, by a measurable alteration in the number of available sites and affinity for Ca^{2+} binding. The theoretical consequences of these changes have been discussed (15, 19, 28, 43), and conceptual models were proposed to bridge the gap between molecular biology and neurophysiologic function. The respiratory enzymes are repeating units of proteins and are integral parts of the mitochondrial membrane. When all of these units are predominantly in the same conformation there is a 1 : 1 correspondence between the conformation of the proteins and the configuration of the membrane. Conformational change involves a whole train of interrelated changes and any one of these can be used as the basis of a conformational probe; for example, changes in the titer of sulphydryl groups, in ion binding or proton concentration, or changes in configuration demonstratable by light scattering or electron microscopy. Using this last-named technique, Hackenbrock (20) first reported a reversible ultrastructural change in isolated rat liver mitochondria which varied with the metabolic energy state of the mitochondria as defined by Chance and Williams (3). This was described as a transition from an orthodox (essentially uncoupled, typical of configuration seen in anesthetized preparations) to a condensed (capable of coupled oxidative phosphorylation) ultrastructural state. (State 4, resting to state 3, active = orthodox to condensed.) Inhibitors of electron transport prevent the condensed to orthodox transition; uncouplers, such as dinitrophenol prevent the orthodox to condensed transitions instituted by ADP. These findings provided the early basis for the conformational coupling hypothesis of oxidative phosphorylation of Green (19), who maintains that the integrity of energy transduction resides in the membrane. Other workers later invoked conformational changes at the molecular level as the basis for the ultrastructural changes observed (for example, the chemiosmotic theory)

(32). It is not known whether the morphologic changes are due to the electron transport, i.e., whether they represent the transduction of chemical energy into mechanical energy, or whether alternatively the transformations of the mitochondrial cristae and matrix are the results of the movements of ions and water.

Repeatedly in the past, attempts to provide evidence toward a unified theory of the relationship between the state of anesthesia and the metabolic rate or oxygen consumption have failed. Recent studies continue to offer no support. Comparison of the anesthetic concentrations required to produce functional changes in the whole body, in tissue slices, cell cultures, or mitochondria with the concentrations required to produce measurable biochemical changes in the same preparations (25) indicates that the anesthetic concentrations are by several orders of magnitude lower than those required for the blockage of electron transport. The implication, however, that alteration of cellular respiration is not involved as a consequence of clinical anesthesia may not be justified at this early stage of the understanding of energy relationships. The model system (the mitochondrial electron-transport system) used in testing is itself still something of a "black box" and Wainio (46) in 1971 summarized the state of the art: "...after years of advances our knowledge of the respiratory chain is still rudimentary...even the sequences of electron carriers remains somewhat uncertain, while the hypotheses of oxidative phosphorylation are no more than guesses." Furthermore, alternative pathways in quantitatively "minor" organ systems could be easily masked in overall body reactions.

Until recently, with interest focused primarily on defining the pathways of intermediary metabolism it appeared that the enzymatic mechanism of a given metabolic transformation was similar if not identical in widely different organisms. Such simplification of this mechanism would allow similar generalizations concerning metabolic inhibitors. It is now evident that there are differences in some of the properties among enzymes obtained from different cell types which catalyze the same reactions ("heteroenzymes"). There is also evidence for the existence of multiple forms of enzymes in the same cells which catalyze a given reaction by the same mechanism ("isoenzymes") or by different mechanisms ("isoalloenzymes"). The existence of additional pathways has been evident in studies where the primary electron pathway has been blocked by phenothiazine (16) or by steroids (48). Studies of the evolutionary development of succinate dehydrogenase (41) in relation to the physiologic needs of the organism are another case in point.

It is conceivable that the predominant effect of anesthetics is mediated via indirect action on the electron-transport system. There may be, for example, a reversal of the Pasteur effect, an anesthetic-induced intracellular "Crabtree effect." Anesthetic capability to alter the rate (V_{max}) of transformation of substrates without any changes in their affinity to respiratory enzymes was demonstrated by Miller (31). The tissue or organ site of action of the anesthetic may be important; critical reduction in energy production or enzyme utilization in only a small group of neurons may be responsible for narcosis without significantly altering overall metabolic activity of brain segments. It is conceivable that the glial–neuronal relationship may be disturbed. Whether or not mitochondria from these two cell populations are biochemically heterogenous and capable of individual responses to anesthetics is not known. The meticulous studies of Matschinsky (29) demonstrate beyond doubt that the capacities of energy-yielding and energy-consuming processes of dissimilar neuronal material are closely determined by the specific environment and the functions of the nervous elements. Variable patterns of energy metabolism signal unique structural features. Recent investigations have demonstrated significant differences in blood flow, and presumably in metabolism not only in differing regional areas of the brain, but also in the same areas under differing functional situations, i.e., between mental rest and mental arithmetic.

Such studies as the relationship between Ca^{2+} uptake, oxidative phosphorylation, and heart function, the relationships between alkalosis and increased oxygen consumption, and studies of the protective effect of acidosis against hypoxia further suggest that electron transport and oxidative phosphorylation may be the common denominator between anesthesia and the clinically observed myocardial effects, and between anesthesia and its protective role in hemorrhagic shock and hypoxia.

References

1. Bruni, A., and Racker, E., Resolution and reconstitution of the mitochondrial electron transport system. Reconstitution of the succinate-ubiquinone reductase. *J. Biol. Chem.* **243**, 962 (1968).
2. Brunner, E. A., and Passonneau, J. V., The effect of inhalational anesthetic agents on brain metabolite levels. *In* "Cellular Biology and Toxicity of Anesthetics" (B. R. Fink, ed.), p. 39. Williams & Wilkins, Baltimore, Maryland, 1972.
3. Chance, B., and Williams, G. R., The respiratory chain and oxidative phosphorylation. *Adv. Enzymol.* **17**, 65 (1956).
4. Chance, B., Hollunger, G., and Hagihara, B.: Inhibition of energy transfer at the pyridine nucleotide-flavin site. *Biochem. Biophys. Res. Commun.* **8** 180, (1962).
5. Chance, B., and Hollunger, G., Inhibition of electron and energy transfer in mitochondria. Effects of amytal, thiopental, rotenone, progesterone, and methylene glycol. *J. Biol. Chem.* **238**, 418 (1963).
6. Cohen, P. J., Marshall, B. E., Harris, J. E., Lecky, J. H., and Rosner, B. S., Halothane-induced changes in mitochondrial oxygen uptake and respiratory control. *Fed. Proc., Fed. Am. Soc. Exp. Biol.* **27**, 705 (1968).
7. Cohen, P. J., and Marshall, B. E., Effects of halothane on respiratory control and oxygen consumption of rat liver mitochondria. *In* "Toxicity of Anesthetics" (B. R. Fink, ed.), p. 24. Williams & Wilkins, Baltimore, Maryland, 1968.
8. Cohen, P. J., Marshall, B. E., and Lecky, J. H., Effects of halothane on mitochondrial oxygen uptake. *Anesthesiology* **30**, 337 (1969).

9. Cohen, P. J., and McIntyre, R., The effects of general anesthesia on respiratory control and oxygen consumption of rat liver mitochondria. *In* "Cellular Biology and Toxicity of Anesthetics" (B. R. Fink, ed.), p. 109. Williams & Wilkins, Baltimore, Maryland, 1972.

10. Cowger, M. L., Labbe, R. F., and Mackler, B., Relation of barbiturate structure to DPNH oxidase inhibition. *Arch. Biochem. Biophys.* **96**, 583 (1962).

11. Cunningham, W. P., Crane, F. L., and Sottocasa, G. L., Sequential release of nicotinamide-adenine dinucleotide dehydrogenase activity by a series of physical treatments. *Biochim. Biophys. Acta* **110**, 265 (1965).

12. Eger, E. I., II, Brandstater, B., Saidman, L. J. Regan, M. J., Severinghaus, J. W., and Munson, E. S., Equipotent alveolar concentrations of methoxyflurane, diethyl ether, fluroxene, cyclopropane, xenon, and nitrous oxide in the dog. *Anesthesiology* **26**, 771 (1965).

13. Ernster, L., Jelling, O., Low, H., and Lindberg, O., Alternate pathways of mitochondrial DPNH oxidation studied with amytal. *Exp. Cell Res. Suppl.* **3**, 124 (1955).

14. Ernster, L., Dallner, G., and Azzone, G. F., Differential effects of rotenone and amytal on mitochondrial electron and energy transfer. *J. Biol. Chem.* **238**, 1114 (1963).

15. Eyring, H., Woodbury, J. W., and D'Arrigo, J. S., A molecular mechanism of general anesthesia. *Anesthesiology* **38**, 415 (1973).

16. Gallagher, C. H., Koch, J. H., and Mann, D. M., The effect of phenothiazine on the metabolism of liver mitochondria. *Biochem. Pharmacol.* **14**, 789 (1965).

17. Gatz, E. E., and Jones, J. R., Effects of six general anesthetics upon mitochondrial phosphorylation. *Fed. Proc., Fed. Am. Soc. Exp. Biol.* **28**, 356 (1969).

18. George, P., The fitness of oxygen. *Oxidases Relat. Redox Syst., Proc. Symp., 1964* p. 3 (1965).

19. Green, D. E., and Baum, H., "Energy and the Mitochondrion." Academic Press, New York, 1969.

20. Hackenbrock, C. R., Ultra-structural basis for metabolically linked mechanical activity in mitochondria. II. Electron transport-linked ultrastructural transformation in mitochondria. *J. Cell Biol.* **37**, 345 (1968).

21. Harris, R. A., Munroe, J., Farmer, B., Kim, K. C., and Jenkins, P., Action of halothane upon mitochondrial respiration. *Arch. Biochem. Biophys.* **142**, 435 (1971).

22. Hatefi, Y., Haavik, A. G., and Jurtshuk, P., DPNH-cytochrome c reductase. *Biochim. Biophys. Acta* **52**, 106 (1961).

23. Hatefi, Y., Haavik, A. G., and Griffiths, D. E., Preparation and property of mitochondrial DPNH-coenzyme Q reductase. *J. Biol. Chem.* **237**, 1676 (1962).

24. Hatefi, Y., Flavoproteins of the electron transport system and the site of action of amytal, rotenone, and piericidin A. *Proc. Natl. Acad. Sci. U. S. A.* **60**, 733 (1968).

25. Quastel, J. H., Effects of anesthetics, depressants, and tranquilizers on cerebral metabolism. In Metabolic Inhibitors. Hochster, R. M., and Quastel, J. H., Editors, Academic Press, New York and London, 1963, Vol II, p. 517.

26. Horgan, D. J., Singer, T. P., and Casida, J. E., Binding sites of rotenone, piercidin A and amytal in the respiratory chain. *J. Biol. Chem.* **243**, 834 (1968).

26 a. Keilin, D., On cytochrome, a respiratory pigment common to animals, yeast and higher plants. Proc. Royal Soc., London, B, 98:312, 1925.

27. Green, D. E., Wharton, D. C., Tzagoloff, A., Rieske, J. S., and Brierley, G. P., The mitochondrial electron transfer chain. In Oxidases and Related Redox Systems. King, T. E., Mason, H. S., and Morrison, M., Editors: John Wiley and Sons, New York, London, Sydney. 1965, Vol 2, p. 1032.

28. Krnjevic, K., Excitable membranes and anesthetics. *In* "Cellular Biology and Toxicity of Anesthetics" (B. R. Fink, ed.), p. 3. Williams & Wilkins, Baltimore, Maryland, 1972.

29. Matchinsky, F. M., Energy metabolism of the microscopic structures of the cochlea, the retina, and the cerebellum. *Adv. Biochem. Psychopharmacol.* **2**, 217 (1970).

30. Miller, R. N., and Hunter, F. E., The effect of halothane on electron transport, oxidative phosphorylation and swelling in isolated rat liver mitochondria. *Mol. Pharmacol.* **6**, 67 (1970).

31. Miller, R. N., and Hunter, F. E., Jr., Is halothane a true uncoupler of oxidative phosphorylation? *Anesthesiology* **35**, 256 (1971).
32. Mitchell, P., Coupling of phosphorylation to electron and hydrogen transfer by a chemiosmotic type of mechanism. *Nature (London)* **191**, 144 (1961).
33. Nahrwold, M. L., and Cohen, P. J., Additive effects of nitrous oxide and halothane on respiration of rat liver mitochondria. *Fed. Proc., Fed. Am. Soc. Exp. Biol.* **32**, 789 (abstr.) (1973).
34. Nahrwold, M. L., and Cohen, P. J., The effects of forane and fluroxene on mitochondrial respiration. *Anesthesiology* **38**, 437 (1973).
35. Paradise, R. R., and Ko, K. C., The effect of fructose on halothane depressed rat atria. *Anesthesiology* **32**, 124 (1970).
36. Pumphrey, A. M., and Redfearn, E. R., An inhibition by amytal of succinate oxidation in tightly coupled mitochondria. *Biochem. Biophys. Res. Commun.* **8**, 92 (1962).
37. Redfearn, E. R., and King, T. E., Mitochondrial NADH dehydrogenase and NADH oxidase from heart muscle. *Nature (London)* **202**, 1313 (1964).
38. Saidman, L. J., Eger, E. I., II, Munson, E. S., Babad, A. A., and Muallem, M., Minimum alveolar concentrations of methoxyflurane, halothane, ether, and cyclopropane in man. *Anesthesiology* **28**, 994 (1967).
39. Saktor, B., Packer, L., and Estabrook, R. W., Respiratory activity of brain mitochondria. *Arch. Biochem. Biophys.* **80**, 68 (1959).
40. Singer, T. P., Horgan, D. J., and Casida, J. E., Reactions of rotenone, piericidin A, and barbiturates with components of the respiratory chain. *Flavins Flavoproteins Proc. Conf., 2nd, 1967* p. 192 (1968).
41. Singer, T. P., Comparative biochemistry of succinate dehydrogenase. *Oxidases Relat. Redox Syst., Proc. Symp., 1964* p. 448 (1965).
42. Spiegel, H. E., and Wainio, W. W., Some features of barbiturate interaction and inhibition of NADH-cytochrome c oxidoreductase in respiring systems. *J. Pharmacol. Exp. Ther.* **165**, 23 (1969).
43. Taylor, C. A., Williams, C. H., Wakabayashi, T., Valdiva, E., Harris, R. A., and Green, D. E., The effect of halothane on energized configurational changes in heart mitochondria in situ. *In* "Cellular Biology and Toxicity of Anesthetics" (B. R. Fink, ed.), p. 117. Williams & Wilkins, Baltimore, Maryland, 1972.
44. Tyler, D. D., Gonze, J., and Estabrook, R. W., Observations on the inhibitor sensitivity of the choline oxidase system. *Arch. Biochem. Biophys.* **115**, 373 (1966).
45. Ueda, I., and Kamaya, H., Kinetic and thermodynamic aspects of the mechanism of general anesthesia in a model system of firefly luminescence in vitro. *Anesthesiology* **38**, 425 (1973).
46. Wainio, W. W., "The Mammalian Mitochondrial Respiratory Chain." Academic Press, New York, 1970.
47. Yagi, K., ed., "Flavins and Flavoproteins." Univ. Park Press, Baltimore, Maryland, 1968.
48. Yielding, K. L., Tomkins, G. M., Munday, J. S., and Cowley, I. J., The effect of steroids on electronic transport. *J. Biol. Chem.* **235**, 3413 (1960).

Chapter 13

Creatine Phosphokinase

Elemer K. Zsigmond

Properties and Enzymologic Characteristics

Enzymologic Classification

Creatine phosphokinase (CPK) (ATP:creatine phosphotransferase) (EC 2.7.3.2*) belongs to a group of enzymes called *transferases* (EC 2...), more specifically to the *kinases* (EC 2.7.3.) which transfer phosphorous containing groups (81). These enzymes are responsible for the majority of transphosphorylation reactions involving ATP and are essential to the maintenance of normal energy transfer from one system to the other (187). There is an energy-rich bond on both sides of the equation and therefore these reactions are readily reversible, as shown below (144).

Creatine + ATP $\rightleftharpoons$ creatine phosphate + ADP

Molecular Weight

All mammalian CPK have a molecular weight of about 80,000 (73, 142, 186). The molecular weight of CPK in the dystrophic mouse muscle is also 80,000 (114).

Specificity

Creatine is the physiologic substrate of CPK (187). Kuby *et al.* (143) found that creatine cannot be replaced by creatinine, *l*-arginine, *d*-arginine, or *l*-histidine. It can be replaced, however, by *N*-ethylglycocyamine and glycocyamine (71, 142–144, 245). The energy donor of the reaction ATP cannot be

*EC 2.7.3.2. = Classification of International Commission on Enzymes.

replaced by ADP, adenosine-2-phosphate, adenosine-3'-phosphate, adeno-sine-5'-phosphate, or inosinetriphosphate. Similarly, ADP cannot be replaced by ATP in the reverse reaction. CPK also possesses slight adenosine-triphos-phatase (ATPase) (EC 3.6.1.3.) activity (219).

Cofactors

The cofactors are Ca^{2+}, Mg^{2+}, and Mn^{2+} which facilitate the reactions in one or other direction (143, 144).

Inhibitors

Several adenosine phosphate compounds, and NaCl, *l*-thyroxine, *l*-triio-dothyronine, malonate, and Zn^{2+} and some inorganic anions (143, 188) inhibit the reaction. The inhibitory effect of NaCl is due to the increased Cl^- concentration (111). 1-Fluoro-2,4-dinitrobenzene which reacts with the sulf-hydryl groups, converts CPK into an S-dinitrophenylated-CPK (157). 2-Mercaptoethanol restores CPK activity by thiolysis (157). Iodine also inhibits CPK by converting the cysteinyl SH-groups into sulfenyliodide which in turn can be reversed by 2-mercaptoethanol or dithiothreitol (251) [Cleland reagent (DTT)]. In addition, many anions such as HCO_3^-, HCO_2^-, NO_3^-, NO_2^-, Br^-, F^-, SO_4^{2-}, HPO_4^-, ClO_4^-, and BF_4^- inhibit CPK (173).

Protein Structure

Rabbit muscle CPK contains 13 percent glutamic acid, 12 percent aspartic acid, 12 percent lysin, 10 percent leucine, 7 percent arginine, 7 percent valine, and all the other essential amino acids (88, 142). In dystrophic skeletal muscle, cysteine is replaced by glutamic acid (114). The amino acid spectrum of human CPK is not known.

Active Centers

The active center of CPK contains a nitrogenous group which requires Mg^{2+} ion as cofactor. Sulfhydryl groups present in the enzyme are also essential to its function. Therefore, cysteine, glutathione, and DTT can reactivate the inhibited enzyme (85, 100, 157, 191).

Measurement of Activity

Enzymologic Basis

The measurement of CPK activity is based on either the forward or reverse reaction. The forward reaction leads to the synthesis of creatine-phosphate (CP) and the reverse reaction splits CP to creatine and phosphate. In the

European countries, the methods based on the forward reaction are the most popular (15, 41, 76, 77, 85, 153, 193, 212, 245, 269), while in the United States the reverse reaction is used most frequently (110, 115, 209, 220, 247). The two most commonly employed spectrophotometric methods utilize two auxiliary enzyme reactions. In Rosalki's method (209) first CP is converted by CPK to creatine and phosphate and the phosphate is taken up by ADP. In the next step, requiring hexokinase (HK), the ATP formed phosphorylates glucose into glucose 6-phosphate (G-6-P) while ATP is converted to ADP. Finally, G-6-P reacts in the presence of G-6-P-dehydrogenase (G-6-PDH) with nicotinamide–adenine dinucleotide phosphate (NADP) and forms 6-phosphogluconate, reduced NADP, and H^+:

$$CP + ADP \xrightarrow{\text{CPK}} \text{creatine} + ATP$$

$$\text{Glucose} + ATP \xrightarrow{\text{HK}} \text{G-6-P} + ADP$$

$$\text{G-6-P} + NADP \xrightarrow{\text{G-6-PDH}} \text{6-phosphogluconate} + NADPH + H^+$$

The amount of NADPH produced can be measured with an ultraviolet (UV) spectrophotometer at 340 nm. The reagents are available in prepackaged kits from the manufacturer.* The activity is expressed in milliunits (mU). One milliunit equals 1 nmole of substrate converted in 1 min at 25°C. The normal range for each laboratory should be given, since experimental conditions may modify the results. The cysteine "activated" CPK activities are always higher than those determined without activators.

The other commonly employed method of Tanzer and Gilvarg (245) is based on the measurements of ADP synthesis with the utilization of two auxiliary reactions. First, creatine is converted into CP while ATP is reduced to ADP. ADP in the presence of phosphoenolpyruvate (PEP) is converted back to ATP by pyruvate kinase (PK). The net effect of ATP generation is to shift the equilibrium reaction to the right. The third reaction is the indicator reaction. Pyruvate is converted to lactate by LDH in the presence of reduced NAD.

$$\text{Creatine} + ATP \rightarrow CP + ADP$$

$$ADP + PEP \xrightarrow{\text{PK}} ATP + \text{pyruvate}$$

$$\text{Pyruvate} + NADH_2 \xrightarrow{\text{LDH}} \text{lactate} + NAD^+$$

Since all the reactions are stoichiometric, the disappearance of 1 μmole of $NADH_2$ indicates that 1 μmole of creatine has been phosphorylated. Two

---

*Kit available from Calbiochem, 3625 Medford Street, Los Angeles, California 90063.

test-kits* based on this system are commercially available. The pyruvate can also be determined by phenylhydrazon (189).

There are many other methods employed for the measurement of CPK activity (17, 38, 100, 115, 127, 245). One of these, "the spot test," requires very little equipment. Therefore, it may be adapted to mass screening of large populations (127).

Preparation and Preservation of Samples

Clotted blood serum is usually used since fluoride, citrate, heparin, and ethylenediaminetetraacetic acid (EDTA) may inhibit CPK activity. CPK is an unstable enzyme; it must be stabilized to obtain accurate assessment of its activity. This can be accomplished by the addition of either reduced gluta-thione, cysteine, or DTT to the sample (85, 100, 157, 191, 251, 269). Unless the activity is determined within 1 hr, a stabilizing agent must be used and the serum or muscle samples must be stored frozen to preserve activity.

Storage

As already mentioned, the CPK activity of sera stored at room temperature declines rapidly (44, 75, 191, 209, 266). Escher and Zimmerman (75) reported that CPK activity of sera stored for 24 hr at $-5°$ to $-10°C$ remained unaltered. Their results were confirmed by others (100, 170, 202), who found no significant decrease of CPK activity in sera stored for 12 to 48 hr at temperatures below $0°C$. In contrast, in samples kept at $4°C$, CPK activity decreases 55, 81, and 90 percent, respectively after 12, 24, and 48 hr. Others, however, observed only 10–35 percent reduction in activity in sera kept at $+4°C$ for 24 hr (213). Sera treated with thioglycolic acid, however, lost only 12 percent activity when kept for 5 hr at $25°C$ or for 8 days at $4°C$ (100). Therefore, it is essential that blood samples obtained for the determination of CPK activity be immediately centrifuged and 1 mg/ml cysteine be added to the separated serum, quick frozen in an acetone Dry Ice mixture and shipped in Dry Ice.

Reproducibility

Szasz et al. (241) showed that the accuracy of serum CPK measurements is the same as with other enzymes (2–7 percent) when the activity exceeds 20 mU/ml. Below this level, the coefficient of variation was two to three times greater. The day-to-day variation in activity (7–15 percent) of sera, however, is somewhat greater than the accuracy of the parallel determinations.

*Test kits available from Boehringer Mannheim Corp., 219 East 44th Street, New York, New York 10017 or Worthington Biochem. Corp., Freehold, New Jersey.

Incubation Time

The studies of Graig *et al.* (95) indicate that the optimal incubation time is less than 5 min in sera with high activity. The optimal incubation time is much longer (up to 90 min) in sera with low activity (95).

Dilution Effect

Graig *et al.* (95) called attention to the fact that dilution with distilled water or buffers caused a marked increase in the CPK activity of highly active sera, regardless of whether the forward or reverse reaction was used for the determination. Therefore, they recommended that dilution in highly active sera should be performed with heat-inactivated serum. Their results were corroborated by Hess *et al.* (109) and by Szasz *et al.* (241). The inactivation of serum samples should be carried out in a waterbath of 56°C for 30 min (95).

Temperature Effect

As with all enzyme reactions, within certain limits, the higher the reaction temperature, the faster the reaction rate (Arrhenius' Law). According to Szasz *et al.* (241), the temperature conversion factors are:

From 25°C to 30°C $F = 1.40 \pm 0.12$

From 25°C to 37°C $F = 2.01 \pm 0.28$

The average rise of CPK activity is 8 percent/°C. In order to avoid error in the interpretation of activity, these conversion factors should be used if the reaction rate was not determined at 25°C.

Hemolysis and Other Possible Sources of Errors

Hemolyzed sera should not be used for CPK determinations, since there is a linear increase in CPK activity with increasing free hemoglobin levels. The increase is about 5–10 mU/ml for each 100 mg% hemoglobin present in the serum (34). While in sera with low activity, hemolysis may cause considerable error in CPK activity determinations, this error is of little clinical importance in high activity sera. Especially with the reverse reaction, both washed and hemolyzed red cells markedly increased CPK activity (34). Although red cells have negligible CPK activity (223), adenylatekinase [(AK), myokinase, (EC 2.7.4.3.)], and G-6-PDH which are present in high concentrations (51, 268) may distort the results of CPK determinations by altering either ATP or ADP levels. AK catalyzes the following reaction:

$$2ADP \rightleftarrows ATP + AMP$$

Although high concentrations of AMP inhibit the forward reaction, it is not known whether or not this inhibition is adequate in the presence of high concentrations of AK. Although some commercially available kits incorporate AMP into the reaction mixture, and thereby eliminate the problem of hemolysis, the best is to prevent hemolysis in sera.

Isoenzymes: Their Tissue Distribution and Function

Types of Isoenzymes

It has long been recognized that although enzymes from different organs and different species may have identical substrate specificity and catalyze identical reactions, they may, nevertheless, differ markedly in their physical or chemical properties. These are called the isoenzymes of the specific enzyme. These isoenzymes of CPK can be identified by electrophoresis, chromatography, immunoelectrophoresis, solubility determinations, and by variation in their sensitivity to enzyme inhibitors.

Analysis of the molecular structure of CPK revealed that CPK is a dimeric enzyme. The two monomers may be identical or different. One type of monomer (M) predominates in the muscle and the other (B) in the nerve tissue. Therefore, three possible combinations of CPK monomers exist: MM, CPK_3, skeletal muscle-type or cathodal-enzyme; BB, CPK_1, brain-type or anodal-enzyme; and MB, CPK_2, cardiac muscle-type, or hybrid enzyme of CPK (52, 72, 73, 79) (see Figs. 13-1 and 13-2).

Figure 13.1 CPK isoenzyme fractions of normal human tissues (283). (Reproduced with the permission of *Anesthesia and Analgesia*).

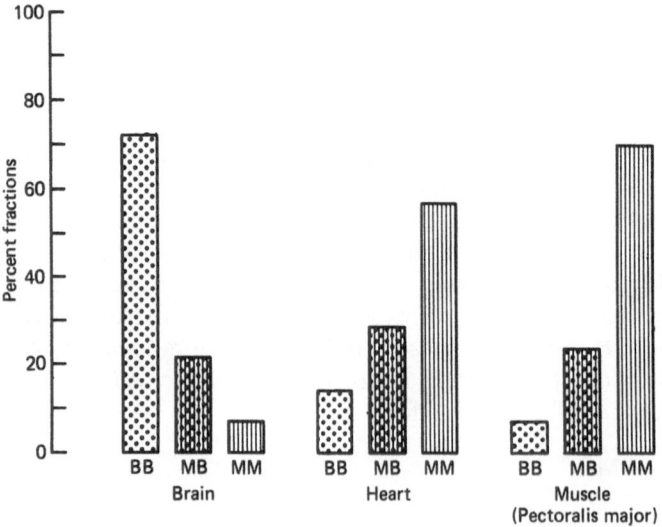

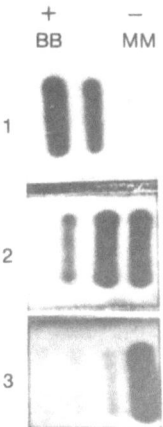

Figure 13.2 Normal tissue CPK isoenzyme patterns. (1) Brain isoenzyme pattern; (2) heart muscle isoenzyme pattern; (3) skeletal muscle pattern (287). (Reproduced with the permission of Der Anaethetists).

The Brain-Type Isoenzyme. From the ontogenetic point of view this enzyme is the earliest form of CPK in all tissues of mammals and avians (79). MB- and MM-isoenzymes appear later in the course of ontogenic development. In mature species BB is the predominant isoenzyme in all tissues except the skeletal and cardiac muscle (255).

The Heart Muscle-Type Isoenzyme. This isoenzyme is the characteristic CPK of mammalian heart muscle. Despite this, MM or skeletal muscle type CPK is present in greater quantity than the MB variety in the mammalian heart muscle. In contrast to this mammalian pattern, chicken has only BB-isoenzyme in both embryonic and mature heart muscle. This indicates that the BB-isoenzyme of the chicken heart undergoes no ontogenetic changes (30, 52, 73, 79, 252). The functional significance of the ontogenetic alteration of the tissue specific isoenzyme distribution is not well known (252).

The Skeletal Muscle-Type Isoenzyme. This isoenzyme is predominant in the skeletal muscle of adult humans and other mammals. MB-isoenzyme also may be present in smaller quantities (287). The human fetal skeletal muscle up to 6 weeks contains primarily BB-CPK (30, 93, 207) and not MM. In the newborn, especially in premature babies, the MB > MM is the normal pattern (30, 93, 207). At the age of 4 years, the normal adult pattern is fully developed with the predominance of MM-homodimer. The ratio of MB to MM is always less than 0.31 in normal adult skeletal muscle (93, 155). It is interesting that both Goto *et al.* (93) and Magalhaes (155) found that only the paravertebral muscles, in contrast to other skeletal muscles studied, usually contain all three CPK isoenzymes. The differing isoenzyme patterns found in normal adult human skeletal muscles are shown in Fig. 13-3.

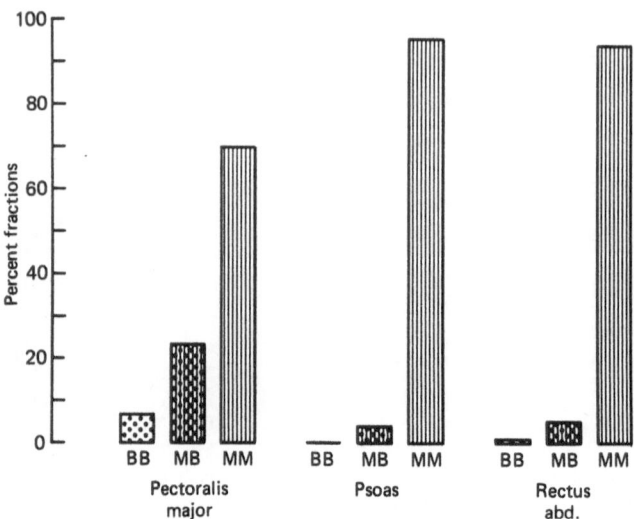

Figure 13.3 Variation in CPK isoenzyme fractions in normal human muscles (W. H. Starkweather and E. K. Zsigmond, unpublished data).

The amount of CPK present in skeletal muscles is usually a good index of the degree of muscle specialization. The more highly specialized, e.g., the faster the muscle, the higher the CPK activity. Cholinesterase activity was also found to be higher in muscles with more specialized functions (232). When the activity pattern of muscle changes from fetal to adult type, the synthesis of MM–CPK reaches its peak and the synthesis of BB–CPK is suppressed (129). Turner and Eppenberger (252, 253), in *in vitro* studies, found that the transition from BB- to MM-CPK occurs at the start of development of the structural characteristic of the muscle fiber and this change coincides with increased myosin synthesis (74). Furthermore, they have demonstrated in chicken embryonic cultures that this is blocked when the fusion of the myoblast is prevented by the use of media containing low Ca^{2+} concentrations (253). This temporal relationship of myoblast fusion to the BB- to MM-CPK isoenzyme transition has been repeatedly confirmed by other investigations (40, 176, 278). The activity of certain other enzymes in muscle (e.g., myokinase, phosphorylase) also increases concomitant with this change in the CPK isoenzyme pattern (226, 275).

MM-CPK was found to be identical (254) with the so-called M-protein (175), or M-band protein (145) previously isolated from the M-line region of the skeletal muscle myofibril. This M-protein binds to myosin filaments (145, 276) but not to actin (5). The MM-CPK bound to the myofibril comprises about one-fourth of the total muscle CPK (175). At high ionic strength the CPK-myofibril complex is associated, while at lower ionic strength it becomes dissociated. Baskin and Deamer (8) found that about 1 percent muscle CPK activity is in the sarcoplasmic reticulum, a structure which is vital to the normal relaxation process. The function of CPK bound at the M-line is questionable. According to Yagi and Mase, CPK increases myosin ATPase activity (276). In the mitochondria, the role of CPK is the phosphorylation of creatine. The sarcoplasmic reticular CPK is probably necessary for the function of sodium pump, essential for the relaxation process (8). Since MM-CPK is associated with the cross-striated organelles of the skeletal and of the heart muscle, it was proposed (74, 252) that the levels of CPK isoenzymes may be regulated in different tissues according to the types of actin and myosin present. In contrast, although actin and myosin are also present in platelets (1, 168), brain (12), and fibroblasts (1), the CPK activity is due solely to BB-CPK in these structures (40, 64, 253).

The best characterization of the isoenzyme pattern in the normal adult skeletal muscle was proposed by Magalhaes (155). According to him, the MM-CPK is predominant, MB-CPK is less than 25 percent of the total activity and the BB-CPK is absent.

Isoenzymes of Various Tissues. *Serum.* Unless the blood serum CPK activity is greater than twice the highest normal range, the identification of isoenzymes is impossible with previously available techniques. In sera with low CPK activity, however, a tenfold concentration of the serum permits the identification of the isoenzymes. The most common isoenzyme of the serum is the MM-CPK. Even in patients with brain damage (64), schizophrenia (162), and pulmonary embolism (249), the only isoenzyme present is the MM-CPK. There are only a few conditions in which MB-CPK is predominant. These include myocardial infarction (44, 75, 76, 85, 115, 153, 266), malignant hyperpyrexia (4, 287), myotonia congenita (93, 155), and neurogenic atrophy (227). Malignant hyperthermia is the only condition in which BB-isoenzyme was clearly demonstrated in the sera of patients (4, 287) and their relatives (287).

Cerebrospinal Fluid. BB-CPK can be found in the cerebrospinal fluid after stroke, head injury, and meningitis (64). The elevated serum CPK levels in these patients, however, are due to increase of the MM variant (249).

Other Tissues. The CPK activity of organs, other than muscles and heart, is caused solely by the presence of BB-isoenzymes (41, 73).

Methods for the Determination of Isoenzymes

Three different methods are available at present for the identification of various isoenzymes: electrophoresis (28, 59, 258), column chromatography (169), and immunochemistry (27). The latter is solely a research tool at present and is not discussed in this review.

Electrophoresis. *Separation of Isoenzymes.* For the electrophoretic technique different support media such as cellulose–acetate (150), agarose–gel (238, 267), and polyacrylamide–gel (4) are utilized. Some of the medical instrument suppliers have made available gel-strips and electrophoretic chambers in kits which are specially adapted to their electric power supply. The different media and electrophoretic conditions (4, 59, 208, 258)—in contrast to γ-globulins (92)—have little influence on the mobility of CPK isoenzymes. Therefore, the location of the CPK isoenzyme bands is almost identical in the three different gels. Madsen (154) observed two additional electrophoretic bands (X and Y bands) of CPK isoenzymes in human tissues with agarose–gel electrophoresis. These two additional fractions, however, were only active in the presence of cysteine in the reaction mixture. Omission of cysteine from the developing mixture caused disappearance of the X and Y bands. The most reliable separation of CPK isoenzymes so far was achieved by the method of Anido *et al.* (4). However, the method of Starkweather *et al.* (238) for LDH electrophoresis adapted to CPK determination, also results in adequate separation of BB and MB-CPK from MM-CPK in plasma and muscle homogenates of malignant hyperpyrexic patients and their relatives (239).

Preparation of Serum and Tissue Extracts. The frozen or fresh serum activity is first measured to determine whether the activity of serum is high enough for the electrophoretic separation of the CPK isoenzymes. Sera with activity less than twice the upper normal range are concentrated tenfold by ultrafiltration. The tissue extracts are prepared by homogenizing the specimen in four- to tenfold volume of 12 percent sucrose in an ice bath. The supernatant is used for CPK electrophoresis.

The essential activation of the serum or tissue CPK is accomplished by incubating the specimen for 30 min at room temperature in *l*-cysteine hydrochloride (191), glutathione (94), or DTT (235). The rate of migration of CPK isoenzymes, or in other words, the position of the bands is also influenced by the thiol compounds. Therefore, it is essential to know the specific thiol reagent used when comparing the findings of various authors (154).

Other factors such as temperature, pH, and the type of buffer used for incubation, the electrophoretic system, the voltage of the electrical current, and the duration of electrophoresis also influence the separation of CPK isoenzymes. Consistent findings of three major bands, however, were obtained by all the techniques so far utilized (4, 59, 154, 227, 235, 238, 258, 267). In order to avoid errors, the proper controls should include blanks without creatine-phosphate and thiol reagents, comparative electrophoresis of serum proteins and inhibition of myokinase of muscle and adenylatekinase of red cells by AMP in the incubation medium (92, 154).

Identification of the Activity. Basically, two methods may be used for the identification of CPK activity: (a) staining method; and (b) fluorescence method. Both methods are based on the reverse reaction shown in the equation on page 281. The staining method requires two additional reactions:

$NADPH_2$ + phenazine methosulfate (PMS)→NADP + reduced PMS

Reduced PMS + nitrobluetetrazolium (NBT)→PMS + reduced NBT (color)

Because of the light sensitivity of the reaction, this method should be carried out in the dark. Another problem with this technique is the occurrence of nonspecific staining (32, 94, 235, 250, 258). The fluorescent method eliminates those steps of the staining reaction which may result in unspecific staining (42, 206, 235, 258). With this method, CPK isoenzyme bands can be separated in samples with low CPK activity.

Many commercial kits and electrophoretic systems are available, which can be easily adapted to CPK isoenzyme determinations. At the present time, one of the major problems is the lack of standardization of the techniques. This makes difficult the comparison of the findings of various authors obtained with different methods.

Chromatography. The preparation of gels for electrophoresis is time consuming and the use of four additional reactions may result in errors. Therefore, Mercer (169) devised an ion-exchange column chromatographic technique for the separation of isoenzymes of CPK.

Those unfamiliar with chromatography in general are referred to the reviews of Flodin (80) and Mercer (169). In the case of ion-exchange chromatography, the column possesses exchangeable ions. In Mercer's method for CPK isoenzyme separation (169), the chloride is the exchangeable anion. The major advantages of chromatography are: simplicity; insensitivity to composition of the eluent and to temperature; stability of the enzyme structure; and economy. The recovery rate of this method (169) is between 80 and 103 percent in sera containing different ratios of added MB-CPK and MM-CPK. This technique is easily adaptable to routine isoenzyme screening on a large scale.

Factors That Alter Creatine Phosphokinase Activity

Physiologic Factors

Age. According to Sabater and Villalba (216) the serum CPK level in the umbilical cord blood of neonates is higher than that of adults. Similar findings were made by Wharton *et al.* (272) who observed a marked rise in CPK activity of neonates in the first 24 hr following birth, decreasing to somewhat higher than adult levels on the fifth day. Okinaka *et al.* (191) found that children up to 5 or 6 years have lower CPK levels than adults. Feraudi *et al.* (78) reported the highest CPK levels in the 12 to 21-year age group and a decline thereafter. Meltzer (167) observed that CPK levels of males tended to peak between ages of 10–19 and 40–49 years. Others (109, 241, 247), however, have found no difference in serum CPK levels in various age groups.

Sex. It is generally agreed that males have significantly higher CPK activity than females (115, 189, 191, 196, 216, 247). Meltzer (167) confirmed this sex difference between males and females in both whites and blacks. Differences in muscular activity, muscle bulk (115), and hormonal influences on the skeletal muscles may explain these differences (189).

Race. Blacks, males and females alike, have higher mean serum CPK levels than the corresponding sexes of whites (167). Meltzer (167) contributes this finding to the relatively larger muscle bulk of blacks. However, the possibility of nutritional factors and greater physical activity prior to hospital admission cannot be ruled out.

Body Weight and Height. No correlation of serum CPK with body weight or height has been established.

Pregnancy. Blyth and Hughes (18) observed lower than normal CPK levels in patients with Duchenne muscular dystrophy during pregnancy. Others (67, 136) reported increased CPK levels in late pregnancy and postpartum in normal women. Recently Emery and Pascasion (67) found that during the first 20 weeks of pregnancy, the CPK activity in serum is significantly lower than in nonpregnants or during late pregnancy.

Physical Fitness and Exercise. Several studies (29, 86, 87, 99, 101, 190, 198, 259) demonstrated only moderate and short-lasting rise in serum CPK after moderate exercise (100 W during a 6 min period) involving limited number of muscles (bicycle-ergometer). CPK activity, however, was not followed long enough after exercise, therefore, further elevations of CPK activity might have been missed. Review of these reports suggests that the exercise-induced rise in serum CPK and its extent depends on several factors: the type, severity, and duration of exercise, and the extent of muscle involvement, the physical fitness of the exercising individual. Indeed, Keif et al. (126) showed that gymnastic exercise caused a considerable rise in serum CPK and that the degree of elevation was directly proportional to the state of training. In well-trained test persons, the enzyme level rise was minimal while in un-trained persons it could be as high as 20 times the mean normal value. The rise was maximal at 12 hr, but elevated levels were still found at 36 hr following heavy exercise in untrained individuals, while only minimal rise occurred in the latter group following ergometric exercise. Their findings were corroborated by Ledwick (151) who also showed that in those individuals who had high initial CPK levels, positive exercise ECG, and lower level of physical fitness, the elevation of CPK activity was greater than in fit individuals with no ECG abnormalities and normal initial CPK levels. Furthermore, it was demonstrated (19) that subthalamic stimulation, mimicking cardiovascular effects of exercise, caused enzyme elevation of serum CPK activity similar to that induced by exercise. It is probable that catecholamines, local muscle hypoxia, hypercarbia, and acidosis might be responsible for the release of CPK from the muscle cell membrane. It is also conceivable that untrained persons have some disuse atrophy of the skeletal musculature, and therefore their muscles are more prone to release CPK than normal muscles. These assumptions are corroborated by the findings of Nutall and Jones (190), that after a 3 to 5-week period of physical training, the postexercise peak CPK levels were only 10 percent of the pretraining peaks measured in untrained individuals. Unaccustomed heavy exercise (53-mile walk in 15 hr), however, can cause up to 20- to 25-fold elevation of serum CPK levels even in trained individuals (99). No myocardial MB-isoenzyme release could be demonstrated, however, even after a 26-mile run (210). These findings in normal individuals with no myopathies were also confirmed in patients with Duchenne muscular dystrophy (183), in acutely psychotic patients (163), and also in patients with motor neuron disease (271).

Griffiths (99) found that in the general population, some seemingly healthy "enzyme-labile" males have elevated serum CPK values. Emery and Spikesman (68) confirmed this finding. Six out of 18 healthy subjects over 45 years of age had elevated serum CPK which could not be explained. Examination of the relatives indicated normal levels. It is likely that these individuals may have a subclinical myopathy, since malignant hyperpyrexic myopathic probands also have elevated serum CPK at rest without any clinical symptoms (55, 120, 281). Emery and Spikesman (69) studied 10 "enzyme-labile" individuals and their relatives for survival after general anesthesia. Although 34 of the 233 "enzyme-labile" individuals and relatives underwent anesthesia, no deaths were encountered. The degree of CPK elevation and the type of anesthesia and muscle relaxant used were not reported in this communication. Since the duration of the exposure to the triggering anesthetic agents and degree of muscle involvement determines the morbidity and mortality in malignant hyperthermia, the absence of mortality in this small group of operated probands and relatives does not guarantee that the triggering agents can always be used with impunity in these "enzyme-labile" individuals. It is better to avoid the triggering agent in these individuals and to err on the safe side.

Pathologic Conditions

Diseases of the Nervous System. *Lesions of the CNS and Neuropathies.* Trauma to the brain, vascular thrombosis, and embolization, stroke, and meningitis cause about a 40–80 percent rise in serum CPK (64, 141, 166, 222).

It would be expected that the BB-CPK and not the MM-CPK would cause most of the elevation of serum CPK activity. Contrary to this expectation, however, only the MM-CPK was found in the serum, although the BB-CPK was simultaneously found in the spinal fluid (64). There was no correlation between serum and cerebrospinal fluid CPK activity on one hand and between the severity of disease and the degree of CPK elevation in either serum or spinal fluid on the other (64). Dubo *et al.* (64) also reported that in stroke patients the elevation reaches its peak only on the second to the sixth day. He suggested that some metabolic derangement of the skeletal muscle may be responsible for this finding. The most likely explanation for the finding of MM-CPK in the serum of patients with upper or lower neuron injuries is that the normal trophic influence on the muscle is disturbed and the resultant myopathy alters the muscle membrane permeability and allows the leakage of CPK. The studies of Cooperman (47) demonstrated the increased permeability and the greater lability of the muscle membrane in patients with neurologic disorders, manifested by the greater succinylcholine-induced potassium release in these patients than in normals.

Tetanus. Tetanus patients have elevated CPK levels. Rabbits with experimentally induced tetanus showed marked elevation of serum MM-CPK isoenzyme. Therefore, it is likely that the MM-CPK is also responsible for the serum CPK elevation in men (131).

Myasthenia Gravis. Hess *et al.* (109) found only one out of 13 myasthenic patients with elevated serum CPK activity. The rest of them were normal. Others (135, 231) were also unable to demonstrate any differences in the CPK activity of normal and myasthenic subjects.

Cerebrospinal Fluid in Neurologic Disease. Lisak and Graig (152), using a very sensitive method, found a mean CPK activity value of 1.9 mU/ml in normal man, which is not detectable with routine methods. In patients with brain tumors, Guillaine-Barré Syndrome, epilepsy, lumbar discs, disseminated sclerosis, and spinal cord lesions, elevated spinal fluid CPK levels were reported (107, 229). Elevated spinal fluid levels of CPK were also found in 30 percent of psychotic patients (257). Since BB-CPK does not pass the blood–brain barrier (229), the increased CPK level suggests that the brain cells leak out an abnormal amount of CPK in psychotic patients. It is of interest that spinal fluid CPK remains elevated only for a short period. This suggests that only a limited amount of CPK is available for release from the damaged brain cells.

Acute Psychoses. The serum CPK is elevated in many untreated acutely psychotic patients, (e.g., schizophrenics, manic–depressives, simple depressives) (162, 164). Patients with chronic psychiatric conditions, however, have normal serum CPK levels. The onset of psychosis is heralded by a five- to ten-fold rise in serum CPK levels which fall after treatment with 2-chloropromazine (162). Sleep deprivation (146), catatonia, and acute psychosis with severe agitation results in excessive muscle activity, which may be the cause of the serum CPK elevation (166). As mentioned earlier, Meltzer and Moline (163) observed that after brief exercise the serum CPK becomes more markedly elevated in psychotic patients than in normals. This suggests that psychotic patients may suffer from a subclinical myopathy, which results in the release of CPK from muscle. Indeed, Engel and Meltzer (70) and Meltzer and Moline (163, 164) demonstrated microscopic abnormalities in muscle biopsies of psychotic patients. Since Vale *et al.* (257) also found elevated spinal fluid CPK levels, evidently CPK is simultaneously released from both the brain and skeletal muscles in psychotic patients.

Diseases of the Heart Muscle. *Myocardial Infarction.* Pioneering studies of Karmen *et al.* (125) showed that characteristic serum enzyme elevations occur in myocardial infarction. Since then, many studies demonstrated the diagnostic value of elevated serum CPK in patients with acute myocardial infarction (4, 43, 45, 46, 62, 63, 84, 109, 137, 169, 230, 263, 280). The classical studies of Forster and Escher (84) on the correlation of CPK elevation with the time of onset, extent, and severity of myocardial infarction indicate that, in contrast to the slow rise in level of other enzymes, such as LDH and SGOT, the CPK rise occurred promptly, within 3–4 hr after the onset of the initial signs and symptoms of the infarction. This was not surprising since CPK is more specific for the heart muscle than SGOT and LDH and is more readily released than the latter. The CPK levels in uncomplicated infarctions reach their peak in 24–36 hr and return to normal levels by the third or fourth day. These findings were repeatedly confirmed by others (43, 45, 46, 75, 109, 137, 233, 249). Forster and Escher (84) also noted elevation of serum CPK in three out of five patients following angina pectoris attacks. SGOT and LDH levels remained unchanged in these patients. They assumed that microinfarcts might have been responsible for this rise. Since they did not identify MB-CPK in serum, it is not unlikely that MM-CPK released from the skeletal muscle might have caused it. The finding of Ledwick (151), that unfit individuals with positive exercise ECG may show elevated serum CPK levels following even moderate exercise, further stresses the need for studies of the CPK isoenzymes in patients with angina. The magnitude of the serum CPK elevation and its time course may be helpful for the differentiation of angina from infarction.

A recent study on the prognostic value of serum CPK levels in myocardial infarction by Coodley (46) is the most thorough and concise report on this subject. He studied the CPK activity in 125 consecutive patients daily during the first 4 days following acute myocardial infarction. He found the following correlations: (1) the greater the CPK rise, the higher the mortality. Levels above six times the normal were associated with 56 percent mortality; (2) the greater the CPK rise, the higher the incidence of arrhythmias. Levels above four times normal were associated with an 82 percent incidence of ventricular arrhythmias; (3) Persistent and marked rise above four to five times the normal level was associated with a 10 percent incidence of cardiogenic shock. Conversely, none of the patients, who had less than fourfold elevations of the serum CPK levels developed congestive failure or cardiogenic shock. These findings were corroborated by Kluge (133) who observed a mortality rate of 6 percent in those with CPK levels of five times the normal and of 50 percent in those with a tenfold elevation of the CPK activity. The possibility of the correct estimation of the infarct size and prognosis on the basis of serum CPK determinations was also confirmed by Sobel (233, 234).

Some degree of cardiac muscle damage estimation can be made from the time course and the degree of elevation of serum CPK activity in patients with clinical signs and symptoms and/or ECG evidence of myocardial infarct. In doubtful cases, however, the identification of MB-CPK isoenzyme is required for differential diagnosis (4, 169). When attempting to make a differential diagnosis it should be remembered that MB isoenzyme may also be present in patients with carbon monoxide poisoning (4) in myopathies (93, 155), and in probands with malignant hyperthermic myopathy and their relatives (4, 239, 287). These conditions, however, are associated with different clinical signs and symptoms than myocardial infarction. The MB band may only be present in the sera for a short period. This may explain why Konttinen and Somer (138) found no MB-CPK in the sera of two out of the 21 patients with known myocardial infarction. Anido et al. (4) emphasized that although the presence of MB-CPK may be indicative of myocardial damage, it is not pathognomonic of myocardial infarction. The recently introduced column chromatographic method of Mercer (169) will greatly facilitate the routine determination of serum CPK isoenzymes for cardiac diagnosis.

Elevated levels of unspecified serum CPK were observed in experimental pulmonary embolism (106). In five patients with pulmonary embolism, the MM-CPK isoenzyme was only found to be elevated in the serum (139). This latter finding may help in the differential diagnosis between pulmonary embolism and myocardial infarct.

Trauma to the Heart. Since the presence of the MB-CPK band usually indicates cardiac muscle trauma, the degree of myocardial damage during coronary artery surgery can be adequately assessed by isoenzyme analysis (37, 91, 192, 211). An early sign of rejection following cardiac transplantation is a marked elevation of serum CPK activity (6, 224). A moderate rise in serum CPK activity may also occur following cardioversion (58, 132, 158).

Other Cardiac Diseases. Elevated serum CPK levels were found in patients with pericarditis, myocarditis, dissecting aneurysm (109), and following coronary angiography (48, 171). Michie *et al.* (171) found that uncomplicated cardiac catheterization itself causes elevation of serum CPK. Therefore, it does not necessarily indicate complications associated with the catheterization.

Diseases of the Skeletal Muscle. *Muscle Trauma.* Intramuscular injection may cause significant rise in serum CPK (165). In a 1-year survey of all serum CPK determinations in a general hospital, Reece (203) found that in 10 percent of the cases no cause could be found for elevation of serum CPK activity. He found that intramuscular injections, contusion, or necrosis of muscles caused 150 to 300 percent average increase of serum CPK activity. Intramuscularly injected meperidine and other analgesics (9, 165), chlorpromazine (165), digoxin (97), lidocaine (280), insulin (39), penicillin (108), ampicillin (134), and diuretics (9) may also cause marked rise in CPK activity (203).

Surgical trauma to the skeletal muscle causes only slight elevation of serum CPK activity (177). Two groups of patients studied received either neurolept or halothane anesthesia with succinylcholine, but no electrocautery was used during surgery. The members of the third group were anesthetized with either technique, but electrocautery was used during surgery. There was no difference in the serum CPK activities between the two anesthetized groups operated on without electrocautery. In the group in which electrocautery was employed there was a severalfold increase of CPK activities during the first 7 postoperative hr. The magnitude of the rise equaled that observed in patients with acute myocardial infarct. Burn injury to the skeletal muscle was found to cause tenfold elevation in serum CPK activity for 2 weeks after injury (289).

Muscle Atrophy. On the basis of serum CPK activities or of isoenzyme patterns, at the present time it is not possible to differentiate between neurogenic and myogenic skeletal muscle atrophy (141). In a survey of isoenzymes of biopsied human muscles, Magalhaes (155) and Shapira *et al.* (227) found BB-CPK in some patients with both myogenic and neurogenic atrophies, also found in some patients with myotonia congenita (93). This isoenzyme pattern of atrophic muscle resembles that of the embryonic muscle (227). Since the same pattern was observed in malignant hyperthermic (MH) myopathy (239), it is important to emphasize that an abnormal embryonic type LDH-isoenzyme pattern was also found in atrophic muscles (277). In contrast, no such LDH-isoenzyme pattern was found in malignant hyperthermic muscles (287).

Myotonia Congenita. Elevation of serum CPK activity is commonly associated with myotonia congenita (36, 50, 93, 155). Goto *et al.* (93) also observed abnormal CPK isoenzyme patterns with elevated MB bands in myotonia dystrophica patients.

Muscular Dystrophies. Elevation of serum CPK activity was reported in almost all types of muscular dystrophies (49, 54, 116, 181). Abnormal CPK values were found in 71 percent of patients with Duchenne muscular dystrophy (228). Others (21, 115, 181) found that the elevation of CPK activity was more useful than that of aldolase activity for the diagnosis of this condition. No biochemical explanation is available at the present time for the continuous leakage of CPK from diseased muscle. The striking elevation of serum CPK activity especially at the early stages of the disease, when histopathologic findings are scanty, suggest that abnormal membrane permeability may be an early and prominent defect. It is possible to diagnose, with the help of serum CPK determinations, Duchenne's myopathy in 90 percent of patients and in 75 percent of the carriers (115, 181, 228). Furthermore, it was reported (181) that the mean CPK activity was almost 30 times greater than normal in the early stages and approximately seven times greater than normal in the late stages of this disease. In contrast, other muscle enzymes were only elevated two- to sixfold. Marked drop in CPK activity was observed in Duchenne's dystrophy patients after age 10, and in carriers after age 20. Therefore it is important to identify female carriers of this disease in early childhood, when the CPK activity is still high (181). It is of interest that the frequency and degree of elevated serum CPK activity were similar in Becker (benign) and Duchenne (rapidly progressing) muscular dystrophies, despite the dissimilar rate of progression of the two diseases. In contrast, the elevation in CPK activity is greater in rapidly progressing denervation disorders (115, 181, 228). The degree of elevation of serum CPK activity paralleled the progression and severity of the pathology in the other myopathies, such as limb–girdle dystrophy, facioscapulohumeral dystrophy, congenital myopathies, and infantile spinal muscular dystrophy. In the polymyositis patients, aldolase was the muscle enzyme most frequently elevated (21).

In muscular dystrophies, CPK isoenzyme variants in biopsied or autopsied muscles are helpful in the differential diagnosis. Those interested in the details of the CPK isoenzyme studies carried out in various myopathies are refered to the following publications: 7, 31, 33, 50, 90, 93, 94, 105, 114, 155, 174, 214, 227, 248.

From the anesthesiologist's point of view, it is of great interest that in MH the CPK isoenzyme patterns in serum and muscle are quite different from those present in normals or in patients with various myopathies. In normals there is only MM isoenzyme present in the serum and in the muscle, the MB variant is less than 30 percent of the total CPK. In myopathies, variable amounts of MB may be present in the sera and its concentrations may also be increased in the muscle (31, 54, 93, 155, 227). However, BB isoenzyme was not encountered in myopathies in serum (4) and seldom in muscle (93, 227). In contrast, in MH significant amounts of not only MB, but also BB are present in both serum and muscles (4, 284). It is of interest that in the human fetus up to the end of the sixth week the CPK isoenzyme pattern of the skeletal muscle (221, 227) is similar to that of MH muscle (287).

Interstitial Myofibrositis. A chronic inflammation of fibrous tissues of the fascial aponeurosis and of the interstitial tissues of muscle causes no significant elevation of serum CPK activity (118).

Influenza Myositis. CPK activities exceeding twice the normal range were found in the sera of 14 out of 21 patients with B and A_2 virus influenza infection (172).

Polymyositis. Marked elevations of CPK activity were found in some polymyositis patients (262, 274).

Dermatomyositis. Elevated levels of serum CPK were found in dermatomyositis (53, 140). CPK activity further increased after whole body massage used for the therapy of this condition (20).

Other Myopathies. In hypertrophia musculorum vera (200), in motor neuron diseases (270), in acromegalic-myopathy (159, 160), in idiopathic scoliosis (90), in chronic spinal muscular atrophy (184), and in exertion induced cramps (225), elevated CPK activities were reported.

Myoglobinuria and Serum Creatine Phosphokinase. Myoglobinuria and myoglobinemia were reported in patients with various forms of myositis (13, 123), and in *Herbicola lathyri* septicemia (237). In all these conditions marked elevation of the serum CPK activity accompanied the myoglobinuria.

Alcoholic Myopathy. Elevated serum CPK activities were found in patients with acute alcoholism, especially in those with delirium tremens (149, 182). The elevation was due to the MM isoenzyme and no CPK was found in spinal fluid (221).

Iatrogenic Factors

The Effect of Drugs on Creatine Phosphokinase. The trauma of intramuscular injection can cause moderate elevation of serum CPK activity (165). This is discussed in the section on *muscle trauma*.

Subcutaneously injected epinephrine, intravenous (i.v.) aminocaproic acid or clofibrate, peroral gluthethimide, and imipramine were reported to cause elevation of serum CPK activity (131). Protriptyline overdose (98), heroin addiction (205), and 2, 4-dichlorophenoxyacetic acid poisoning (16) were also reported to increase serum CPK activity. Vorburger *et al.* (264) showed that the i.m. injection of 5 to 10 mg diazepam, or 30 mg pentazocin caused elevation of serum CPK activity which reached its peak at 24 hr and returned to normal after 3 days. Radiotherapy to large areas of the skeletal musculature caused no elevation of CPK but irradiation of the heart may cause rise in serum CPK activity. This may reflect radiation induced heart-muscle damage (180).

Surgery. *Surgical Muscle Trauma.* Sorenson *et al.* (236) first reported elevation of serum CPK activity following operations. This was confirmed in patients who underwent gastric resection, cholecystectomy, or operations on the colon (279). The elevation reached its peak at 24 hr and returned to normal 4–5 days postoperatively. Zelder (279) claimed that the elevated CPK activity is related to the influence of the increased release of hormones (e.g., catecholamines, steroids) due to the stress of surgery on the musculature rather than to direct muscle damage. Hobson *et al.* (112, 113) studied four groups of 20 patients, each undergoing surgery associated with increasing degrees of muscle trauma. Insignificant mean elevation was observed in the breast biopsy group, but significant elevations occurred in the herniorraphy, cholecystectomy, and miscellaneous other laparotomy groups. Maximal elevation again occurred at 24 hr postoperatively and lasted 4–5 days. The elevation after breast biopsy did not exceed 100 percent while 10–12 times normal values were seen in the other groups. Dixon *et al.* (60) compared the elevation of serum CPK activities following cardiac and other thoracic and abdominal operations and confirmed that the rise correlated with the degree of muscle trauma. There was no statistically significant difference between the three groups in the degree and duration of increased CPK activity. The time course was identical with those reported by others (112, 113, 279). Transurethral resection of the bladder neck, but not of the prostate, was also associated with elevation of CPK activity in the serum (35, 156).

Myocardial Injury during Surgery and Anesthesia.. Hultgren *et al.* (117) showed that patients undergoing coronary-bypass surgery of either the Vineberg type or saphenous aortocoronary bypass, and showing ECG changes indicative of ischemic injury, had higher postoperative CPK activities than those with no ECG changes. It should be remembered, however, that blunt thoracic trauma (201) which is frequently associated with closed chest cardiac massage may also cause elevated serum CPK activity.

Differential Diagnosis of Muscle and Heart Muscle Trauma. The identifications of tissue-specific CPK isoenzymes can aid in the differential diagnosis of muscle and heart muscle trauma (4, 169). Oldham *et al.* (192) determined serum MB-CPK intra- and postoperatively in 39 patients undergoing coronary bypass surgery. Twenty-three patients had no MB-CPK in their sera; in 10 patients MB-CPK was transiently elevated for a mean of 5.2 hr and in thirteen patients for an average of 37.5 hr. Persistent presence of MB-CPK was encountered more frequently in patients with severe coronary disease and in those with triple-grafts and in cases of excessively long cardiopulmonary bypass. Dixon *et al.* (61) found that there was a 100 percent correlation between the presence of MB-CPK in the sera and myocardial infarction confirmed by ECG.

Anesthetic Agents. *Inhalational Anesthetics.* Airaksinen and Tammisto (2) and Tammisto and Airaksinen (242, 243) studied the influence of succinylcholine (Sch) on serum CPK with various induction sequences. Elevation of serum CPK activity after the use of Sch was less frequently encountered with thiopental than with either ether or methoxyflurane induction. In ophthalmic operations, the incidence of increased CPK activity was seven times greater when Sch was used with halothane and N_2O-O_2 than when it was used with N_2O-O_2 alone. The degree of elevation of serum CPK activity paralleled the incidence and degree of visible fasciculations. There was no correlation with the incidence and severity of the postoperative muscle pain. Innes and Stromme (119) observed no significant elevation of serum CPK activity in children induced with halothane and ethyl ether alone. Significant increases occurred, however, in those induced with barbiturate-Sch and diethylether-Sch sequences. The highest rise of CPK activity occurred with the halothane-Sch sequence. When halothane was given following Sch, a lesser elevation of serum CPK activity occurred than when halothane preceded Sch. It is probable that Sch triggers the rise of serum CPK activity encountered in conjunction with anesthesia (119). Of all anesthetic agents, halothane seems to enhance the Sch-induced release of CPK from the muscle most markedly (119, 124, 217, 242, 243, 246). It is of interest that Tammisto *et al.* (242) observed that children with strabismus may be prone to muscle injury induced by Sch, since in malignant hyperpyrexic families the incidence of skeletomuscular abnormalities is higher than in the general population (23, 24, 55, 246, 273, 281, 282, 286).

Muscle Relaxants. Of all anesthetic agents and adjuvant drugs, Sch is most likely to release muscle CPK. Airaksinen and Tammisto (2), Tammisto and Airaksinen (242, 243), and Ventafridda *et al.* (261), showed that intermittent, *repeated* doses of Sch, especially in combination with halothane, caused the highest incidence and degree of elevated serum CPK activity. Patients induced without Sch had no elevation, while a *single dose* of Sch caused less elevation of CPK activity than repeated doses. Increased CPK activity was usually accompanied by postoperative muscle pain. Hyperpotassemia was also associated with increased CPK activity and postoperative muscle pain. Palecchi *et al.* (195), Ventafridda *et al.* (261), and Tammisto and Airaksinen (242) all observed the potentiating effect of halothane on the Sch-induced increase in serum CPK activity probably due to muscle damage. That muscle damage may occur during Sch-induced muscle fasciculations is also evidenced by the myoglobinuria that follows Sch administration (2, 11, 261) and the rise in CPK activity that is induced by the administration of antagonists of non-depolarizing muscle relaxants, (e.g., edrophonium, neostigmine) in cats (243).

Both the muscle damage manifested in the increased serum CPK activity and muscle pain can be effectively prevented by the administration of small doses of d-tubocurarine prior to Sch (244). Ventafridda *et al.* (261) found that intravenous local anesthetics also prevent increased CPK activity since they prevent Sch induced fasciculations, as shown by Usubiaga *et al.* (256). For this reason the routine use of small doses of d-tubocurarine prior to the administration of Sch was recommended by Mayrhofer as early as 1959 (161).

Malignant Hyperthermia and Creatine Phosphokinase

Genetics, Symptomatology, Therapy

Malignant Hyperthermia (MH) is a syndrome characterized by rapid rise in temperature triggered by anesthetic agents and/or muscle relaxants. The rise may occur either during or after anesthesia and may be so fulminating that it rightfully deserves the terms alternatively used: malignant hyperpyrexia or fulminant hyperthermia. Individuals with an autosomal, dominantly inherited genetic myopathy are the targets. It is not an X-linked anomaly such as the Duchenne-type muscular dystrophy. The incidence of MH is greater in males than in females. Young individuals are especially prone to MH, but cases were reported in patients up to the seventh decade (23, 24). Its true incidence (1 : 50,000 to 1 : 15,000 claimed) is not known, since many individuals who may have this genetic anomaly are never exposed to anesthesia. Furthermore, until recently, even severe reactions and deaths due to MH were not reported. "Forme frustes" may be so mild that they go unnoticed (284). Rigid and nonrigid varieties were differentiated by Britt and Kalow (24, 25). The nonrigid variety, however, may represent patients in whom the muscle rigidity might have gone unnoticed, or patients with such a severe muscle atrophy that the rigidity was not visible (284, 285). As to the triggering agents, all the potent inhalational anesthetics and muscle relaxants were implicated (24, 25). N_2O supplemented with Innovar (24, 25), diazepam, morphine (282), ketamine (265), and rapidly acting barbiturates (25) were proven to be safe in MH survivors. Britt (25) claimed that d-tubocurarine and lidocaine might also act as triggering agents (288), but Harrison (104) was unable to induce or increase the incidence of MH by d-tubocurarine administration in strains of pigs, which develop a syndrome similar or identical to human MH when exposed to Sch and halothane. Indeed, these two agents were involved in most cases of MH recently reported (25). Until 1969, the mortality of the fully developed MH was reported to be about 70 percent (22, 24). Beldavs *et al.*

(10) introduced intravenous procaine in the therapy of MH in 1971. Procaine also blocks the MH syndrome triggered by either halothane or Sch in pigs (103).

Malignant Hyperthermic Myopathy and Creatine Phosphokinase

Inhalation anesthetics and/or Sch and decamethonium may cause in MH patients, muscle fasciculation and sustained contracture, hyperpnea, tachycardia, arrhythmia, cyanosis, sweating, pallor, decreased P_{aO_2}, elevated P_{aCO_2}, metabolic acidosis, hyperphosphatemia, hyperglycemia, hyperkalemia and hypercalcemia, followed later by hypocalcemia and hypokalemia. These events are associated with the elevation of serum CPK activity (3, 54, 55, 57, 130, 199, 239, 240, 246, 281, 285). This may occur prior to or during the rapid temperature rise. Denborough et al. (57) reported 240-fold elevation of serum CPK activity in a patient during a hyperthermic episode. Associated with the elevated CPK activity, myoglobinuria (often causing fatal renal failure) was also reported (3, 14, 22–25, 89, 194, 215, 240). Myalgia and muscle weakness develops in the survivors (24, 25).

High serum CPK activity was found not only in surviving MH probands, but also in many of their relatives (56, 121, 239, 240, 281, 287). The autosomal, dominant pattern of increased serum CPK activity is shown in Fig. 13-4. Since then, numercus reports (14, 25, 128, 147, 179, 204) confirmed these findings. Moulds and Denborough (179) recently demonstrated that serum CPK is a reliable screening test for *MH families*. The observation of high serum CPK activity in an *individual*, however, is in itself not sufficient for the diagnosis of MH.

In order to determine the source of the elevated serum CPK activity and the isoenzyme variants, Zsigmond et al. carried out electrophoretic studies on the sera of several probands and their relatives (283, 286, 287). Surprisingly the MB and BB isoenzymes were also found to be responsible to a considerable extent for the elevated serum CPK activity (see Figs. 13-5 and 13-6). MB and BB isoenzymes were also found in the biopsied muscles of the probands and their relatives (286). Recently, Anido et al. (4) corroborated these findings. Peters et al. (199), however, could not confirm the presence of MB or BB band in the sera of MH probands.

Malignant Hyperthermic Myopathy and Other Myopathies

As already mentioned, in the normal adult pectoralis muscle and serum, MM is the predominant CPK isoenzyme. In MH families BB-CPK and/or MB-CPK may be dominant in the sera and muscles. A high concentration of BB-CPK is a consistent finding in the fetal muscle (30, 93, 207). Abnormal isoenzyme patterns with elevated MB/MM ratios were also described in various forms of myopathies in a survey of 1730 muscle biopsies by Demos (54). Since the CPK isoenzyme pattern in myopathies resembles that found

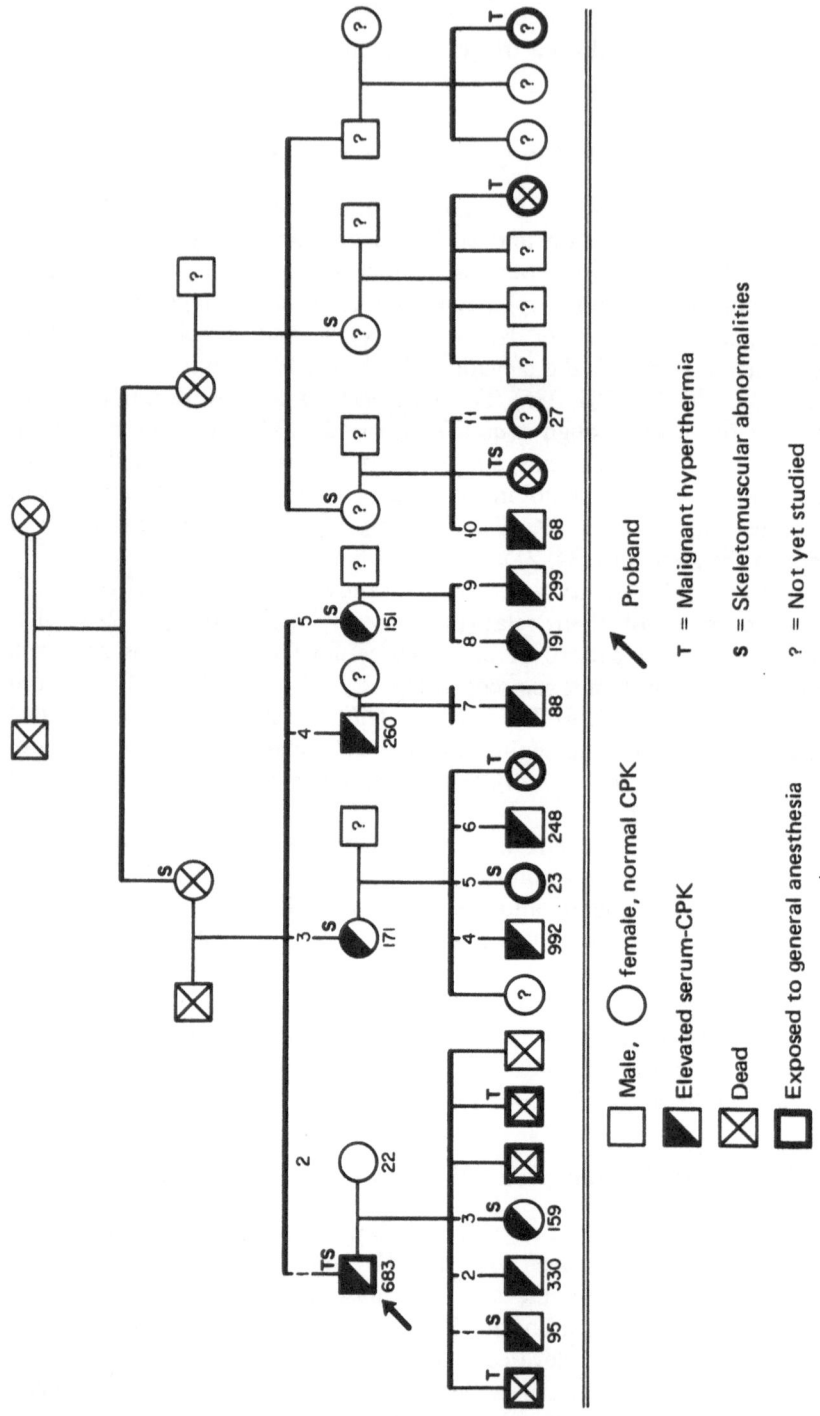

Figure 13-4 Abnormally elevated serum CPK and death in a family with malignant hyperpyrexia and consanguinity. Note that the serum CPK elevation follows a dominant pattern of inheritance (284). (Reproduced with the permission of *Anesthesia and Analgesia*).

Male, ▢ ◯ female, normal CPK

◨ Elevated serum-CPK

⊠ Dead

▢ Exposed to general anesthesia

↑ Proband

T = Malignant hyperthermia

s = Skeletomuscular abnormalities

? = Not yet studied

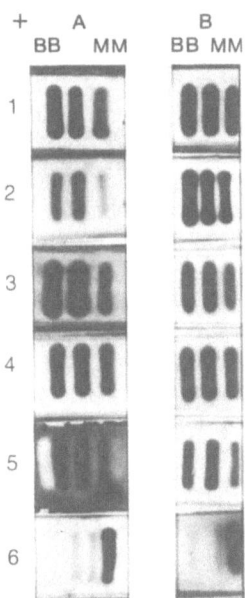

Figure 13-5   Tissue-CPK isoenzyme patterns in the family presented in Fig. 13-4. (1) A-1 and B-1: Serum and muscle CPK isoenzymes of the proband; III/1. (2) A-2 and B-2: Corresponding isoenzymes of son, IV/1. (3) A-3 and B-3: Corresponding isoenzymes of son, IV/2. (4) A-4 and B-4: Corresponding isoenzymes of daughter, IV/3. (5) A-5: 10× conc. serum with added brain homogenate; B-5; brain homogenate. (6) A-6: 10× conc. normal serum; B-6: muscle homogenate from biopsied muscle of normals (287).

Figure 13-6   CPK isoenzyme monomer fractions of the skeletal muscle homogenates of malignant hyperthermic patients as compared to normal brain and muscle homogenates (284). (Reproduced with the permission of *Anesthesia and Analgesia*.)

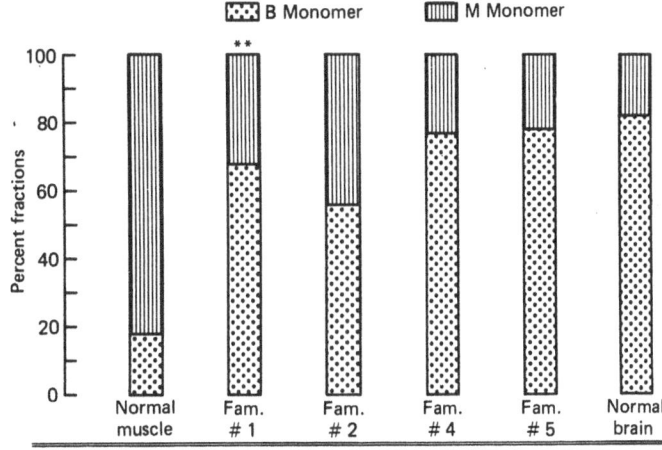

during late fetal development, they suggested that the arrest of the ontogenesis of CPK may be the cause of some myopathies (30, 72, 93, 94, 155, 207, 221). Similarly, it may be assumed that in MH the ontogenesis of CPK was arrested at an even earlier stage, before the disappearance of the BB isoenzyme of CPK (287).

Another alternative explanation for the abnormal CPK isoenzyme pattern in MH probands and relatives is that the BB-CPK is continually released from a high number of degenerating and regenerating nerve fibers present in the abnormal skeletal muscle. The neurophysiologic and neuropathologic studies of La Cour et al. (148) favor the latter assumption. They found hyperexcitability of the muscle fibers by electromyography. They attributed this to continuous denervation which was confirmed by histologic studies of muscle biopsies which showed degeneration and regeneration of the intramuscular nerve fibers. There was an increase in the number of endplates and pronounced sprouting of the nerve terminals. Their findings cast some doubt on the assumption that the BB-isoenzyme originates in the skeletal muscle fiber (284).

Malignant Hyperthermic Myopathy and Myotonias

Sch may cause sustained contracture, similar to that found in MH, in myotonia congenita patients (82, 83). Myotonia congenita patients do not as a rule, develop hyperthermia following the development of muscle contracture caused by Sch. Some known myotonic patients, however, developed characteristic MH during anesthesia (130, 218). In view of the above, Sch and other depolarizing relaxants should not be administered to patients with myotonia congenita or other myopathies (287).

Malignant Hyperthermic Myopathy and Denervation Myopathies

Sch and other depolarizing relaxants produce muscular fasciculations and/or tonic–clonic contractions of the muscles of normal subjects (197). The incidence and severity of these side effects may be reduced by the *slow* intravenous injection of *moderate* (0.4 to 0.6 mg/kg) doses of Sch (82, 83). In subjects with central or peripheral motor neuron damage, depolarizing relaxants may produce sustained contracture instead of relaxation (83, 96, 122, 124, 277), similar to that observed in avians and amphibians (82). The severe fasciculations in normal subjects and the contracture in the presence of motor neuron injuries may cause rhabdomyolysis accompanied by elevation of serum CPK activity, myoglobinemia, and hyperkalemia (119, 195, 242, 243, 261). It is of interest that La Cour et al. (148) and Bernhardt and Schiller (14) found some histopathologic and electromyographic changes, characteristic of denervation myopathies.

Age Distribution of Malignant Hyperthermic Myopathy

The age distribution of death from MH (23, 26) and that of death observed in myopathic patients (49) follows similar patterns. The mean age of death from MH was 21 years (273).

Consanguinity and Malignant Hyperthermic Myopathy

Consanguineous marriages of second cousins were reported in several large MH families (24, 285). Similarly, in pigs specifically bred for higher meat production, inbreeding results in an abnormal skeletal muscle development which makes them susceptible to MH under stress, anesthesia, and surgery (102, 185, 260).

Possible Etiology and Pathogenesis of Malignant Hyperthermia

The etiology of MH is unknown. It is possible that the triggering mechanism causes the abnormal release of (25) or reuptake of Ca^{2+} to the sarcoplasmic reticulum (281). This interferes with the physiologic contraction–relaxation process dependent on the reversible association of the released Ca^{2+} with the troponin–tropomyosin complex, essential for the formation of the contractile actomyosin. If Ca^{2+} is released in excessive amounts from genetically defective membranes or its reuptake is inhibited, the actomyosin may remain in an active state for prolonged periods. This is manifested in the contracture of the muscle. This in turn will result in increased heat and acid production and the reduction of the muscle ATP levels (103). This will cause a pathologic increase of the permeability of the muscle membrane resulting in the leakage of CPK, myoglobin, K^+ and phosphate (25, 281).

Recently Baskin and Deamer (8) suggested that the relatively small portion of muscle CPK (about 1 percent) present in the sarcoplasmic reticulum is an essential source of ATP necessary for the reuptake of Ca^{2+}. Indeed, Kalow and Hards (124), and Moulds and Denborough (178) and Britt et al. (26) found that the Ca^{2+} uptake was inhibited in vitro by halothane and/or Sch in biopsied MH muscle. This suggests that the defect in MH muscle may be in the sarcoplasmic reticulum.

Diagnostic Tests for Malignant Hyperthermia

Ellis et al. (65, 66) recently described a test which has great value in the differential diagnosis of MH. They exposed the biopsied muscles in vitro to halothane and/or Sch. In contrast to the normal muscle, contracture developed in MH muscle. Moulds and Denborough (179) and Ellis et al. (65) also observed that the muscle contracture and the histopathologic findings correlated well. Furthermore, Moulds and Denborough (179) found that the elevated serum CPK in MH families correlated well with the observed *in vitro* contracture of their biopsied muscles. These authors, therefore, suggested that serum CPK determinations should be used for the initial screening of suspected MH probands and their families. In doubtful cases, muscle biopsy and the *in vitro* contracture-test of Moulds and Denborough (179) should be used for the differential diagnosis.

Summary

The value of serum CPK and CPK isoenzymes in the detection of MH probands and the identification of the affected families no longer can be questioned. One must be cognizant, however, of the many factors which may alter serum CPK activity or CPK-isoenzyme pattern. It is hoped that in the future, simpler and less expensive CPK isoenzyme methods suitable for mass laboratory screening will be available to permit routine preoperative screening of all patients in whom the use of potent inhalation anesthetics and depolarizing muscle relaxants is planned. If regional or balanced anesthesia is selected, preoperative screening is not necessary, since these agents and techniques are well tolerated by MH patients.

The routine preoperative serum CPK level will aid the practicing anesthesiologist in the following ways:

(1) Elevated serum CPK activity and abnormal, MB/MM or BB/MM isoenzyme ratios will help to identify *MH patients* prior to induction.
(2) Patients who developed *myocardial infarction* preoperatively would be recognized before induction.
(3) Patients with *myotonias* who would develop muscle contracture following Sch could be recognized preoperatively.
(4) In patients with *central* or *peripherial neuropathies*, the degree of muscle damage could be estimated and if indicated, the use of Sch avoided.
(5) In *traumatized patients*, the degree of muscle trauma could be assessed before induction of anesthesia and Sch not used.
(6) Preoperative baseline CPK values would be helpful in the assessment of the degree of skeletal *muscle or cardiac muscle trauma that could have occurred during anesthesia and surgery*, and the CPK isoenzymes would help differentiate between the two types of muscle damage.
(7) Elevated serum CPK levels would call attention to the presence of *unrecognized muscular dystrophies* which may alter the response to anesthetic agents and muscle relaxants.
(8) Routine serum CPK tests would lead to *new discoveries of patients with MH, various muscular dystrophies, and cardiomyopathies*. Based on this recognition, proper selection of anesthetics could be made.

Acknowledgments

I am indebted to Dr. Francis F. Foldes, the editor of this volume, for his editorial help and suggestions in the preparation of this chapter, to Miss Catherine Papes for her devoted work of typing, and to Sarla P. Kothary, M. B. B. S. for the proofreading of the manuscript.

References

1. Adelstein, R. S., and Conti, M. A., The characterization of contractile proteins from platelets and fibroblasts. *Cold Spring Harbor Symp. Quant. Biol.* **37**, 599 (1972).
2. Airaksinen, M., and Tammisto, T., Myoglobinuria after intermittent administration of succinylcholine during halothane anesthesia. *Clin. Pharmacol. Ther.* **7**, 583 (1966).
3. Aldrete, J. A., Padfield, A., Solomon, C. C., and Rubright, M. W., Possible predictive tests for malignant hyperthermia during anesthesia. *J. Am. Med. Assoc.* **215**, 1465 (1971).
4. Anido, V., Conn, R. B., Jr., Mengoli, H. F., and Anido, G., Diagnostic efficacy of myocardial creatine phosphokinase using polyacrylamide disk-gel electrophoresis. *Am. J. Clin. Pathol.* **61**, 599 (1974).
5. Arnold, H., and Pette, D., Binding of glycolytic enzymes to structure proteins of the muscle. *Eur. J. Biochem.* **1**, 163 (1968).
6. Bachmann, O., and Largiader, F., Serum enzymes during cardiac allograft rejection. *Z. Gesamte Exp. Med. Einschl. Exp. Chir.* **153**, 162 (1970).
7. Banerji, A. P., Khopkar, P. P., Deshpande, D. H., and Desai, A. D., Study of creatine phosphokinase isoenzymes of serum and cerebrospinal fluid in a patient with Duchenne muscular dystrophy. *Clin. Chim. Acta* **43**, 431 (1973).
8. Baskin, R. J., and Deamer, D. W., A membrane-bound creatine phosphokinase in fragmented sarcoplasmic reticulum. *J. Biol. Chem.* **245**, 1345 (1970).
9. Batsakis, J. G., Preston, J. A., Briere, R. O., and Giesen, P. C., Iatrogenic aberrations of serum enzyme activity. *Clin. Biochem.* **2**, 125 (1968).
10. Beldavs, J., Small, V., Cooper, D. A., and Britt, B. A., Postoperative malignant hyperthermia: A case report. *Can. Anaesth. Soc. J.* **18**, 202 (1971).
11. Bennike, K. A., and Jarnum, S., Myoglobinuria with acute renal failure possibly induced by suxamethonium. *Br. J. Anaesth.* **36**, 730 (1964).
12. Berl, S., Puszkin, S., and Nicklas, W. J., Actomyosin-like protein in brain. *Science* **179**, 441 (1973).
13. Berlin, B. S., Simon, N. M., and Bovner, R. N., Myoglobinuria precipitated by viral infection. *J. Am. Med. Assoc.* **227**, 1414 (1974).
14. Bernhardt, D., and Schiller, H., Malignant hyperthermia under general anaesthesia. Abnormal histochemical and electron-microscopic muscle findings in combination with pathological serum-CPK-values evidencing the existence of primary myopathy. *Anaesthesist* **22**, 367 (1973).
15. Bernt, E., and Bergmeyer, H.-U., Creatine phosphokinase. *In* "Methoden der enzymatischen Analyse" (H.-U. Bergmeyer, ed.), 1st ed., p. 859. Verlag Chemie, Weinheim, 1962.
16. Berwick, P., 2,4-dichlorophenoxyacetic acid poisoning in man. Some interesting clinical and laboratory findings. *J. Am. Med. Assoc.* **214**, 1114 (1970).
17. Beutler, E., and Baluda, B., Simplified determination of blood adenosine-triphosphate using the fire-fly system. *Blood* **23**, 688 (1964).
18. Blyth, H., and Hughes, B. P., Pregnancy and serum-CPK levels in potential carriers of "severe" X-linked muscular dystrophy. *Lancet* **1**, 855 (1971).

19. Bolter, C. P., and Critz, J. B., Plasma enzyme activities after stimulation of the cardiovascular responses to exercise by subthalamic stimulation. *Fed. Proc., Fed. Am. Soc. Exp. Biol.* **29**, 266 (1970).

20. Bork, K., Korting, G. W., and Faust, G., Action of some serum enzymes following whole-body muscle massage. Contribution to the problem of physical therapy in dermatomyositis. *Arch. Dermatol. Forsch.* **240**, 342 (1971).

21. Bray, G. M., and Ferrendelli, J. A., Serum creatine phosphokinase in muscle disease: An evaluation of two methods of determinations and comparison with serum aldolase. *Neurology* **18**, 480 (1968).

22. Britt, B. A., and Kalow, W., Hyperrigidity and hyperthermia associated with anesthesia. *Ann. N.Y. Acad. Sci.* **151**, 947 (1968).

23. Britt, B. A., Locher, W. G., and Kalow, W., Hereditary aspects of malignant hyperthermia. *Can. Anaesth. Soc. J.* **16**, 89 (1969).

24. Britt, B. A., and Kalow, W., Malignant hyperpyrexia: A statistical review. *Can. Anaesth. Soc. J.* **17**, 293 (1970).

25. Britt, B. A., Recent advances in malignant hyperthermia. *Anesth. Analg. (Cleveland)* **51**, 841 (1972).

26. Britt, B. A., Kalow, W., Gordon, A., Humphrey, J. G., and Newcastle, N. B., Malignant hyperthermia: An investigation of five patients. *Can. Anaesth. Soc. J.* **20**, 431 (1973).

27. Bulcke, J. A., and Sherwin, A. L., Organ specificity of creatine phosphokinase muscle isoenzyme. *Immunochemistry* **6**, 681 (1969).

28. Burger, A., Richterich, R., and Aebi, H., Die Heterogenität der Kreatin-Kinase. *Biochem. Z.* **339**, 305 (1964).

29. Calvy, G. L., Cady, L. D., Mutson, A., Nierman, J., and Gertler, M. M., Serum lipids and enzymes: Their levels after high-caloric, high-fat intake and vigorous exercise regimen in Marine Corps recruit personnel. *J. Am. Med. Assoc.* **183**, 1 (1963).

30. Cao, A., de Virgilis, S., and Falorni, A., The ontogeny of creatine-kinase isoenzymes. *Biol. Neonat.* **13**, 375 (1968).

31. Cao, A., de Virgilis, S., de Marco, A., and Coppa, G., Gli isoenzimi della creatinfosfochinasi del muscolo nelle distrofie muscolari progressive. *Riv. Clin. Pediatr.* **81**, 504 (1968).

32. Cao, A., de Virgilis, S., Lippi, C., and Trabalza, N., Creatine kinase isoenzymes in serum of children with neurological disorders. *Clin. Chim. Acta* **23**, 475 (1969).

33. Cao, A., de Virgilis, S., Lippi, C., and Coppa, G., Serum and muscle creatine phosphokinase isoenzymes and serum aspartate aminotransferase isoenzymes in progressive muscular dystrophy. *Enzyme* **12**, 49 (1971).

34. Caraway, W. T., Chemical and diagnostic specificity of laboratory tests. Effects of hemolysis, lipemia, anticoagulants, medications, contaminants, and other variables. *Am. J. Clin. Pathol.* **37**, 445 (1962).

35. Cattolica, E. V., Effect of transurethral surgery upon the serum enzyme creatine phosphokinase. *J. Urol.* **106**, 262 (1971).

36. Caughey, J. E., Relationship of dystrophia myotonica (myotonic dystrophy) and myotonia congenita (Thomsen's Disease). *Neurology* **8**, 469 (1958).

37. Chalençon, J. P., Losman, J., Leclerc, P., Mathon, C., Froment, R., Dalloz, C., Froment, A., Age, C., Didier-Laurent, J. F., Du Grès, B., Revillard, J. P., Betuel, H., Fleurette, J., Collobert, L., Bertagnold, J. G., Perrin, A., and Loire, R., Case of cardiac transplantation for fully developed myocardiopathy. *Arch. Mal. Coeur Vaiss.* **62**, 780 (1969).

38. Cho, A. K., Haslett, W. L., and Jenden, D. J., A titrimetric method for the determination of creatine phosphokinase. *Biochem. J.* **75**, 115 (1960).

39. Cohen, L., CPK test: Effect of intramuscular injection in myocardial infarction. *J. Am. Med. Assoc.* **219**, 625 (1972).

40. Coleman, J. R., and Coleman, A. W., Muscle differentiation and macromolecular synthesis. *J. Cell. Physiol.* **72**, Suppl. 1, 19 (1968).

41. Colombo, J. P., Richterich, R., and Rossi, E., Serum-Kreatin Phosphokinase; Bestimmung und diagnostische Bedeutung. *Klin. Wochenschr.* **40**, 37 (1962).
42. Conn, R. B., Jr., and Anido, V., Creatine phosphokinase determinations by the fluorescent ninhydrin reaction. *Am. J. Clin. Pathol.* **46**, 177 (1966).
43. Coodley, E. L., Use of enzymes in cardiac diagnosis. *Angiology* **16**, 209 (1965).
44. Coodley, E. L., Current status of enzyme diagnosis in cardiovascular disease. *Am. J. Med. Sci.* **252**, 633 (1966).
45. Coodley, E. L., Enzymes and isoenzymes in myocardial infarction. *Cardiovasc. Clin.* **1**, 140 (1970).
46. Coodley, E. L., Prognostic value of enzymes in myocardial infarction. *J. Am. Med. Assoc.* **225**, 597 (1973).
47. Cooperman, L. H., Succinylcholine-induced hyperkalemia in neuromuscular disease. *J. Am. Med. Assoc.* **213**, 1867 (1970).
48. Crowley, L. V., and Alton, M., A comparison of four methods of measuring CPK. *Am. J. Clin. Pathol.* **53**, 948 (1970).
49. Danowski, T. S., Sabeh, G., Vester, J. W., Alley, R. A., Robbins, T. J., Tsai, C. T., Pazirandeh, M., and Sekaran, K., Serum CPK in muscular dystrophy and myotonia dystrophica. *Metab., Clin. Exp.* **17**, 808 (1968).
50. Danowski, T. S., Wissinger, H. A., Hohmann, T. C., Gerneth, J. A., Folkers, K., Vester, J. W., and Fisher, E. R., Tabulation of findings in the muscular dystrophies and in myotonia dystrophica. *Arch. Phys. Med. Rehabil.* **52**, 193 (1971).
51. Dawson, D. M., Eppenberger, H. M., and Kaplan, N. O., Creatine-kinase: Evidence for a dimeric structure. *Biophys. Biochem. Res. Commun.* **21**, 346 (1965).
52. Dawson, D. M., Eppenberger, H. M., and Kaplan, N. O., The comparative enzymology of creatine-kinase. II. Physical and chemical properties. *J. Biol. Chem.* **242**, 210 (1967).
53. Debreczeni, M., and Ladanyi, E., Determination of creatine-phosphokinase in patients with scleroderma and dermatomyositis. *Hautarzt* **21**, 81 (1970).
54. Demos, J., La détection des porteuses saines du trait myopathique dans la myopathie humaine. *Ann. Genet.* **12**, 191 (1969).
55. Denborough, M. A., Ebeling, P., King, J. O., and Zapf, P. W., Myopathy and malignant hyperpyrexia. *Lancet* **1**, 1138 (1970).
56. Denborough, M. A., Hird, F. R., King, J. O., Marginson, M. A., Mitchelson, K. R., Naylor, W. G., Rex, M. A., Zapf, P. W., and Condron, R. J., Mitochondrial and other studies in Australian Landrace Pigs affected with malignant hyperthermia. *In* "Malignant Hyperthermia" (A. H. Gordon, B. H. Britt, and W. Kalow, eds.), p. 229. Thomas, Springfield, Illinois, 1973.
57. Denborough, M. A., Forster, J. F. A., Hudson, M. C., Carter, N. G., and Zapf, P. W., Biochemical changes in malignant hyperpyrexia. *Lancet* **1**, 1137 (1970).
58. Derweduwen, H., Enderle, J., De Geest, H., Polis, O., Vancrombreuco, J. C., and Joossens, J. V., Direct-current shock in the treatment of cardiac arrhythmias. A clinical study. *Acta Cardiol.* **24**, 205 (1969).
59. Deul, D. H., and Van Breemen, J. F. L., Electrophoresis of creatine kinase from various organs. *Clin. Chim. Acta* **10**, 276 (1964).
60. Dixon, S. H., Jr., Fuchs, J. C., and Ebert, P. A., Changes in serum creatine phosphokinase activity following thoracic, cardiac and abdominal operations. *Arch. Surg. (Chicago)* **103**, 66 (1971).
61. Dixon, S. H., Jr., Limbird, L. E., Roe, C. R., Wagner, G. S., Oldham, H. N., and Sabiston, D. C., Recognition of postoperative acute myocardial infarction. Application of isoenzyme techniques. *Circulation* **137**, Suppl. III, 47 (1973).
62. Dreyfus, J. C., Schapira, G., Resnais, J., and Scebat, L., Serum creatine kinase in the diagnosis of myocardial infarct. *Rev. Fr. Etud. Clin. Biol.* **5**, 386 (1960).
63. Dreyfus, J. C., Schapira, G., and Demos, J., Study of serum creatine kinase in myopathic patients and their families. *Rev. Fr. Etud. Clin. Biol.* **5**, 384 (1960).

64. Dubo, H., Park, D. C., Pennington, R. J. T., Kalbag, R. M., and Walton, J. N., Serum creatine-kinase in cases of stroke, head injury and meningitis. *Lancet* **2**, 743 (1967).

65. Ellis, F. R., Keaney, N. P., Harriman, D. G. F., Sumner, D. W., Kyei-Mensah, K., Tyrrell, J. H., Hargreaves, J. B., Parikh, R. K., and Mulrooney, P. L., Screening for malignant hyperpyrexia. *Br. Med. J.* **3**, 559 (1972).

66. Ellis, F. R., Keaney, N. P., and Harriman, D. G. F., Histopathological and neuropharmacological aspects of malignant hyperpyrexia. *Proc. R. Soc. Med.* **66**, 12 (1973).

67. Emery, A. E. H., and Pascasion, F. M., The effect of pregnancy on the concentration of creatine kinase in serum, skeletal muscle and myometrium. *Am. J. Obstet. Gynecol.* **91**, 18 (1965).

68. Emery, A. E. H., and Spikesman, A. M., Evidence against the existence of subclinical form of X-linked Duchenne muscular dystrophy. *J. Neurol. Sci.* **10**, 523 (1970).

69. Emery, A. E. H., and Spikesman, A. M., Serum creatine kinase levels, correspondence. *Br. Med. J.* **2**, 790 (1970).

70. Engel, W. K., and Meltzer, H. Y., Histochemical abnormalities of skeletal muscle in patients with acute psychoses. *Science* **168**, 273 (1970).

71. Ennor, A. H., Rosenberg, H., and Armstrong, M. D., Specificity of creatine phosphokinase. *Nature (London)* **175**, 120 (1955).

72. Eppenberger, H. M., Eppenberger, M., Richterich, R., and Aebi, H., The ontogeny of creatine kinase isozymes. *Dev. Biol.* **10**, 1 (1964).

73. Eppenberger, H. M., Dawson, D. M., and Kaplan, N. O., The comparative enzymology of creatine kinase. I. Isolation and characterization from chicken and rabbit tissues. *J. Biol. Chem.* **242**, 204 (1967).

74. Eppenberger, M. E., Eppenberger, H. M., and Kaplan, N. O., Evolution of creatine kinase. *Nature (London)* **214**, 239 (1967).

75. Escher, J., and Zimmerman, H. J., Creatine phosphokinase in disease. *Am. J. Med. Sci.* **253**, 272 (1967).

76. Evers, C., Die Bedeutung der Kreatinphosphokinase (CPK) für die Diagnostik des Herzinfarktes und der Myokarditis. *Med. Klin. (Munich)* **58**, 1260 (1963).

77. Feraudi, M., and Harm, K., Ein optischer Test zur Bestimmung der Kreatin-Phosphokinase. *Z. Klin. Chem. Klin. Biochem.* **5**, 270 (1967).

78. Feraudi, M., Harm, K., and Schmidt, H., Statistiche Untersuchungen zur Frage der Normalwerte und der Serum-Kreatin-Phosphokinase. *Enzymol. Biol. Clin.* **9**, 338 (1968).

79. Fitzsimmons, J. A. E., Studies on the creatine-kinase of the invertebrate: Morphysa sanguinea. Ph.D. Thesis, University of Queensland, Brisbane, Australia (1971).

80. Flodin, P., Dextran gels and their application in gel filtration. Uppsala, U.S. Patent 3,208,994 (1962).

81. Florkin, M., and Stotz, E., eds., "Comprehensive Biochemistry," Vol. 13. Elsevier, Amsterdam, 1965.

82. Foldes, F. F., "Muscle Relaxants in Anesthesiology." Thomas, Springfield, Illinois, 1957.

83. Foldes, F. F., Factors altering effects of muscle relaxants. *Anesthesiology* **20**, 474 (1959).

84. Forster, G., and Escher, J., Die Kreatinphosphokinase in der Diagnostik von Herzinfarkt und Myopathien. *Helv. Med. Acta* **28**, 513 (1961).

85. Forster, G., Zur Enzymdiagnostik von Herzinfarkt und Myopathien. *Schweiz. Med. Wochenschr.* **97**, 329 (1967).

86. Fowler, W. M., Jr., Chowdhury, S. R., Pearson, C. M., Gardner, G. W., and Bratton, R., Changes in serum enzyme levels after exercise in trained and untrained subjects. *J. Appl. Physiol.* **17**, 943 (1962).

87. Fowler, W. M., Jr., Gardner, G. W., Kazerunian, H. H., and Lauvstad, W. A., The effect of exercise on serum enzymes. *Arch. Phys. Med. Rehabil.* **49**, 554 (1968).

88. Friedberg, F., The aminoacid composition of adenosine triphosphate-creatine transphosphorylase. *Arch. Biochem. Biophys.* **61**, 263 (1956).

89. Gibson, J. A., and Gardiner, D. M., Malignant hypertonic hyperpyrexia syndrome. *Can. Anaesth. Soc. J.* **16**, 106 (1969).

90. Girlando, V., On the constant positivity of 2 plasmatic enzyme tests, specific for myopathy, found in 92 subjects with idiopathic scoliosis. *Minerva Ortop.* **22**, 288 (1971).

91. Goldman, B. S., Trimble, A. S., Sheverini, M. A., Teasdale, S. J., Silver, M. D., and Elliott, G. E., Functional and metabolic effects of anoxic cardiac arrest. *Ann. Thorac. Surg.* **11**, 122 (1971).

92. Gordon, A. H., Electrophoresis of proteins in polyacrylamide and starch gels. *In* "Laboratory Techniques in Biochemistry and Molecular Biology" (T. S. Work and E. Work, eds.), Vol. 1, p. 119. North-Holland Publ., Amsterdam, 1970.

93. Goto, I., Nagamine, M., and Katsuki, S., Creatine-phosphokinase isoenzymes in muscles. *Arch. Neurol. (Chicago)* **20**, 422 (1969).

94. Goto, I., and Katsuki, S., Creatine phosphokinase isoenzymes in pathological human serum. *Clin. Chim. Acta* **30**, 795 (1970).

95. Graig, F. A., Smith, J. C., and Foldes, F. F., Effect of dilution on the activity of serum creatine phosphokinase. *Clin. Chim. Acta* **15**, 107 (1967).

96. Granit, R., Skögland, S., and Thesleff, S., Activation of muscle spindles by succinylcholine and decamethonium. The effects of curare. *Acta Physiol. Scand.* **28**, 134 (1953).

97. Greenblatt, D. J., Duhme, D. W., and Koch-Weser, J., Pain and CPK elevation after intramuscular digoxin. *N. Engl. J. Med.* **288**, 689 (1973).

98. Greenblatt, D. J., Koch-Weser, J., and Schader, R. I., Multiple complications and death following protryptilene overdose. *J. Am. Med. Assoc.* **229**, 554 (1974).

99. Griffiths, P. D., Serum levels of ATP: Creatine phosphotransferase (creatine kinase). The normal range and effect of muscular activity. *Clin. Chim. Acta* **13**, 413 (1966).

100. Häcker, M. R., Krüger, E., and Augustin, H. W., Über eine einfache Methode zur Bestimmung der Kreatinkinase (ATP: Kreatin-Phosphotransferase, EC.2.7.3.2.) im Serum. *Z. Med. Labortech.* **8**, 259 (1967).

101. Halonen, P. I., and Konttinen, A., Effects of physical exercise on some enzymes in the serum. *Nature (London)* **193**, 942 (1962).

102. Harrison, G. G., Saunders, S. J., Biebuyck, J. F., Hickman, R., Dent, D. M., Weaver, V., and Terblanche, J., Anaesthetic-induced malignant hyperpyrexia and a method for its prediction. *Br. J. Anaesth.* **41**, 844 (1969).

103. Harrison, G. G., Anesthesia induced malignant hyperpyrexia: A suggested method of treatment. *Br. Med. J.* **3**, 454 (1971).

104. Harrison, G. G., The effect of procaine and curare on the initiation of anesthetic-induced malignant hyperpyrexia. *In* "Malignant Hyperthermia" (A. Gordon, B. A. Britt, and W. Kalow, eds.), p. 271. Thomas, Springfield, Illinois, 1973.

105. Harvey-Sklar, S., and Wigand, J. S., Creatine phosphokinase as indicator of adenosine triphosphate activity. Letter to the Editor. *J. Am. Med. Assoc.* **226**, 1464 (1973).

106. Henry, P. D., Bloor, C. M., and Sobel, B. E., Increased serum CPK activity in experimental pulmonary embolism. *Am. J. Cardiol.* **26**, 151 (1970).

107. Herschkovitz, N., and Cummings, J. N., Creatine kinase in the cerebrospinal fluid. *J. Neurol., Neurosurg. Psychiatry* **27**, 247 (1964).

108. Hess, J. W., and MacDonald, R. P., Serum creatine phosphokinase activity: A new diagnostic aid in myocardial and skeletal muscle disease. *J. Mich. State Med. Soc.* **62**, 1095 (1963).

109. Hess, J. W., MacDonald, R. P., Frederick, R. J., Jones, R. N., Neely, J., and Gross, D., Serum creatine phosphokinase (CPK) activity in disorders of heart and skeletal muscle. *Ann. Intern. Med.* **61**, 1015 (1964).

110. Hess, J. W., Murdock, K. J., and Natho, G. J. W., Creatine phosphokinase. A spectrophotometric method with improved sensitivity. *Am. J. Clin. Pathol.* **50**, 89 (1968).

111. Heyde, E., and Morrison, J. F., Studies on the inhibition of ATP: creatine phosphotransferase by NaCl. *Biochim. Biophys. Acta* **212**, 288 (1970).

112. Hobson, R. W., Conant, C., Fleming, A., Mahoney, W. D., and Baugh, J. H., Postoperative serum enzyme patterns. *Mil. Med.* **136**, 624 (1971).

113. Hobson, R. W., Conant, C., Mahoney, W. D., and Baugh, J. H., Serum creatine phos-

phokinase. Analysis of postoperative changes. *Am. J. Surg.* **124**, 625 (1972).

114. Hooton, B. T., and Watts, D. C., Adenosine 5-triphosphate-creatine phosphotransferase from dystrophic mouse skeletal muscle. A genetic lesion associated with the catalytic-site thiol groups. *Biochem. J.* **100**, 637 (1966).

115. Hughes, B. P., A method for the estimation of serum creatine kinase and its use in comparing creatine kinase to aldolase activity in normal and pathologic sera. *Clin. Chim. Acta* **7**, 597 (1962).

116. Hughes, B. P., Creatine phosphokinase in facioscapulohumeral muscular dystrophy. *Br. Med. J.* **3**, 464 (1971).

117. Hultgren, H. N., Miyagama, U., Buck, W., and Angell, W. W., Ischemic myocardial injury during coronary artery surgery. *Am. Heart J.* **82**, 624 (1971).

118. Ibrahim, G. A., Awad, E. A., and Kottke, F. J., Interstitial myofibrositis: Serum and muscle enzymes and lactate dehydrogenase-isoenzymes. *Arch. Phys. Med. Rehabil.* **55**, 23 (1974).

119. Innes, R. K. R., and Stromme, J. H., Rise in serum creatine phosphokinase associated with agents used in anaesthesia. *Br. J. Anaesth.* **45**, 185 (1973).

120. Isaacs, H., and Barlow, M. B., Malignant hyperpyrexia during anesthesia; possible association with subclinical myopathy. *Br. Med. J.* **1**, 275 (1970).

121. James, O. F., Hyperpyrexia and hypertonia in anaesthesia. *Med. J. Aust.* **1**, 1154 (1970).

122. Jarcho, L. W., Berman, B., Eyzaguirre, C., and Lilienthal, J. L., Curarization of denervated muscle. *Ann. N. Y. Acad. Sci.* **54**, 337 (1951).

123. Kagen, L. J., Myoglobinemia and myoglobinuria in patients with myositis. *Arthritis Rheum.* **14**, 457 (1971).

124. Kalow, W., and Hards, J., A species-specific action of halothane on human skeletal muscle. *Pap. Int. Congr. Pharmacol., 4th, 1969* Abstracts, p. 423 (1970).

125. Karmen, A., Wroblewski, F., and La Due, J. S., Transaminase activity in human blood. *J. Clin. Invest.* **34**, 126 (1955).

126. Keif, W., Klein, B., and Möller, E., Enzyme Bewegungen unter Körperlicher Belastung bei Trainierten und Untrainierten Probanden. *Med. Klin. (Munich)* **67**, 195 (1972).

127. Kelly, S., Copeland, W., and Smith, R. O., A fluorescent spot test for creatine kinase. *Clin. Chim. Acta* **21**, 431 (1968).

128. Kelstrup, J., Haase, J., Jorni, J., Reske-Nielsen, E., and Hanel, H. K., Malignant hyperthermia in a family. *Acta Anaesthesiol. Scand.* **17**, 283 (1973).

129. Kendrich-Jones, J. and Perry, S. V., The enzymes of adenine nucleotide metabolism in developing skeletal muscle. *Biochem. J.* **103**, 207 (1967).

130. King, J. O., Denborough, M. A., and Zapf, P. W., Survey of 18 malignant hyperpyrexic families. *Lancet* **1**, 365 (1972).

131. King, J. O., and Zapf, P. W., A review of the value of creatine phosphokinase estimations in clinical medicine. *Med. J. Aust.* **1**, 699 (1972).

132. Kironova, I. U. P., Gracheva, G. V., and Mishurova, V. P., Evaluation of the effect of electric defibrillation on the myocardium from the serum creatine kinase level. *Kardiologia* **11**, 74 (1971).

133. Kluge, W. F., Prognostic value of serum creatine phosphokinase levels in myocardial infarction. *Northwest Med.* **68**, 847 (1969).

134. Knirsch, A. K., and Gralla, E. J., Abnormal serum transaminase levels after parenteral ampicillin and carbenicillin administration. *N. Engl. J. Med.* **282**, 1081 (1970).

135. Kolins, J., and Gilroy, J., Serum enzyme levels in patients with myasthenia gravis after aerobic and ischemic exercise. *J. Neurol., Neurosurg. Psychiatry* **35**, 34 (1972).

136. Konttinen, A., and Pyörälä, T., Serum enzyme activity in late pregnancy, at delivery, and during puerperium. *Scand. J. Clin. Lab. Invest.* **15**, 429 (1963).

137. Konttinen, A., and Halonen, P. I., Serum creatine phosphokinase and hydroxybutyrate dehydrogenase activities compared with glutamic-oxaloacetic transaminase and lactic

dehydrogenase in myocardial infarction. *Cardiol. Prat.* **43**, 56 (1965).

138. Konttinen, A., and Somer, H., Determination of serum creatine kinase isoenzymes in myocardial infarction. *Am. J. Cardiol.* **29**, 870 (1972).

139. Konttinen, A., Somer, H., and Auvinen, S., Serum enzymes and isoenzymes. *Arch. Intern. Med.* **133**, 243 (1974).

140. Korting, G. W., Weber, G., and Werle, H., Enzympathologische Beobachtungen bei Dermatomyositis. *Hautarzt* **13**, 485 (1962).

141. Koufen, H., and Consbruch, V., Die Serum Kreatin-Phosphokinase (CPK) Activität bei Amyotropher Lateralsklerose (ALS) und anderen neurogenen Muskelatrophien unter Berüchsichtigung differatial-diagnostischer Aspekte. *Nervenarzt* **41**, 599 (1970).

142. Kuby, S. A., Noda, L., and Lardy, H. A., Adenosinetriphosphate-creatine transphosphory-lase. I. Isolation of the crystalline enzyme from rabbit muscle. *J. Biol. Chem.* **209**, 191 (1954).

143. Kuby, S. A., Noda, L., and Lardy, H. A., Adenosine triphosphate-creatine transphos-phorylase. III. Kinetic studies. *J. Biol. Chem.* **210**, 65 (1954).

144. Kuby, S. A., and Noltmann, E. A., ATP-creatine transphosphorylase. *In* "The Enzymes" (P. D. Boyer, H. Lardy, and K. Myräck, eds.), 2nd ed., Academic Press, Vol. 6, p. 515. New York, 1962.

145. Kundrat, E., and Pepe, F. A., The M-band studies with fluorescent antibody staining. *J. Cell Biol.* **48**, 340 (1971).

146. Kupfer, D. J., Meltzer, H. Y., Wyatt, R. J., and Snyder, F., Serum enzyme changes during sleep deprivation. *Nature (London)* **228**, 768 (1970).

147. Kyei-Mensah, K., Lockwood, R., Tyrrell, J. H., and Willett, I. H., Malignant hyper-pyrexia: A study of a family. Case report. *Br. J. Anaesth.* **43**, 811 (1971).

148. La Cour, D., Juul-Jensen, P., and Reske-Nielsen, E., Central and peripheral mechanisms in malignant hyperthermia. *In* "Malignant Hyperthermia" (A. Gordon, B. A. Britt, and W. Kalow, eds.), p. 380. Thomas, Springfield, Illinois, 1973.

149. Lafair, J. S., and Myerson, R. M., Alcoholic myopathy with special reference to the significance of creatine phosphokinase. *Ann. Intern. Med.* **122**, 417 (1968).

150. Lawrence, S. H., Melnick, P. J., and Weimer, H. E., A species comparison of serum proteins and enzymes by starch gel electrophoresis. *Proc. Soc. Exp. Biol. Med.* **105**, 572 (1960).

151. Ledwick, J. R., Changes in serum creatine phosphokinase during submaximal exercise testing. *Can. Med. Assoc. J.* **109**, 273 (1973).

152. Lisak, R. P., and Graig, F. A., Lack of diagnostic value of CPK assay in spinal fluid. *J. Am. Med. Assoc.* **199**, 750 (1967).

153. Maassen, J. H., and Broy, H., Kritische Bemerkungen zur Enzymediagnostik beim Herzinfarkt. *Muench. Med. Wochenschr.* **104**, 2497 (1962).

154. Madsen, A., Creatine phosphokinase isoenzymes in human tissue with special reference to brain extracts. *Clin. Chim. Acta* **36**, 17 (1972).

155. Magalhaes, A. S., Isoenzymes de la créatine-phosphokinase de biopsies musculaires humaines. *Acta Neurol. Belg.* **70**, 471 (1970).

156. Mahon, F. B., Myangman, U., and Madsen, P. O., Serum enzyme determinations after transurethral resection of the prostate. *J. Urol.* **107**, 88 (1972).

157. Mahowald, T. A., and Agodoa, L., Thiolysis of S-dinitrophenylated creatine kinase with restoration of enzymatic activity. *Biochem. Biophys. Res. Commun.* **37**, 576 (1969).

158. Mandecki, T., Giec, L., and Kargul, W., Serum enzyme activities after cardioversion. *Br. Heart J.* **32**, 600 (1970).

159. Mastaglia, F. L., Barwich, D. D., and Hall, R., Myopathy in acromegaly. *Lancet* **2**, 907 (1970).

160. Mastaglia, F. L., and Walton, J. N., Histological and histochemical changes in skeletal muscle from cases of chronic juvenile and early adult spinal muscular atrophy (the

Kugelberg-Welander Syndrome). *J. Neurol. Sci.* **12**, 15 (1971).

161. Mayrhofer, O., Die Wirksamkeit von d-Tubocurarine zur Verhütung der Muskelschmerzen nach Succinylcholina. *Anaesthesist* **4**, 313 (1959).

162. Meltzer, H. Y., Muscle enzyme release in acute psychoses. *Arch. Gen. Psychiatry* **21**, 102 (1969).

163. Meltzer, H. Y., and Moline, R. A., Plasma enzyme activity after exercise. *Arch. Gen. Psychiatry* **22**, 390 (1970).

164. Meltzer, H. Y., and Moline, R. A., Muscle abnormalities in acute psychoses. *Arch. Gen. Psychiatry* **23**, 481 (1970).

165. Meltzer, H. Y., Mrozak, S., and Boyer, M., Effect of intramuscular injections on serum creatine phosphokinase activity. *Am. J. Med. Sci.* **259**, 42 (1970).

166. Meltzer, H. Y., Serum creatine phosphokinase activity in newly admitted psychiatric patients. *Arch. Gen. Psychiatry* **24**, 568 (1971).

167. Meltzer, H. Y., Factors affecting serum creatin-kinase levels in the general population: The role of race, activity and age. *Clin. Chim. Acta* **33**, 165 (1971).

168. Meltzer, H. Y., and Guschuren, A., Type I (brain type) creatine phosphokinase (CPK) activity in rat platelets. *Life Sci.* **11**, 121 (1972).

169. Mercer, D. W., Separation of tissue and serum creatine kinase isoenzymes by ion-exchange column chromatography. *Clin. Chem.* **20**, 36 (1974).

170. Michie, D. D., Booth, R. W., Conley, M. A., and McQuire, H. J., The effect of quick freezing on the preservation of selected serum enzyme activities. *Am. J. Clin. Pathol.* **52**, 329 (1969).

171. Michie, D. D., Conley, M. A., Carretta, R. F., and Booth, R. W., Serum enzyme changes following cardiac catheterizations with and without selective coronary arteriography. *Am. J. Med. Sci.* **260**, 11 (1970).

172. Middleton, P. J., Alexander, R. M., and Symanski, M. T., Severe myositis during recovery from influenza. *Lancet* **2**, 533 (1970).

173. Milner-White, E. J., and Watts, D. C., Inhibition of adenosine 5-triphosphate creatine-phosphotransferase by substrate anion complexes. *Biochem. J.* **122**, 727 (1971).

174. Miyazaki, K., Toyoda, R., Tomino, H., Yoshimatsu, M., Saijo, K., Katsunuma, N., and Fujino, A., Electrophoretic pattern of creatine kinase isoenzymes during development or in the case of skeletal muscular dystrophy. *Chem. Abstr., Biochem. Sect.* **64**, 1152c (1966).

175. Morimoto, U., and Harrington, W. F., Isolation and physical chemical properties of M-line protein from skeletal muscle. *J. Biol. Chem.* **247**, 3052 (1972).

176. Morris, G. E., Cooke, A., and Cole, R. J., Isoenzymes of creatine phosphokinase during myogenesis in vitro. *Exp. Cell Res.* **74**, 582 (1972).

177. Mostert, J. W., Trudnowski, R. J., Hobika, G. H., and Moore, R., Effect of electrocautery on serum creatine phosphokinase in surgical patients. *J. Surg. Oncol.* **2**, 189 (1970).

178. Moulds, R. F. W., and Denborough, M. A., Biochemical basis of malignant hyperpyrexia. *Br. Med. J.* **2**, 241 (1974).

179. Moulds, R. F. W., and Denborough, M. A., Identification of susceptibility to malignant hyperpyrexia. *Br. Med. J.* **2**, 245 (1974).

180. Muggia, F. M., Ghossein, N. A., and Hanok, A., Creatine phosphokinase and other serum enzymes during radiotherapy. Comparison of cardiac vs noncardiac irradiation. *J. Am. Med. Assoc.* **211**, 1345 (1970).

181. Munsat, T. L., Baloh, R., Pearson, C. M., and Fowler, W., Serum enzyme alterations in neuromuscular disorders. *J. Am. Med. Assoc.* **226**, 1536 (1973).

182. Myerson, R. M., and Lafair, J. S., Alcoholic muscle disease. *Med. Clin. North Am.* **54**, 723 (1970).

183. Nakane, K., Change of serum creatine phosphokinase activity after exercise in Duchenne type of progressive muscular dystrophy. *Nagoya Med. J.* **17**, 203 (1972).

184. Namba, T., Aberfeld, D. C., and Grob, D., Chronic proximal spinal muscular atrophy. *J. Neurol. Sci.* **11**, 401 (1970).

185. Nelson, T., Porcine stress syndrome. *In* "Malignant Hyperthermia" (A. Gordon, B. A. Britt, and W. Kalow, eds.), p. 191. Thomas, Springfield, Illinois 1973.

186. Noda, L., Kuby, S. A., and Lardy, H. A., Adenosine triphosphate-creatine transphosphorylase. II. Homogenicity and physiochemical properties. *J. Biol. Chem.* **209**, 203 (1954).

187. Noda, L., Kuby, S. A., and Lardy, H. A., ATP-creatine transphosphorylase. *Methods Enzymol.* **2**, 605 (1955).

188. Noda, L., Nihei, T., and Moralis, M. F., The enzymatic activity and inhibition of adenosine 5-triphosphate creatine transphosphorylase. *J. Biol. Chem.* **235**, 2830 (1960).

189. Nuttall, F. Q., and Wedin, D. S., A simple rapid colorimetric method for determination of creatine-kinase activity. *J. Lab. Clin. Med.* **68**, 324 (1966).

190. Nuttall, F. Q., and Jones, B., Creatine kinase and glutamic oxaloacetic transaminase activity in serum: Kinetics of change with exercise and the effect of physical conditioning. *J. Lab. Clin. Med.* **71**, 847 (1968).

191. Okinaka, S., Sugita, H., Momoi, H., Toyokura, Y., Watanabe, T., Ebashi, F., and Ebashi, S., Cysteine-stimulated serum creatine kinase in health and disease. *J. Lab. Clin. Med.* **64**, 229 (1964).

192. Oldham, H. N., Roe, C. R., Young, W. G., Jr., and Dixon, S. H., Jr., Intraoperative detection of myocardial damage during coronary artery surgery by plasma creatine phosphokinase isoenzyme analysis. *Surgery* **74**, 917 (1973).

193. Oliver, J. T., A spectrophotometric method for the determination of creatine phosphokinase and myokinase. *Biochem. J.* **61**, 116 (1955).

194. Oppermann, C., Podlesch, I., and Purschke, R., Malignant hyperthermia during general anesthesia with rigor, myoglobinuria and disturbance of the blood coagulation mechanism. *Anaesthesist* **20**, 315 (1971).

195. Palecchi, A. E., Dall'Orso, F., and Fossa, S., La sofferenza muscolare secondaria alla somministrazione di succinilcolina e anestetici diversi valuata in base all'incremento serico delle attivitá aldolasica e creatinfosfochinasica. *Acta Anaesthesiol.* **19**, 1113 (1968).

196. Paterson, Y., and Lawrence, E. F., Factors affecting serum creatine phosphokinase levels in normal adult females. *Clin. Chim. Acta* **42**, 131 (1972).

197. Paton, W. D. M., The effects of muscle relaxants other than muscle relaxation. *Anesthesiology* **20**, 454 (1959).

198. Pearce, J. M. S., Pennington, R. J. T., and Walton, J. N., Serum enzymes in muscle disease. Part I. *J. Neurol., Neurosurg. Psychiatry* **27**, 1 (1964).

199. Peter, H. J., Zapf, J., Froesch, E. R., Bogenmann, E., Eppenberger, H., Bernhard, K., and Hossli, G., Kreatin-phosphokinase und ihre Isoenzyme im Serum von Patienten mit maligner Hyperthermie. In: Maligne Hyperthermie, Acupunctur, Biomedizinische Technik und Abdominelle Intensiwe Therapie. Anesthesiologie und Wiederbelebung No. 91. Bergmann, H. und Blauhut, B. (eds.). Springer Verlag, Berlin, Heidelberg, New York, 1975, p. 47.

200. Poch, G. F., Sica, E. P., Taratuto, A., and Weinstein, I. H., Hypertrophia musculorum vera. Study of a family. *J. Neurol. Sci.* **12**, 53 (1971).

201. Pomerantz, M., Delgado, F., and Eiseman, B., Unsuspected depressed cardiac output following blunt thoracic and abdominal trauma. *Surgery* **70**, 865 (1971).

202. Preston, J. A., Batsakis, J. G., Briere, R. O., and Taylor, R. V., Serum creatine phosphokinase: A clinical and laboratory evaluation. *Am. J. Clin. Pathol.* **44**, 71 (1965).

203. Reece, R. L., Creatine phosphokinase (CPK) as part of a screening profile: A one year survey. *Minn. Med.* **57**, 58 (1974).

204. Reinecke, R. D., Fatal fever with anesthesia. *Am. J. Ophthalmol.* **70**, 858 (1970).

205. Richter, R. W., Challenor, Y. B., Pearson, J., Kagen, L. J., Hamilton, L. L., and Ramsey, W. H., Acute myoglobinuria associated with heroin addiction. *J. Am. Med. Assoc.* **216**, 1172 (1971).

206. Roe, C. R., Limibird, L. E., Wagner, G. S., and Nerenberg, S. T., Combined isoenzyme

analysis in the diagnosis of myocardial injury: Application of electrophoretic methods for the detection and quantitation of the CPK-MB isoenzyme. *J. Lab. Clin. Med.* **80**, 577 (1972).

207. Roget, J., Beaudoing, A., Jobert, J., and Bost, M., Créatine phosphokinase sérique au cours des états d'anoxie du nouveau-né. *Pediatrie* **24**, 163 (1969).

208. Rosalki, S. B., Creatine-phosphokinase isoenzymes. *Nature (London)* **207**, 414 (1965).

209. Rosalki, S. B., An improved procedure for serum creatine phosphokinase determination. *J. Lab. Clin. Med.* **69**, 696 (1967).

210. Rose, L. I., Bousser, J. E., and Cooper, K. H., Serum enzymes after marathon running. *J. Appl. Physiol.* **29**, 355 (1970).

211. Rose, M. R., Glassman, E., Isom, O. W., and Spencer, F. C., Electrocardiographic and serum enzyme changes of myocardial infarction after coronary artery bypass surgery. *Am. J. Cardiol.* **33**, 215 (1974).

212. Rotthauwe, H. W., Zurukzoglu-Sklarvounou, S., and Hamann, H., Untersuchungen zur Methodik der Aktivitätsbestimmung der Kreatinphosphokinase. *Klin. Wochenschr.* **39**, 1269 (1961).

213. Rotthauwe, H. W., and Kowalewski, S., Bestimmung der Aktivität der Serum-Kreatin Phosphokinase. *Z. Klin. Chem. Klin. Biochem.* **5**, 254 (1967).

214. Rotthauwe, H. W., and Kowalewski, S., Kongenitale Muskeldystrophie. *Z. Kinderheilk.* **106**, 131 (1969).

215. Ryan, J. E., and Papper, E. M., Malignant fever during and following anesthesia. *Anesthesiology* **32**, 196 (1970).

216. Sabater, J., and Villalba, M., Creatine phosphokinase (CPK) levels in umbilical cord blood. *Clin. Chim. Acta* **36**, 201 (1972).

217. Sabawala, P. B., and Dillon, J. B., The positive inotropic action of cyclopropane on human intercostal muscle in vitro and its modification by *d*-tubocurarine. *Anesthesiology* **19**, 473 (1958).

218. Saidman, L. J., Harvard, E. S., and Eger, E. I., II, Hyperthermia during anesthesia. *J. Am. Med. Assoc.* **190**, 1029 (1964).

219. Sasa, T., and Noda, L., Adenosine triphosphatase activity of adenosine triphosphate-creatine phosphotransferase. *Biochim. Biophys. Acta* **81**, 270 (1964).

220. Savignano, T., Hanok, A., and Kuo, J. F., Creatine phosphokinase activity. A study of normal and abnormal levels. *Am. J. Clin. Pathol.* **51**, 76 (1969).

221. Schaposnik, F., Salvioli, M. V., Laquens, R., Neuman, M., and Cacciatore, J., La fosfocreatinquinasa como expresion de daño muscular. *Prensa Med. Argent.* **56**, 615 (1969).

222. Schiavone, D. J., and Kaldor, J., Creatine phosphokinase levels and cerebral disease. *Med. J. Aust.* **2**, 790 (1965).

223. Schmidt, F. H., Über die Bestimmung der Myokinase (Adenylatkinase) Aktivität im Serum. *Klin. Wochenschr.* **42**, 476 (1964).

224. Schwartz, K., and Mendonca, M. C., Value of enzymatic surveillance after cardiac transplantation. *Arch. Mal. Coeur Vaiss.* **62**, 767 (1969).

225. Serratrice, G., Roux, H., Aquaron, R., Gastaut, J. L., and Lapousse, J. C., On 8 cases of cramp due to exertion accompanied by tissue enzyme abnormalities. *Rev. Neurol.* **124**, 80 (1971).

226. Shainberg, A., Yagil, G., and Yaffe, D., Alterations of enzyme activities during muscle differentation in vitro. *Dev. Biol.* **25**, 1 (1971).

227. Shapira, F., Dreyfus, J. C., and Allard, D., Les isoenzymes de la créatine kinase et de l'aldolase du muscle foetal et pathologique. *Clin. Chim. Acta* **20**, 439 (1969).

228. Shaw, R. F., Chowdhury, S. R., and Pearson, C. M., Statistical characteristics and normal values of four serum enzymes, glutamic-oxalacetic transaminase, glutamic-pyruvic transaminase, aldolase and creatine-phosphokinase. *Enzymol. Biol. Clin.* **6**, 10 (1966).

229. Sherwin, A. L., Norris, J. W., and Bulcke, J. A., Spinal fluid creatine kinase in neurologic disease. *Neurology* **19**, 993 (1969).

230. Shirey, E. K., Proudfit, W. L., and Sones, F. M., Serum enzyme and electrocardiographic changes after coronary artery surgery. *Chest* **57**, 122 (1970).

231. Simpson, J. A., The biochemistry of myasthenia gravis. *In* "Progressive Muskeldystrophie Myotonia-Myasthenia" (E. Kuhn, ed.), p. 339. Springer-Verlag, Berlin and New York, 1966.

232. Smith, J. C., Foldes, V. M., and Foldes, F. F., Distribution of cholinesterase in normal human muscle. *Can. J. Biochem. Physiol.* **41**, 1713 (1963).

233. Sobel, B. E., Estimation of infarct size in man and its relation to prognosis. *Circulation* **46**, 640 (1972).

234. Sobel, B. E., Serum creatine phosphokinase and myocardial infarction. *J. Am. Med. Assoc.* **229**, 201 (1974).

235. Somer, H., and Konttinen, A., Demonstration of serum creatine kinase isoenzymes by fluorescence technique. *Clin. Chim. Acta* **40**, 133 (1972).

236. Sorenson, N. S., Dahl, J. A., and Aastrup, J. E., Creatine phosphokinase and operation. *Ugeskr. Laeg.* **127**, 1431 (1965).

237. Soule, T. I., and Cunningham, G. R., Herbicola Lathyri septicemia, myoglobinuria and acute renal failure. *J. Am. Med. Assoc.* **223**, 1265 (1973).

238. Starkweather, W. H., Spencer, H. H., Schwartz, E. L., and Schoch, H. K., The electrophoretic separation of lactate dehydrogenase isoenzymes and their evaluation in clinical medicine. *J. Lab. Clin. Med.* **67**, 329 (1966).

239. Starkweather, W. H., Zsigmond, E. K., Duboff, G. S., and Flynn, K., Creatine phosphokinase isoenzyme patterns in a malignant hyperthermic family. *In* "Malignant Hyperthermia" (A. Gordon, B. A. Britt, and W. Kalow, eds.), p. 339. Thomas, Springfield, Illinois, 1973.

240. Steers, A. J. W., Tallack, J. A., and Thompson, D. E. A., Fulminating hyperpyrexia during anaesthesia in a member of a myopathic family. *Br. Med. J.* **2**, 341 (1970).

241. Szasz, G., Busch, E. W., and Farohs, H. B., Serum-Kreatinkinase. I. Methodische Erfahrungen und Normalwerte mit einem neuen handelsüblichen test. *Dtsch. Med. Wochenschr.* **15**, 829 (1970).

242. Tammisto, T., and Airaksinen, M., Increase of creatine kinase activity in serum as sign of muscular injury caused by intermittently administered suxamethonium during halothane anesthesia. *Br. J. Anaesth.* **38**, 510 (1966).

243. Tammisto, T., and Airaksinen, M., Effect of some anesthetics on the increase of serum creatine kinase activity in cats by depolarizing agents. *Ann. Med. Exp. Biol. Fenn.* **44**, 404 (1966).

244. Tammisto, T., Leikkonen, D., and Airaksinen, M., The inhibitory effect of *d*-tubocurarine on the increase of serum creatine kinase activity produced by intermittent suxamethonium administration during halothane anesthesia. *Acta Anaesthesiol. Scand.* **11**, 333 (1967).

245. Tanzer, M. L., and Gilvarg, C., Creatine and creatine kinase measurement. *J. Biol. Chem.* **234**, 3201 (1959).

246. Thompson, D. E. A., and Tallack, J. A., Co-existent muscle disease and malignant hyperpyrexia. *In* "Malignant Hyperthermia" (A. Gordon, B. A. Britt, and W. Kalow, eds.), p. 309. Thomas, Springfield, Illinois, 1973.

247. Thomson, W. H. S., Determination and statistical analysis of the normal ranges for five serum enzymes. *Clin. Chim. Acta* **21**, 469 (1968).

248. Thomson, W. H. S., Serum enzyme studies in acquired disease of skeletal muscle. *Clin. Chim. Acta* **35**, 193 (1971).

249. Thürmer, J., and Lübcke, P., Aktivierte Serum-Kreatin-Kinase. *Dtsch. Med. Wochenschr.* **98**, 1568 (1973).

250. Trainer, T. D., and Gruenig, D., A rapid method for the analysis of creatine phosphokinase isoenzymes. *Clin. Chim. Acta* **21**, 151 (1968).

251. Trundle, D., and Cunningham, L. W., Iodine oxidation of the sulfhydyl groups of creatine-kinase. *Biochemistry* **8**, 1919 (1969).

252. Turner, D. C., and Eppenberger, H. M., Developmental changes in creatine kinase and aldolase isoenzymes and their possible function in association with the contractile elements. *Enzyme* **15**, 224 (1973).

253. Turner, D. C., Maier, V., and Eppenberger, H. M., Creatine kinase and aldolase isoenzyme transitions in culture of chick skeletal muscle cells. Unpublished observations.

254. Turner, D. C., Walliman, T., and Eppenberger, H. M., A protein that binds specifically to the M-line of skeletal muscle is identified as the muscle form of creatine kinase. *Proc. Natl. Acad. Sci. U.S.A.* **70**, 702 (1973).

255. Turner, D. C., and Eppenberger, H. M., unpublished observations.

256. Usubiaga, J. E., Wikinski, J. A., Wikinski, R. L., and Usubiaga, L. E., Prevention of succinylcholine fasciculation with local anesthetics. *Anesthesiology* **2**, 3 (1965).

257. Vale, S., Espejel, A., Calcaneo, F., Ocampo, J., and Diaz-de-Leon, J., Creatine phosphokinase; Increased activity in the spinal fluid of psychotic patients. *Arch. Neurol. (Chicago)* **30**, 103 (1974).

258. Van der Veen, K. J., and Willebrands, A. F., Isoenzymes of creatine phosphokinase in tissue extracts and in normal and pathologic sera. *Clin. Chim. Acta* **13**, 312 (1966).

259. Vejjajva, A., and Teasdale, G. M., Serum creatine kinase and physical exercise. *Br. Med. J.* **7**, 1653 (1965).

260. Venable, J. H., Skeletal muscle structure in Poland China Pigs suffering from malignant hyperthermia. *In* "Malignant Hyperthermia" (A. Gordon, B. A. Britt, and W. Kalow, eds.), p. 208. Thomas, Springfield, Illinois, 1973.

261. Ventafridda, V., Terno, G., and Sciancalepore, G., Compartamento della creatin-fosfochinasi e del potassio serico in rapporto alla mialgia de succinilcolina. *Min. Anestesiol.* **36**, 200 (1970).

262. Vignos, P. J., Jr., and Goldwyn, J., Evaluation of laboratory tests in diagnosis and management of polymyositis. *Am. J. Med. Sci.* **263**, 291 (1972).

263. Vincent, W. R., and Rapaport, E., Serum creatine phosphokinase in the diagnosis of acute myocardial infarction. *Am. J. Cardiol.* **15**, 17 (1965).

264. Vorburger, C., Fässler, B., and Köhl, P., Serum-Kreatinphosphokinase and Intramuskuläre Injektion. *Schweiz. Med. Wochenschr.* **103**, 927 (1973).

265. Wadhwa, R. K., and Tantisira, B., Parotidectomy in a patient with a family history of hyperthermia. *Anesthesiology* **40**, 191 (1974).

266. Warburton, F. G., Bernstein, A., and Wright, A. C., Serum creatine phosphokinase estimations in myocardial infarction. *Br. Heart J.* **27**, 740 (1965).

267. Weimer, R. J., Van Saude, M., Karcher, D., Loewenthal, A., and Van der Heim, H., A modified technique for direct staining with nitroblue-tetrazolonium of LDH upon agar gel electrophoresis. *Clin. Chim. Acta* **7**, 750 (1962).

268. Weinreich, J., Busch, D., Gottstein, U., Schaefer, J., and Rohr, J., Über zwei neuen Falle von hereditärer-nichtsphärocytärer hämolytischer Anämie bei Glucose-6-phosphat-dehydrogenase-Defekt in einer norddeutschen Familie. I. Mitteilung. *Klin. Wochenschr.* **46**, 146 (1968).

269. Weismann, U., Colombo, J. P., Adam, A., and Richterich, R., Determination of cysteine activated creatinine kinase in serum. *Enzymol. Biol. Clin.* **7**, 266 (1966).

270. Welch, K. M. A., and Goldberg, D. M., Serum creatine phosphokinase in motor neuron disease. *Neurology* **22**, 697 (1972).

271. Welch, K. M. A., and Goldberg, D. M., Response of serum enzymes and other biochemical constituents to strenuous exercise in control subjects and patients with motor nerve disease. *J. Neurol. Sci.* **19**, 225 (1973).

272. Wharton, B. A., Bassi, U., Gough, G., and Williams, A., Clinical value of plasma creatine kinase and uric acid levels during first week of life. *Arch. Dis. Child.* **46**, 356 (1971).

273. Wilson. R. D., Dent, T. E., Traber, D. L., McCoy, N. R., and Allen, C. R., Malignant hyperpyrexia with anesthesia. *J. Am. Med. Assoc.* **202**, 183 (1967).

274. Wolf, E., and Magora, A., Serum analysis in the diagnosis and follow up of acute polymyositis. *Harefuah* **84**, 534 (1973).

275. Yaffe, D., and Dym, H., Gene expression during differentiation of contractile muscle fibers. *Cold Spring Harbor Symp. Quant. Biol.* **37**, 543 (1972).

276. Yagi, K., and Mase, R., Coupled reaction of creatine kinase and myosin A-adenosine triphosphatase. *J. Biol. Chem.* **237**, 397 (1962).

277. Zaimis, E. J., Action of decamethonium on normal and denervated mammalian muscle. *J. Physiol. (London)* **112**, 176 (1951).

278. Zalin, R., Creatine-kinase activity in cultures of differentiating myoblasts. *Biochem. J.* **130**, 79 (1972).

279. Zelder, O., Postoperative Serumenzymeveränderungen nach Abdominaleingriffen. *Chirurg* **41**, 278 (1970).

280. Zener, J. C., and Harrison, D. C., Serum enzyme values following intramuscular administration of lidocaine. *Arch. Intern. Med.* **134**, 48 (1974).

281. Zsigmond, E. K., Starkweather, W. H., Duboff, G. S., and Flynn, K., Genetic abnormality of muscle creatine phosphokinase in a family with malignant hyperpyrexia. *Adv. Anesthesiol. Resuscitation* **2**, 1419 (1972).

282. Zsigmond, E. K., Comment in Case History, Number 65. Malignant hyperthermia with subsequent uneventful general anesthesia. *Anesth. Analg. (Cleveland)* **50**, 1104 (1971).

283. Zsigmond, E. K., Clinical application of the lactic dehydrogenase zymogram in anesthesia. Guest discussion. *Anesth. Analg. (Cleveland)* **50**, 1102 (1971).

284. Zsigmond, E. K., Starkweather, W. H., Duboff, G. S., and Flynn, K., A genetic abnormality of muscle creatine-phosphokinase in a family with malignant hyperpyrexia. *Rev. Bras. Anesthesiol.* **21**, 265 (1971).

285. Zsigmond, E. K., Starkweather, W. H., Duboff, G. S., and Flynn, K., Elevated serum-creatine phosphokinase activity in a family with malignant hyperpyrexia. *Anesth. Analg. (Cleveland)* **51**, 220 (1972).

286. Zsigmond, E. K., Starkweather, W. H., Duboff, G. S., and Flynn, K., Abnormal creatine-phosphokinase isoenzyme pattern in families with malignant hyperpyrexia. *Anesth. Analg. (Cleveland)* **51**, 827 (1972).

287. Zsigmond, E. K., and Starkweather, W. H., Abnormal serum and muscle creatine phosphokinase (CPK) isoenzyme pattern in a family with malignant hyperpyrexia. *Anaesthesist* **22**, 16 (1973).

288. Zsigmond, E. K., Comment on a pertinent basic science problem. *In* "Malignant Hyperthermia" (A. Gordon, B. A. Britt and W. Kalow, eds.), p. 188. Thomas, Springfield, Illinois, 1973.

289. Zsigmond, E. K., and Fabian, S. F., Serum-butyryl cholinesterase and CPK activities in burn patients. To be published.

Author Index

Magalhaes, A. S. *13*, 155(285, 287, 294, 295, 296, 304)
Maggio, A. *9*, 137(210)
Magnusson, T. *10*, 17(230, 243)
Magora, A. *13*, 274(297)
Magyar, K. *10*, 59(240)
Mahin, D. T. *8*, 12(132)
Mahler, H. R. *3*, 21(31)
Mahon, F. B. *13*, 156(298)
Mahoney, W. D. *13*, 112(298), 113(298)
Mahowald, T. A. *13*, 157(280, 282)
Maickel, R. P. *8*, 53(142), 247(142)
Maier, E. H. *8*, 324(134)
Maier, V. *13*, 253(286, 287)
Main, A. R. *8*, 325(128)
Mair, G. A. *1*, 9(11)
Malacinski, G. M. *2*, 45(24)
Malagodi, M. H. *9*, 133(188)
Malmström, B. G. *8*, 326(108, 109)
Malthe-Sørensen, D. *8*, 327(94), 328(94)
Mandecki, T. *13*, 158(294)
Manian, A. A. *9*, 118(171, 172)
Mann, D. M. *12*, 16(270, 275)
Mann, S. P. *8*, 271(93)
Manner, G. *8*, 329(93), 330(101)
Mannering, G. J. *9*, 5(170, 183), 48a(207), 119(180, 183), 160(190), 171(183), 174(183)
Manoilov, S. E. *6*, 20(63)
Mansfield, G. *8*, 318a(135)
Marchelle, M. *10*, 24(229), 78a(229)
Marginson, M. A. *13*, 56(301)
Mark, L. C. *9*, 7(179), 18(192), 24a(171, 175), 120(169, 171, 172, 173, 174), 121(183), 122(183), 139(192)
Markert, C. L. *8*, 331(110)
Marniemi, J. *9*, 123(189)
Marshall, B. E. *12*, 6(272, 273), 7(272, 273), 8(272, 273)
Martens, S. *9*, 81(194)
Martin, R. *8*, 50(137)
Martinez, G. *1*, 20(7)
Marton, A. V. *8*, 86(112, 115)
Mase, R. *13*, 276(286, 287)
Mason, A. D., Jr. *8*, 200(134)
Mason, S. G. *6*, 6(68)
Massey, V. *3*, 1(33, 40)
Massoulie, J. *8*, 332(105)
Mastaglia, F. L. *13*, 159(297), 160(297)
Matchinsky, F. M. *12*, 29(276)
Mather, L. E. *8*, 419(142)
Mathews, H. B. *9*, 173(209)
Mathias, A. P. *4*, 17(47)
Mathon, C. *13*, 37(294)
Matthews, W. *4*, 28(46)
Maurer, H. M. *9*, 124(186)
Maxwell, J. D. *9*, 91(185)
May, S. C. *8*, 390a(105)
Mayer, I. *8*, 220(134)
Mayfield, D. E. *9*, 180(187)

Mayrhofer, O. *8*, 333(134, 141); *13*, 161(300)
Mazia, D. *6*, 16(67)
Mazze, R. I. *9*, 125(188)
McAfee, D. A. *11*, 20(259)
McCaman, R. E. *8*, 193(132)
McCance, R. A. *8*, 334(132, 142)
McCarville, W. J. *8*, 171b(105)
McComb, R B. *8*, 298(110)
McConn, J. D. *2*, 44(24)
McConnell, H. M. *1*, 22(16)
McCoy, N. R. *13*, 273(299, 305)
McCrea, B. E. *1*, 34(5)
McCreery, D. *8*, 381(93)
McDowell, F. *9*, 106(191), 107(96)
McHorse, T. S. *9*, 101a(206)
McIntyre, R. *12*, 9(272, 273)
McIsaac, R. J. *8*, 323(103)
McLaren, A. D. *3*, 22(39); *6*, 17(69)
McMahon, R. E. *9*, 126(183)
McNall, P. G. *8*, 139(111, 140), 140(144), 140a(123, 144), 155(123, 127, 137, 138), 158(138), 159a(140)
McQuire, H. J. *13*, 170(282)
Mears, G. E. F. *2*, 54(24)
Meffin, P. *9*, 127(202)
Meffin, P. J. *9*, 16(207), 202c(207)
Meister, A. *2*, 49(21)
Meister, W. *9*, 13a(206, 207)
Melmon, K. L. *9*, 16(207), 182(206, 207), 183(207), 202c(207)
Melnick, P. J. *13*, 150(288)
Meloche, H. *1*, 20(7)
Melson, H. *9*, 86(187)
Meltzer, H. Y. *13*, 70(293), 146(293), 162(287, 292, 293), 163(291, 293), 164(292, 293), 165(295, 297), 166(291, 293), 167(290), 168(287)
Melville, K. I. *9*, 128(189)
Mendel, B. *8*, 335(102, 111), 335a(111)
Mendonca, M. C. *13*, 224(294)
Mengle, D. C. *8*, 336(130)
Mengoli, H. F. *13*, 4(287, 288, 293, 294, 296, 298, 301, 303)
Menon, T. *11*, 49(258)
Menten, M. L. *3*, 24(33)
Mercer, D. W. *13*, 169(287, 289, 293, 294, 298)
Merker, H. J. *9*, 153(179)
Merrill, G. G. *8*, 337(146)
Metzger, H. *1*, 60(8)
Metzger, H. P. *8*, 338(118)
Meyer, H. *8*, 10(111)
Meyer, M. *9*, 116(176)
Michaelis, L. *3*, 23(40), 24(33)
Michaelis, R. *2*, 16(24)
Michaelson, I. A. *8*, 339(103), 444(100)
Michel, H. O. *8*, 243(119), 340(131)
Michel, I. M. *11*, 61(262)
Michelson, M. J. *8*, 250(110)
Michie, D. D. *13*, 170(282), 171(295)

Subject Index